AF782229

Democracy, Ecological Integrity and International Law

Democracy, Ecological Integrity and International Law

Edited by

J. Ronald Engel, Laura Westra and Klaus Bosselmann

CAMBRIDGE SCHOLARS

PUBLISHING

Democracy, Ecological Integrity and International Law,
Edited by J. Ronald Engel, Laura Westra and Klaus Bosselmann

This book first published 2010

Cambridge Scholars Publishing

12 Back Chapman Street, Newcastle upon Tyne, NE6 2XX, UK

British Library Cataloguing in Publication Data
A catalogue record for this book is available from the British Library

Copyright © 2010 by J. Ronald Engel, Laura Westra and Klaus Bosselmann and contributors

All rights for this book reserved. No part of this book may be reproduced, stored in a retrieval system, or transmitted, in any form or by any means, electronic, mechanical, photocopying, recording or otherwise, without the prior permission of the copyright owner.

ISBN (10): 1-4438-1767-8, ISBN (13): 978-1-4438-1767-7

DEDICATION

In memory of Aldo Leopold, whose ethic of ecological integrity, and life-long dedication to wilderness preservation and democratic land citizenship, inspired the founding of the Global Ecological Integrity Group.

TABLE OF CONTENTS

Part II – State Sovereignty, Indigenous Governance and International Law

Part III – Environmental Harms, Property Rights and Responsibilities

Part V – Democracy and Civil Society: Neocolonialism v. Nature Protection

LIST OF FIGURES AND TABLES

Figures

Tables

PREFACE

This book resulted from the symposium "Governance for Sustainability" at the European University Institute in Florence, and the Global Ecological Integrity Group Conference "State Sovereignty, Ecological Integrity and International Law" at the University of Florence. Both events were linked in terms of their timing (June 2009) and by their focus on institutionalizing ecological integrity and sustainability.

We live in an era in which governments have all but lost their capacity to secure a good life and a healthy environment for the peoples of the world, to control financial markets and global economics, or to combat climate change; the prospects for humanity are bleaker than ever. Are we now seeing the demise of the most fundamental institutions of modern society – democracy, the state, the corporation, the global financial system and the rule of law? How radical will conceptual and institutional changes have to be to restore trust in, and functioning of, democratic governance?

This question constituted the basic theme of both conferences at Florence and was the overarching framework for the selection of papers in this published volume. Most especially, we have been concerned for the relation of the principle of ecological integrity in environmental ethics, law and policy to current debates on the meaning and future of "democracy." What this relation involves, and the distinctive position that many contributors to this volume take toward it, is further discussed in the Introduction to Part I.

The general democracy debate of Part I is further developed in relation to other forms of governance (Parts II), rights and responsibilities (Part III), water, food and climate change (Part IV), and civil society and the natural world (Part V). In this way, the book shows how democratic governance, as now practiced and conceptualized, fails to preserve ecological integrity. It also makes the case for ways in which democracy can be reconceived so as to support the integrity of natural processes.

—Ron Engel, Laura Westra and Klaus Bosselmann

PART I –

THE DEMOCRACY DEBATE AND ECOLOGICAL INTEGRITY

PART I INTRODUCTION

J. RONALD ENGEL

In the opening years of the 21st century the question of "governance" has moved to the center of our political, economic, social, personal and religious lives.

The most immediate and obvious reason for this is that the dominant institutions of contemporary societies – international law regimes, state governments, the global financial system and the corporate organization of economic life, educational and cultural institutions of all kinds – have lost their capacity to govern the spheres of human activity for which they are deemed responsible in such a way as to maintain the common good.

Persons throughout the world are experiencing a profound sense of insecurity and there is widespread distrust of the legitimacy and effectiveness of societal leadership and institutions.

Increasingly, this is viewed as a question regarding the adequacy of the prevailing modern ideal of governance – democracy. This question is being pressed from a number of critical angles: whether it is a universal ideal or exclusively Western in form and origin; whether it is necessarily linked to the capitalist organization of economic life; whether it can serve as a meaningful and realistic ideal for the governance of international institutions; how it is used ideologically to advance the interests of the so-called "democratic nations"; whether it finds moral and spiritual support in world religions and indigenous traditions; whether as a matter of internal logic it requires a strong doctrine of human rights; whether under conditions of mass urban society and unprecedented concentrations of military, economic and political power it is more than a nostalgic reminder of the kinds of public participation in governance that once characterized small scale social life; not to speak of spirited debates between advocates of competing conceptions of democracy, from libertarian to liberal to social democratic and more.

The Global Ecological Integrity Group (GEIG) annual conference in Florence, Italy in June 2009 brought still another critical angle to this discussion – the urgent question of the compatibility of democratic forms of governance with the preservation of the ecological integrity of the biosphere and the human rights of persons throughout the world who depend upon the health of the biosphere for their survival and well-being.

For the members of GEIG the all-encompassing contemporary issue of governance must be cast as the question of how to so organize our lives together that not only human life, but all life is secure and has the conditions necessary for its full flourishing.

GEIG is not alone, of course, in raising the question of the compatibility of ecological integrity and democracy, although it does so in an unusually broad global context, both in terms of the diversity of participants and disciplines represented in the discussions, and in terms of the long-standing concerns that are at issue, such as health, economics, and indigenous peoples' rights, and most especially, the prospects for international law. Indeed, a small library of publications on the subject has emerged in recent years which can be readily accessed by consulting a recent book by three of the present volume's authors.[1] In addition, a number of allied inquiries, such as the discussions currently organized around the concept of "Earth Jurisprudence," draw heavily on democratic political and moral categories in their aim of devising new forms of global environmental governance.

In most of these discussions the meaning of all three terms – democracy, ecological integrity, and the relationships between them – are in dispute. However, in the conversation that has characterized GEIG, the intrinsic importance of ecological integrity is assumed to stand on par with the intrinsic importance of human dignity and is accepted as a "given" of inquiry and debate.[2] Although there is always opportunity for refinement and further questioning of the scientific and philosophical foundations of the concept of ecological integrity, primary attention is devoted to the ethical requirements of human governance systems to honor it as a precondition of sustainable social life, universal human rights and international law.

Each of the contributors to this volume shares this assumption and in most cases it takes the form of an integrated intellectual and personal commitment. Its centrality is fully evident in the essays in Part I, although a variety of terms in addition to "ecological integrity" – for example, "strong sustainability" (Bosselmann) and "deep ecology" (Sim) – are employed to express it.

"Democracy," however, is another matter. Amartya Sen, following in the footsteps of Jürgen Habermas and John Rawls, argues in the *The Idea of Justice* that the definitive core of democracy is "the exercise of public reason," and its corollary, "government by discussion," and that so defined democracy is a global phenomenon with multiple cultural roots.[3] Sen is here giving voice to a widespread shift by contemporary political philosophers toward acceptance of the ideal and practice of "deliberative democracy" as the central concern of politics and a primary means for the achievement of social justice. The particular institutional means by which deliberative democracy is implemented, such as balloting, while important, are secondary matters. By this definition (*pace* Habermas) – democracy as both open debate about moral values and principles and the justified uses of power and coercion – GEIG has long served as a center of exemplary democratic practice. In Florence in 2009, however, for the first time in the history of the group, the topic of democratic governance as such, and the question of its role in advancing or inhibiting the conservation of ecological integrity and the rule of international law, became matters of explicit critical discussion.

There was large agreement among participants at Florence that "democracy", as practiced (and indeed in large part defended) in most popular political rhetoric, falls far short of the form of governance required to sustain, much less restore, the healthy functioning of the biosphere. This agreement is evident in the papers selected for inclusion in Part I of this volume, although different flaws in the prevailing forms of democracy are stressed by each of the authors. The strongest indictment of democracy comes from Westra who holds that in current practice it is a totally unprincipled form of governance corrupted by monetary and other partial interests. Engel finds the prevailing forms of democracy "thin" because they take as their primary if not sole moral principle the "consent of the governed" and assume that if the institutionally proper procedures are followed the results must be acceptable. One reason Heinrichs finds contemporary democracies to be deficient is that they are governed by one form or another of a party system which robs voters of the opportunity to choose the full range of values they favor. Dellapenna believes that while contemporary liberal democracy may slow the steady slide of the world toward environmental destruction it is only the best of all the bad alternatives. Bosselmann finds that at its best current democratic theory and practice only go as far as embracing "weak sustainability" – the untenable notion that there can be a "balance" between environmental, social and economic needs. There was considerable criticism by the authors of the dominance of economics in contemporary democratic governance, including any notion that there is necessarily a tight linkage between democracy and so-called free-market capitalism, and the closely associated premise that democratic governance is for the sake of ever expanding economic growth.

The most significant agreement among the authors of the papers in Part I, however, has to do with their generally shared position that an adequate concept of governance must rest upon a comprehensive vision of the good life which includes a strong normative understanding of not only human but all of nature. Given the centrality of ecological integrity as a comprehensive normative concept in GEIG this outcome may not be surprising, but when it is set within the context of current discussions of democracy as represented in the work of such leading international political theorists as Sen, Rawls and Habermas, it marks a decisive departure, and suggests an important contribution the group may make in future discussions of governance, including the relation of nature and democracy, in international political philosophy.

With the exception of Dellapenna, who exposes the power of civil society associations, fragile and short-lived as they proved to be, to provide the space where citizens' concerns for the health of the environment could expand into effective movements to overthrow the authoritarian regimes of Eastern Europe, each of the authors of Part I argues that ethically defensible institutions of governance must not only provide the opportunities for the "exercise of public reason" which will "help" discern the demands of justice, as Sen would have it, but also substantive and universal norms by which to evaluate the moral qualities of those who govern, citizens and leaders alike, and the actual outcomes of "government by discussion." Given the rejection of strong ideals of the common good or any teleological understanding of human nature in modern democratic societies, Westra is reluctant to associate such an ideal with the name "democracy" and Sim does not declare herself on the matter.

The most crucial and determinative issue that thus emerges is the meaning and possibilities of "reason" in democratic practice and theory. Does not the "exercise of public reason" involve – indeed, in principle, require – debate concerning ultimate frames of reference and commitment as well the moral principles, values and virtues that are derived from those worldviews? And further, does not the "exercise of public reason" require as its precondition certain foundational commitments, however revisable, to a comprehensive vision of the good, most especially at this juncture of human history, a vision that affirms the comprehensive normative status of ecological integrity as well as an ontologically grounded normative view of human nature that grounds human moral responsibility?

Most interlocutors in the contemporary democracy debates eschew such claims, or like Sen, reduce them to their liberal democratic minimum of freedom. And even the few in the Western philosophical tradition who do affirm a rightful place for debates concerning worldviews in deliberative democracy, such as Jeffrey Stout or Franklin Gamwell, or hold out the possibility of agreement upon a substantive vision of the good life or common good, such as Martha Nussbaum or Michael Sandel, neglect any concept of the normatively natural that extends beyond human capacities alone.

In truth, in virtually all current debates surrounding the meaning of democracy, and even in most debates surrounding the meaning of environmental ethics and policy, including those that seek to make democracy and sustainability mutually supportive, any notion that the concept of nature or the "natural" may be substantive is immediately dismissed as part of the "rhetoric of reaction," a residual anti-historical and cultural imperalist essentialism.

Yet here we find significant agreement between Westra, Engel, Sim, Heinrichs and Bosselmann regarding the need for explicit concern with ultimate aims, goals and norms, and a substantive naturalistic worldview – ancient or modern – as the foundation of any rationally justified theory of governance, however named. Westra and Engel both argue for natural law as a necessary foundationalist perspective, albeit from different metaphysical positions. Aristotle, Westra argues, provides a basis for reclaiming the classical Greek vision of moral and intellectual virtues and universal principle that can undergird international law. Whitehead, Engel proposes, offers a metaphysical grounding for what he describes as the "thick" moral tradition of democracy. Sim argues for the superiority of the metaphysics of a neo-Confucian Chinese philosopher, Zhu Xi, for purposes of deep ecology ethics on grounds that it affirms the human or moral good is identical with the goodness of everything whereas Aristotle sees humans as morally self-sufficient in separation from the rest of the natural world. Heinrichs proposes an anthropologically grounded "value-based theory of democracy" composed of four elected parliaments, one of which is devoted to public debate and decision regarding the society's governing worldview and spirituality. Bosselmann argues that the "European cosmology" of dualism, anthropocentrism, materialism, atomism, greed, and economism underlies the reductionist theories of democracy that are proving to be so dysfunctional in light of the crisis of governance today, and that it must be replaced with a more holistic worldview that draws upon the "ecological wisdom" inherent in the world's great spiritual traditions.

Bosselmann also does something more. He challenges us by concluding his essay with a proposal for a collaborative intellectual and practical agenda, called "Earth Democracy," which will carry forward the discussion begun here, and hopefully enable the viewpoint shared by the authors of this volume –the view that democracy requires a substantive and comprehensive vision of the integrity of the evolutionary and ecological processes of the natural world – to find acceptance in the mainstream political debates on governance in our time.

Notes

[1] Klaus Bosselmann, Ron Engel, Prue Taylor, *Governance for Sustainability – Issues, Challenges and Successes* (Bonn: IUCN Environmental Law and Policy Series, vol. 70, 2008)

[2] Laura Westra, "Ecological Integrity: Its History, Its Future and the Development of the Global Ecological Integrity Group." In Laura Westra, Klaus Bosselmann, and Richard Westra, eds., *Reconciling Human Existence with Ecological Integrity* (London: Earthscan, 2008), pp. 5–20.

[3] Amartya Sen, *The Idea of Justice.* (Cambridge, Massachusetts: Harvard University Press, ,2009), p. 324

CHAPTER ONE

ECOLOGY AND THE LAW: DEMOCRACY, GLOBALIZATION AND THE GREEK ROOTS OF ECOLOGICAL PROBLEMS

LAURA WESTRA

Introduction

> "He who demands that law should rule may thus be regarded as commanding that God and reason alone should rule; he who commands that a man should rule adds the character of the beast … Law [as the pure voice of God and reason] may thus be defined as 'reason free from all passion'." (Barker 1973: 146)

So says Aristotle emphasizing the rule of law, above the "rule of man", that is the presence of regulatory regimes that are based on "reason alone", or on principles that stand aside from desires, or the choices of individuals. In this chapter I would like to discuss several areas of conflict between what international law demands, and what domestic low, or the actions of powerful, but unprincipled states, or other such agents can achieve, in the light of teaching of Aristotle.

The first thing to note is that Aristotle's respect for law is based on the fact that he views it as an entirely principled enterprise, as he defends as best a state with "good laws well obeyed" (Aristotle, *Politics*). Today instead, municipal laws of each nation are strongly influenced by the politics of their countries, so that greed or the quest for power influence not only their constitutions and legal infrastructure, but also the way they interact with the rest of the international legal community.

Hence, for example, a capitalist nation like the US will strongly support so-called "free trade" agreements, and the power of the World Trade Organization (WTO), but it will not ratify the *Convention on the*

Rights of the Child (all countries of the world are signatories, except the US and Somalia), or the *American Convention on Human Rights*, because of the presence of Article 4 on the "Right to Life"; in addition, it does not comply with a number of other international Conventions, against torture, for instance, or on the treatment of prisoners of war, among others.

In contrast, in general terms, international law is based on principles of natural justice; the writings of well-known publicists; and positive laws and precedents in jurisprudence (Kindred 2000). The emphasis however remains on principles. For instance, Judge Tanaka, considered general principles of law a primary source of international law, because they have the character of *jus rationale* and are "valid through all kinds of human societies" (Kindred 2000: 148; *South West Africa Case* 1996). Here Judge Tanaka defines the source of general principles of Law, as these include universality and rationality, following precisely what Aristotle considers to be *eunomia*: "good laws well obeyed" (Aristole, *Politics*; Barker 1973).

Hence the reasoned approach in international law arises from well-established principles, but also in the reasoning of well-respected scholars, and of judges themselves. This triple guarantee of its foundation on reason and principle is not always present in domestic legal instruments, or in the politically dictated choices of the leaders of individual states, including many ratified covenants. The first thing to note is the connection between the laws of nature and natural law, and its origin in Aristotle.

The Laws of Nature and the Origins of Natural Law: Nature and Laws in Aristotle

Nature is central to Aristotle's argument in the *Politics*. This is routinely accepted by Aristotelian scholars:

> "Aristotle conducted his study of things human in the fields of politics and ethics (and also of logic, poetry and oratory), side by side with a study of things natural (physics, medicine, and general biology)." (Barker 1973: xxviii)

In addition, his "inclination towards the Ionic 'becoming' – the genetic doctrine of *phusis*" (Barker 1973: xxix) ensures that nature will be and remain foundational for all his arguments, form the admiration he evinces for the beauty of perfected forms, to the presence of design in nature as such (Aristotle, *Parts of Animals*). We noted that governance, citizenship and the *polis* itself were discussed with reference to natural standards (of size, of completeness and the like). In the same sense, the constitution of the state will provide its "essence", the explanation of its identity as a "quasi-juridical person" (Barker 1973: 100–101). The constitution is analogous to the natural laws governing physical organisms (Barker 1973).

Like all natural entities, the state has two main ends (in this case, not just one end), for the association it represents. Aristotle starts with the basic "natural impulse" according to which "men desire to live a social life"; the other end is represented by the common interest:

> "The good life is the chief end, both for the community as a whole and for each of us individually. But men also come together, and form and maintain political associations merely for the sake of life …" (Aristotle, *Politics*)

Hence the *essential* nature of a state, the laws that regulate it, exist for the sake of maintaining life, social association and the good life (Barker 1973: chapter III). This simply re-elaborates the theme clearly stated in Book I of the *Politics*, that "every *polis* exists by nature", and that the "nature of things consists in their ends or consummation", as "the end, the final cause is the best" (Barker 1973: 5). The *polis* exists "by nature" and man is meant "by nature" to live in a social environment.

If we consider the modern, liberal democratic state, we find something that is in direct conflict with the Aristotelian view of the state. It does provide association, so it satisfies at least one condition Aristotle finds essential to the nature of the state. But note that the other two "ends" or reasons why men join together in political association are missing or under threat. In glaring contrast with the Aristotelian emphasis on the state's support of the "common good", or the happiness that is based on the "natural end of man" as a moral ideal, in modern times even a token quest for that sort of good has been completely eliminated from present political institutions (Westra 1998).

Nature, the State and Right Governance

> "People form governments for their common defense, security and welfare. The first thing that public officials owe their constituents is protection against natural and man-made hazards." (Gostin 2004: 7–12)

We do not find anything of significance regarding ecology in Aristotle, but we can start by considering what "good laws" should be in the teleological sense of what they must achieve in order to be termed "good". Only then we might be able to judge what Aristotle would possible say about major ecological problems that beset us all today, in varying degrees:

1) climate change and its consequences especially (though not exclusively) in island states, costal areas and the Arctic;
2) globalization and economic oppression; and
3) the plight of ecological refugees.

These are all problems far removed from the reality of life in Aristotle's times; hence we cannot expect to find pat, simple answers in his *corpus*. Rather, we must follow Aristotle's thought through the arguments he proposes, in order to see where what he says might shed light on there intractable problems of a different time.

First, if the state, its laws, and its citizen are understood in terms of what is natural, then we must start with the "nature of the end for which the state exists" (Barker 1973: 118). As one might expect, "the end of the state is not mere life; it is rather a good quality of life" (Barker 1973). Such a *polis* will include "the association of families and villages in a perfect and self-sufficing existence; and such an existence, on our definition, consists in a life of true felicity and goodness" (Barker 1973: 120).

Hence the *polis* or the state, must be organized and governed only in ways that will foster "good actions", such that they represent the excellence of character or its citizens, as its laws are meant to make them just.(Barker 1973: 118) A proper understanding of what constitutes justice is basic here:

> "In democracies, for example, justice is considered to mean equality [in the distribution of offices]. It does mean equality – but equality for those who are equal, not all. In oligarchies, again, inequality is the distribution of office is considered to be just; and indeed it is – but only for those who are unequal, and not for all." (Barker 1973: 117)

An oligarchy is a constitution ruled by the wealthy; a democracy is here intended as one where the poor rule, hence, each is limited on its understanding of what a just constitution must be. In addition to being limited, neither starts from then required understanding in the operative aim of the state, on one hand, and the true meaning of citizenship, on the other. Barker says:

> "The principle of a constitution is its conception of justice and this is the fundamental ground of difference between oligarchy and democracy.

> Democrats hold that if men are equal by birth, they should in justice have equal rights: oligarchs hold that if they are unequal in wealth, they should in justice have unequal rights." (Barker 1973: 116)

Aristotle states the problem clearly: both fail to acknowledge "… the really cardinal factor", i.e. the nature of the end for which the state exists (Barker 1973: 118). The state exists to support and promote the "good life", and – *pace* Nussbaum – the "good life" is the life of moral excellence, with the basic means thereof, not a life that entails the right of "flourishing", given the economic connotations that normally attach to that concept.

Instead, its legal structure must encourage habituation to the moral life by its system of rewards and punishments, as no man is born ethical, but only becomes such through habituation (Owens 1959: 343; Aristotle, *Nicomachean Ethics*, II, 1, 1103b 23–25). Only with men are so habituated, their choices will be morally just, and exhibit that individual excellence that is the *causa causans* of a well governed *polis*:

> "Justice is 'that kind of state of character which makes people dispose to do what is just and makes them act justly and wish for what is just …'. (Aristotle, *Nicomachean Ethics*, V, 1, 1129a 7–9). In a wider sense justice is used to mean virtue in general, and in this meaning it, of course, includes all the virtues." (Owens 1959: 347)

Hence, it is neither "alliance for mutual defense against all injury, or to ease exchange and promote economic intercourse" (Barker 1973: 118), that represent the true ends of the state, but the promotion of individual virtue and collective excellence, indicative of true happiness.

Once we have ascertained what is the nature of the *polis* itself for Aristotle, and its basic aim (Barker 1973: 92), we need to consider the nature of its components: its citizens. Citizens are not such in virtue of being residents in the *polis* or of being entitled to sue and be sued, or of being admitted as resident aliens. Aristotle says "We may lay down that citizens are those who share in the holding of office as so defined", e.g. as office held for an indeterminate period (Barker 1973: 94). Hence, we can conclude that for Aristotle, a citizen is only someone who actively participates in the operation and then governance of *polis* and therefore, one who is dedicated to furthering its aims. All that pertains to the state, from its component parts, to the rational laws that govern it, to its ultimate aim, are in clear conflict, however, with the goals, constitutions, and most of the legal instruments that govern today's Western industrialized "democratic" states. We will consider this problem in the next section, before considering the ecological problems listed, in turn.

The Aims, the Composition and the Laws of the Present Western Affluent States

> "In order to protect the biosphere, we can make provisions for the power of sanctions to be exercised by a specific institution. This was provided for in the Declaration adopted by twenty-four States at The Hague on April 3, 1989. Having taken their motto 'My country is the planet', they studied the idea of authority endowed with super state powers capable of directing policies and responsible not only for controlling the conduct of states, but also for taking sanctions against those who violate the fundamental principles that are the basis or respect for the global environment. Such an approach is not possible without acknowledging that humankind has a right to integrity and, in the end, to survival." (Dupuy 1991: 201–204)

As we turn from Aristotle's time to our own, we note the first, basic difference. Aristotle's *polis*, was not only homogenous insofar as its citizens ethnicity was concerned, but also limited in size (for best governance), and self-sufficient. Today's nation's especially Western affluent states are precisely opposite: multi-ethnic large, and – most of all – unable to survive without steady intercourse with other national entities.

Aside from these vast material differences, there is still a major difference, that is a difference of principle, that is, to some extent unavoidable, given the physical discrepancies: the modern Western country's main *raison d'être* is a twofold one: to protect its citizens and its borders, and to thrive economically, while asserting its own power vis-à-vis other nations. The development and the pursuit of virtue, individual and collective, are not even considered as valid objectives. In fact, for the most part, Western nations tend to view morality as something of a personal choice, possible linked with one's religious beliefs and practice: something to which all have a right to, to be sure, but not something that is the state's concern, except insofar as it is obliged to protect that right of their citizens.

In contrast, some have argued (Gostin 2008), and I concur (Westra 2006), that the State's obligations to protect should include the protection of public health, which is gravely threatened by today's environmental attacks, both external (through climate change and toxic exposures), and internal, through the effects of environmental pollutants, both industrial chemicals (Grandjean and Landrigan 2006; Westra 2007) and endocrine disruptors (Colborn et al 1996; Westra 2007). On that topic, Allyn Taylor notes that:

> "WHO's advancement of national and international public health and supervisory institutions is critical to furthering the realization of the right to health. Encouraging countries to develop specific binding legal obligations with respect to the right to health and publicizing their compliance and non-compliance with those obligations, can powerfully influence states to rethink priorities and redirect national resources to national health care." (Taylor 1992: 310–346)

Nevertheless, by appealing to the WHO, we are already transcending the state as such, and this remains yet another cardinally important point of difference with Aristotelian doctrine: the question of self-sufficiency no longer applies to today's state. Not only other states are needed for trade, alliances, and the like, but pathogens travel from continent to continent, hazardous industrial operations and their effluents reach far and wide beyond the site of their operations, most notably through the effects of climate change (McMichael 1995: 310–317; Patz, 2005: 310–317).

Hence, not only trade and education transcend all borders today, but also do infectious diseases and the vectors that travel together with the movement of persons and airplanes and other conveyances, and as do toxic materials and other pollutants.

Hence the "good law" that Aristotle extols cannot be only municipal (or domestic) law today: it has to be international law, in order to be a "good law", at least based on principles and rationality. Its capacity for enforcement is almost non-existent, unfortunately, so that it is weak on the practical side; on the theoretical side, we find that it encompasses some principles that might have even met Aristotle's approval. I am speaking of a special sort or norms that are non-derogable, unlike conventions and agreements that are often ignored, through lack of compliance by powerful countries.

In contrast, *jus cogens* norms are truly universal, and they impose obligations *erga omnes* on both states and individuals. (Ragazzi 1997: 3; *Barcelona Traction Light and Power Co. Ltd.* [second Phase] 1970: 3) These obligations deal with such terrible crimes as genocide, crimes against humanity, war crimes, racial and gender discrimination: no nation may transgress these rules or may sign on to any treaty that would permit any of these crimes. Even if, *per absurdum*, any of the rulers of a state does not believe in the justice of these rules, and despite the fact there are actions that are forbidden, nevertheless, they are bound not to transgress these norms (Westra 2007: 37).

It would be highly desirable if environmental harms or, as I have termed them, forms of "ecoviolence" (Westra 2004) were to be counted among the gravest crimes. My own move from philosophy to jurisprudence was motivated, for the most part, by my desire to work to establish the connection between environmental/ecological rights and other human rights. The Honorable Christian Weeramantry, former Judge of the International Court of Justice, puts it best in his well known "dissenting opinion" in the *Gabcikovo-Nagymaros* case:

> "Environmental rights are human rights. Treaties that affect human rights cannot be applied in such a manner to constitute a denial of human rights, as understood at the time of their application. A Country cannot endorse activities that are a violation of human rights by the standard of their time, merely because they are take under a Treaty which dates back to a period when such action was not a violation of human rights." (*Gabcikovo-Nagymaros Case* 1997)

Today, these are the only norms based on solid principle. Some other related Conventions share status, *The Convention on Genocide*, the *Convention Against Torture and Cruel and Dehumanizing Treatment of Punishment* are examples, although some powerful Western countries as we all know, are not observing them at this time. Instead, they employ well-placed government lawyers to treat these universally binding treaties, as they would tax laws, attempting to find or even manufacture loopholes (Cole 2006: 627).

That said, questions arise about why even the non-derogable, absolute norms that would ensure the protection of human rights against the most heinous crimes, are not fully respected everywhere, as they represent international law. We need to compare the aims of the nation that join together, thus forming the international community, with what Aristotle considered the state's right aim, the right laws, and the right citizens to share in its governance.

The Aristotelian Difference: *Ergon*, *Eudaimonia* and Community over Individuals and Justice

Much of Aristotle's work in the *Nicomachean Ethics* is devoted to an analysis of what it means to live a "good life" and to achieve *eudaimonia*. The latter is a difficult term to translate: its meaning is not simply "happiness", as it is often translated. Happiness is a psychological state, quite different from the activity of *eudaimonia* (Nussbaum 1986). Translations such as "living a good life for a human being" (Nussbaum

1986: 6) or "human flourishing", (Nussbaum 1986: 6; Cooper 1975) both understood as "living well and doing well" are far more accurate. The reason why that is so, is precisely the reason we are reaching back to Aristotle's thought here: *eudaimonia* only makes sense in relation to the proper understanding of what man does and what man is, his *ergon* ("function") and his excellence. In an earlier section we noted how dangerous it is to propose a public policy that was detached from a defensible, logically coherent "public good," as Plato argued in the *Republic*. Further, since, as Plato also argued, the *polis* is only as good as its citizens, the "good" has to be consonant with man's natural excellence in order to be logically and morally defensible, that is, in order to be that "for the sake of which" everything must be done.

The function argument

Aristotle's "function" argument starts, as is his custom, from observations about the natural world. Everything has a natural excellence, and all that is done or happens naturally is also for the sake of something. All natural species pursue the eternity of being in the way available to them, that is, not individually and separately but through their species' reproduction. The physical "function" that supports this quest for "eternity" is shared with all life, from plants to nonhuman animals. In the well-known discussion in the *De Anima* (Aristotle, *On the Soul*), Aristotle acknowledges our link to all other life, even plants. But what we share with plants is simply being alive; in contrast, we share motion and sentience with nonhuman animals: in both cases their respective function lies in the specific capacities and excellences of each kind.

Therefore, our function and excellence cannot lie in anything that we share with other species; it must lie in that which is particularly ours – our rational nature, shared by neither plants nor other animals, for Aristotle. Hence our function embodies our "difference" and indicates the basis for our excellence, and Aristotle's argument provides the tie between the "good", the "good life", or *eudaimonia* – and our own nature: the good is not the choice of the majority, as such, nor one preference among others. The real good for humans is the only choice that is at once logical, practical, moral, and natural. It represents the only way that "flourishing" is meaningful for humans, our most proper function. Aristotle concludes the *ergon* argument as follows:

> "The human good turns out to be activity of soul in accordance with virtue, and if there is more than one virtue, in accordance with the best and most perfect. (Aristotle, *Nicomachean Ethics*, 1098a 16–18)"

The good – that is, not one among many – is the pursuit of an activity in accordance with virtue; hence, given our nature, "the virtuous activity equated with the human good in the function argument is virtuous activity of the reasoning part of the soul" (Kraut 1989). In contrast with J. L. Ackrill, who believes that virtuous activity of the moral kind that is the best (Ackrill 1972: 17–25), Kraut believes that the good "consists in intellectual activity and nothing else" (Kraut 1989). This apparent conflict may not be as stark as it appears, that is, either intellectual, *or* moral value. Kraut suggests that, in the *Nicomachean Ethics*,

> "Aristotle makes a distinction between perfect happiness and a secondary form of happiness. He says that the former consists in contemplation; but that by exercising the practical virtues, the politician can also be happy." (Kraut 1989)

From the point of view of the environment and the ethics of integrity, the "lesser" sort of happiness provides the most appropriate emphasis. When life and health are present, contemplation is "self-sufficient," according to Aristotle; in fact, that is what makes it superior to the "other" *eudaimonia*, that is, the one that needs external circumstances and other goods to flourish. Hence, minimally, we can argue that life and health are primary in Aristotle, as they are in the principle of integrity (Westra, 1994), although they are not argued for; their primacy is implicit. But the main point that can be drawn from Aristotle's argument is the necessary connection between human nature and the good.

The difference between Aristotle and modern thinkers like Rawls, for instance, is that for Aristotle not only must the good must be universalizable and logically defensible, but that the existence of a universal human nature is beyond question; it represents the basis and justification of the "good" itself.

The meaning and role of *eudaimonia*

If human flourishing means moral and intellectual activity in accordance with reason, and if excellence in this activity represents the highest goal in human life, then we must inquire into he prerequisites (or other "goods") necessary for the attainment of that goal. Martha Nussbaum argues that the good life is extremely "vulnerable" to disaster, and hence other circumstances beyond our moral character may be concomitant with or instrumental to *eudaimonia*, or an intrinsic part of it.

In the previous section we saw that Aristotle relies on a twofold interpretation of *eudaimonia*, as the term encompasses the highest intellectual and moral virtues, although the former is primary, and the latter secondary:

> "If happiness is activity in accordance with virtue it is reasonable that it should be in accordance with the highest virtue; and this will be that of the best thing in us …" (Aristotle, *Nicomachean Ethics*, X, 7, 1117a 12–14)

Aristotle adds "That this activity is contemplative we have already said" (*Nicomachean Ethics*, X, 7, 1177a 17–18). Why is this so? He proposes a twofold answer: first, it is the best (rational) activity, directed at the best objects knowable; second, it is "most continuous". But there is another aspect of this primary form of *eudaimonia* that cannot be ignored, as it is equally important: the contemplative use of reason is self-sufficient, hence less vulnerable to both negative chance happenings and to the lack of other goods. In contrast, Aristotle himself recognizes that this sort of life, devoted purely to the philosophical contemplation of truth, "would be too high for men" as it is the exercise of the "divine" in them, that is, this way of life would allow a part (though admittedly the best one) to take over, in a sense, the entire human capacity for activity.

In contrast the "secondary" form of happiness, is entirely suitable to humans, and, as Aristotle points out, it is here that we find the need for specific circumstances and additional "goods" beyond one's life and moral character.

Nussbaum argues that activity is required for the secondary form of *eudaimonia* as well, as a state of character is not sufficient for it. In that case, however the activity must not be "impeded" (Nussbaum 1986: 343–345). Aristotle says:

> "No activity (*energeia*) is complete if it is impeded; but *eudaimonia* is something complete. For the *eudaimon* person needs the goods of the body and external goods, and goods of luck in addition, so that his activities should not be impeded." (*Nicomachean Ethics*, X, 7, 1153b 16)

Examples of such "impediments" are lack of a good birth, "good looks", children or friends. In that case, we can start by noting the obvious – that is, that life and unimpaired rationality and health are necessary preconditions of "living well" – and then add a few candidates to the broad category of possible "impediments". For instance, environmental conditions that render one too weak to interact with one's friends or, through climate changes and powerful infectious diseases, make one ill and, thus, contagious to such friends; or perhaps the sort of environmental "hormone mimics" (Colborn 1996) discussed in earlier chapters might render one incapable of fulfilling one's human destiny because of the loss of normal reproductive capacities (see also the discussion of this question in Chapter 10).

Such contemporary "impediments" are far more than difficulties and possible reductions of our capacity for full achievement of *eudaimonia*. They constitute instead extreme threats to the very aim of our lives, that of achieving our own excellence. Today we tend to view such problems simply as threats to human health or physical well-being, or even as part of the modern human condition. But, insofar as both health and physical well-being are intrinsic components of the natural capacities of humans, Aristotle would view anything impeding such capacities as far more significant than that. Anyone who interferes with the capacities of a man is guilty of preventing him from pursuing his natural goal, that "for the sake of which" all his activity unfolds. The "impeder" would take away, essentially, his reason for living and the very role he is intended to play within the universe. It is a grave harm that attacks at the same time the universal, essential nature of man, and through it, his final goal. In principle, and *prima facie*, an attack on an individual's microintegrity is at the same time an attack on the function the individual performs within her habitat, and hence it is also an attack on the integrity of the systemic whole.

What about other relational goods? How many or how much is necessary for *eudaimonia*? Aristotle simply says that we must be "sufficiently equipped with other goods" (Kraut 1989; Nussbaum 1986: 343–372). This point deserves careful attention. While other external goods are absolutely necessary, it is not true that the more of these we acquire, the better our virtue, or the closer our achievement of *eudaimonia*. Kraut explains:

> "... we should wish for that level of fortune that enhances our virtuous activity. More generally, the right level for any good other than happiness is the level at which it best contributes to one of the ultimate ends in which happiness consists." (Kraut 1989: 159)

As in our examination of the good in Plato, Aristotle uses his conception of *eudaimonia* as the criterion for the need for external goods in general, but also, for the quantity of external goods required

> "to determine the extent to which music, drawing, gymnastics or any other subject should be studied, politics must look to the ultimate end of human life." (Kraut 1989)

Hence, external goods have a limit (*horos*), as Aristotle is applying the doctrine of the mean to both this topic and to the understanding of virtue. The mean must be sought, between the extremes of a complete lack of external goods and an excess, or a limitless amount, which would be equally wrong (Barker 1973).

The appropriate amount of goods is to be found in the mean between too little and too much; further, it is not defined in terms of preferences but must be limited according to the two criteria we have encountered: the criterion of our rational human nature of its function, and that of the ultimate good, the good of human life. In addition, all other goods can be understood as both intrinsic and instrumental goods in regard to the final goal and the two criteria. Not only are external goods necessary prerequisites of the good; they are also, without doubt, components of it. This interpretation is based on the "inclusivist" conception of happiness in Aristotle (Rist 1989) and, Kraut observes, is not accepted by some commentators (Kraut 1989).

Community over individuals and justice

> "One can even say that those who are most responsible for the well-being of the community bear the clearest resemblance among political leaders, to the unmoved mover. For the divine is, in its own way, a cause of the goodness of the universe and the more responsible one is for the goodness of one's community, the more closely one approximates this condition." (Kraut 1989: 100; Aristotle, *Metaphysics*: 4–10; Barker 1976: 11–25)

Political power and office should be shared equally among equals, but it is certain virtues, a certain excellence, that makes specifically every one of these citizens "equals" who must share power for the greater benefit of the community, hence, who must govern in turn:

> "Justice, construed as a specific virtue, required equal treatment (*Nicomachean Ethics*, V, 1–2), and so when citizens are equally deserving of political power, they must take turns governing and being governed." (Barker 1973: 23–27; Kraut 1989: 98; Aristotle, *Metaphysics*: 11–25)

Hence the first vast difference between Aristotelian governance and today's (mostly) neo-liberal "democracies" emerges clearly. Citizens should participate in the governance of the state for the common good, but only when they are deserving: each person must be habituated to virtue, and be determined to seek it for the whole community of fellow citizens.

Today, in Western "democracies", the holders of public office are determined by a combination of wealth, his own, and that of whoever else can be convinced to add to it, but also by the untiring work of the "orators", that is the spin doctors, the speech writers, the political advisors, who work to ensure the "win" of their candidate. I want to emphasize that I believe the European Union is somewhat better than North American and other countries in this regard, because of the predominance of the transnational aspect that can be found in its governance, and I have so argued in my second doctoral thesis (Westra 2004).

For the most part, in democratic political campaigns, the high-sounding words, the commitment to the "people" (unspecified), the assurances of high moral standards when elected, all come from paid promoters, speech-writers, and others. Essentially then, those who rule today do not need to develop their highest intellectual virtues, contemplating (practicing) the theoretical disciplines, such as mathematics, physics and – most of all – theology (Aristotle, *Metaphysics*: 18–19), or, as Aristotle terms it, first philosophy.

The most important of these is the study of the first cause, but the study of nature (physics) is only somewhat lower than that and although all of these are superior to the choice of practical sciences, the latter are also necessary in order to pursue the political life (Kraut 1989: 74–75). The point is that not everyone is suited to pursue the ideal of happiness involving the highest virtue and the highest subjects of study. Doing one's best in "the realm of theory or the realm of practice" (Kraut 1989: 75) is what is required. The latter, intended as one's choice to pursue the political life permits one to achieve happiness insofar as one can.

But the pursuit of happiness through moral or intellectual activity is basic. And only by making the interests of the community primary, the political ruler is also seeking his own happiness, but, first and foremost, that of his community. We need to keep in mind however, that in order to even attempt a comparison, the community today, must include the community of humankind, wherever they might reside, as a ruler who ignore global implications of his actions in order to promote his own state's advantage, at the expense of any and all others, cannot be moral or just.

Aristotle Reconsidered

> "Throughout Aristotle's writings, and with particular frequency in the biological works, we find the rule that 'Nature does nothing [in] vain', or 'to no purpose'. In a well-known passage of the *De Caelo* [*On the Heavens*]... there is a variation on this: God and nature, Aristotle tells us,

> do nothing in vain (Aristotle, *De Caelo*) ... That is nature, not only in the natures of individual things but an immanent god (not dissimilar from that of *On Philosophy*), must work in a law-like fashion." (Rist 1989: 125)

We need to pause here and review in our mind the three examples of grave environmental harms we have chosen to address. All three demonstrate the hazards to which we are exposed, based on four fundamental *lacunae* in globalized governance:

1) the lack of principles that recognize the basic importance of human nature;
2) the lack of understanding of the interconnectedness and the teleology in the laws or nature, and their relation to human nature itself;
3) the lack of emphasis on limit (*horos*) as a basic requirement of a human life worth living; and
4) a better understanding of true "democracy" (since although it is considered the ideal form of governance today, its weaknesses and flaws stand out in contrast with Aristotle's understanding of its reach).

These four "missing" aspect of governance are also additional to the basic necessity for international or transnational legal instruments, that is, for cosmopolitanism that supports universal principles. We still have laws for individual states, but the problems we are addressing are global in scope, and no state can solve them on its own. Municipal laws must internalize international legal rules within their constitutions, in order for these principles to be operational within the state. But the formulation of these instruments must first be international, and only secondarily, they must be incorporated in municipal law.

All four "lacks" stem from the total lack of appropriate limits to growth and to enrichment, as goals. This fundamental lack results for instance, in climate change, and the ferocious resistance of some of the greatest polluters to the imposition of any "limit" whatever, as this is viewed as a limit to economic growth and corporate enrichment, tells the story. The history of the US resistance for decades to even admitting the existence of climate change, let alone to the acceptance of limits, is a clear example (Westra 2007: chapter 8).

Lack of acceptance of the interconnectedness of natural systems is equally vital to the problems. It is in conflict with reality, and with the pursuit of growth/enrichment that is the ultimate goal of most of our so-called "democratic" governance. Corporate crime, including crimes against humanity and even genocide, runs apace with so-called "development", especially in the case of extractive and mining industries. The latter's operations often spell the demise of the indigenous and local communities who depend on the services of the same ecosystems that are instead exploited for gain, with little or no regard for the consequences (Westra, 2007: chapter 4).

Most of all, no system of governance today fully accepts the existence of definable "human nature", let alone one with the Aristotelian connotations of moral and intellectual excellence. If it is considered at all, today human nature is the subject of self-serving, individual and non-universalizable, except insofar as "choosing" and "consuming" appear to be its defining characteristics, in an increasingly homogenized/globalized world. Unrestrained, unprincipled and limitless choice, therefore, represents the ideal, and the "democratically" chosen leadership that follows this analysis, clearly does not include any of these aspects of justice that Aristotle recommends.

Finally, the lack of both "limit" and principles combines with yet another form of "greed": the quest for power. Hence, not only globally engendered environmental disasters, and the thrust for inappropriate enrichment, combine to render the life of impoverished populations impossibly harsh (especially, but not exclusively in developing countries), but the quest for power is an additional global disaster on its own. From the illegal "annexations" of Palestinian territories, to the corporate-funded "ethnic cleansing" in Sudan and other areas in sub-Saharan Africa, to the forcible transfer of populations from Chagos to the "tar sands" operations in Alberta, Canada, in the territories of the Lubicon Cree, the quest for power and unsustainable growth is widespread.

A serious reconsideration of Aristotelian doctrine in all its aspects, well beyond the brief and incomplete discussion of this chapter, appears to be as necessary to avert global tragedy; it is, sadly, unlikely to happen today.

References

Ackrill, J. L., 1972. "Aristotle on 'Good' and the Categories," in M. Stern et al. eds. *Islamic Philosophy and the Classical Tradition: Essays Presented to Richard Walzer*. Oxford, UK.

Aristotle, *Metaphysics* [*Metaphysica*], tr. Hugh Tredennick, 1933–1935, 2 vols, Loeb Classical Library, Harvard University Press, Cambridge, MA.

—. *Nicomachean Ethics* [*Ethica Nicomachea*], ed. and tr. McKeon, 1992. Modern Library, Random House, Inc., New York

—. *On the Heavens* [*De Caelo*], tr. W. K. C. Guthrie, 1939. Loeb Classical Library, Harvard University Press, Cambridge, MA

—. *On the Soul* [*De Anima*], ed. and tr. McKeon, 1992. Modern Library, Random House, New York

—. *Parts of Animals* [*De Partibus Animalium*], tr. A. L. Peck and E. S. Forster, 1945. Loeb Classical Library, Harvard University Press, Cambridge, MA

—. *Politics* [*Politica*], tr. H. Rackham, 1932. Loeb Classical Library, Harvard University Press, Cambridge, MA

Barcelona Traction Light and Power Co. Ltd. [second phase], Judgment ICJ Reports, 1970.

Barker, Ernest, 1973. *The Politics of Aristotle*, ed. And tr., Oxford University Press, Oxford, UK.

Colborn, Theo, 1996. "Plenary Address" to IAGLR (International Association of Great Lakes Researchers), May 27, Erindale College, Toronto, Ontario.

Colborn, Theo, Dumanoski, Dianne, Myers, John Peterson, *Our Stolen Future*, 1996. Dutton, Penguin Books USA Inc., New York, NY.

Cole, David, 2006. "The idea of Humanity: Human Rights and Immigrants' Rights", *Columbia Human Rights Law Review* 37.

Cooper, John, 1975. *Reason and Human Good in Aristotle*, Cambridge University Press, Cambridge, MA.

Dupuy, Rene Jean, 1991. "Humanity and the Environment", *Colorado Journal of International Environmental Law and Policy* 2.

Gabcikovo-Nagymaros Case [1997] I.C.J. Rep.7, Weeramantry, C.J., Dissenting Opinion.

Gostin, Lawrence O., 2004. "Law and Ethics in Population Health", *Australia and New Zealand Journal of Public Health*, Vol.28, No.1.

—. 2008, *Public Health Law*, 2nd ed., University of California Press, Berkeley, CA.

Grandjean and Landrigan, 2006. "Developmental Neurotoxicity of Industrial Chemicals", *The Lancet*, 8 November.

Kindred, H. M. Michelson, K., Provost, R., McDonald, T. L., de Mistral, A., and Williams, S.A., 2000. *International Law*, 6th ed., Edmond Montgomery Publications, Ottawa, Canada.

Kraut, Richard, 1989. *Aristotle on the Human Good*, Princeton University Press, Princeton, NJ.

McMichael, A J., 1995. "The Health of Persons, Populations and Planets: Epidemiology Comes Full Circle", *Epidemiology and Society*, Epidemiology Resources, London, UK.

Nussbaum, Martha C., 1986. *The Fragility of Goodness* Cambridge University Press, Cambridge, MA.

Owens, Fr. Joseph, 1959. *A History of Ancient Western Philosophy*, Appleton-Century-Crofy, Inc., New York, 343; see EN, II, 1, 1103b.

Patz, Jonathan A., 2005. "Impact of Regional Climate Change in Human Health", *Nature*, vol. 384.

Ragazzi, Maurizio, 1997. *The Concept of International Obligations Erga Omnes,* Oxford University Press, Oxford, UK.

Rist, J. 1989. *The Mind of Aristotle*, University of Toronto Press, Toronto.

South West Africa Case [second phase], Dissenting Opinion, [1996] I.C.J. Rep. 6 at 296.

Taylor, Allyn L., 1992. "Making the World Health Organization Work: A Legal Framework for Universal Access to the conditions for Health", *American Journal of Law and Medicine*, vol.18.

Westra, Laura, 1994. *An Environmental Proposal for Ethics: The Principle of Integrity*, Rowman and Littlefield, Lanham, MD.

—. 1998. *Living in Integrity*, Rowman and Littlefield, Lanham, MD.

—. 2004. *Ecoviolence and the Law*, Transnational Publishers Inc., Ardsley, NY.

—. 2006. *Environmental Justice and the Rights of Unborn and Future Generations*, Earthscan, London, UK.

—. 2007. *Environmental Justice and the Rights of Indigenous Peoples*, Earthscan, London, UK.

CHAPTER TWO

CONTESTING DEMOCRACY

J. RONALD ENGEL

Contesting Democracy

Democracy is almost universally accepted today as the norm of good and legitimate governance. This means that every issue we face – from international relations, to economic globalization, to how we govern each of our domestic societies, our families and our interpersonal relations, to the issues that join those of us who come together each year in the Global Ecological Integrity Group, the issues of sustaining the planet's ecological integrity and the health of its peoples – depends upon how we think about and institutionalize the ideal of democracy and the principles and purposes it entails.

This situation first became apparent in the years immediately following WWII and the founding of the United Nations. In 1946 UNESCO launched two inquiries on the "differences which separate peoples in their views of right, liberty, democracy, and similar moral and political concepts" (McKeon 1951: vi). The first of these studies was published in 1949 in the volume *Human Rights: Comments and Interpretations* (UNESCO 1949); the second was published in 1951 in the volume *Democracy in an Age of Tensions*. The latter study concluded:

> "For the first time in history of the world, no doctrines are advanced as anti-democratic. The accusation of antidemocratic action or attitude is frequently directed against others, but practical politicians and political theorists agree in stressing the democratic element in the institutions they defend and the theories they advocate. This acceptance of democracy as the highest form of political or social organization is the sign of a basic agreement in the ultimate aims of modern social and political institutions." (McKeon 1951: 522)

Forty years later, concluding a study of the history of democracy from 508 BC to AD 1993, the Oxford historian John Dunn confirmed this judgment:

> "Why is democracy today the overwhelmingly dominant, and increasingly the well-nigh exclusive, claimant to set the standard for legitimate political authority? It takes some imaginative concentration to register quite what an extraordinary fact this really is." (Dunn 1993: 239)

In 2005 Princeton political philosopher Patrick Deneen concurred, declaring "democracy is the only regime most living humans now deem worthy of serious consideration, exploration, clarification, articulation, exportation, importation, and finally, faith" (Deneen 2005: 1).

Yet exactly in what this unprecedented agreement consists remains heavily contested. In spite of the millions of pages devoted to its interpretation and defense, and the millions of persons who have sacrificed their lives for it, we have yet to reach a common understanding of democratic governance adequate to the issues of our age or to the majesty and tragedy of the history of humanity on this planet.

I would like to propose that our way through this situation is to recognize that "democracy" is a word that denotes a complex "tradition" of human thought, action and experience, and that the kind of debates in which we are engaged are precisely the kind of debates we would expect of a tradition that is so vital and influential in human affairs.

Alisdair MacIntyre writes:

> "Traditions, when vital, embody continuities of conflict … A living tradition … is an historically extended, socially embodied argument, and an argument precisely in part about the goods which constitute that tradition." (MacIntyre 1981: 222)

This suggests that the tradition of democracy is an historically extended, socially embodied argument about the goods of self-government and the grounds for our belief that these goods are true goods.

As persons concerned for transforming contemporary norms and practices of democratic self-governance in ways that will sustain nature as well as humankind, it is imperative that we engage this argument directly. Our discussion this year at the Global Ecological Integrity Group will hopefully make a contribution to furthering this transformation.

Thin Versus Thick Interpretations of the Democratic Ideal

In my judgment, the primary dividing line in contemporary debates about democracy is the line that separates what I am calling "thin" from "thick" interpretations of the democratic ideal. I distinguish thin from thick interpretations on the basis of the historical depth and conceptual breadth they accord the tradition of democracy, and how they justify the ideal of democracy itself. This usage is different from that of Michael Walzer who identifies thin interpretations of morality with universal and abstract principles and thick interpretations with culturally rich locally established symbolic systems. By my definition, "thin" democracy is culturally and conceptually narrow and bare, while "thick" democracy is culturally and conceptually broad and complex (Walzer 1994). I hold that what Clifford Geertz describes as a "thick description" of culture can apply to the democratic tradition in its unity and diversity (Geertz 1973).

By far the most widely accepted interpretation of democracy today is what I am calling a "thin" interpretation – by which I include what is variously called "procedural democracy," "liberal democracy," "representative democracy," or simply the "democratic process." This interpretation holds that the democratic tradition is a modern secular affair. While it may have had precedents in the direct democracy of the Athenian polis and elsewhere, democracy as we know and practice it is mainly a phenomenon of the post-Westphalian era, and largely the invention of the Enlightenment. In this view, the democratic ideal is a way of bringing free and equal but competitive individuals and groups (or "interests") into cooperative and stable relationships by such devices as constitutionally protected civil and political rights, limited government, basic fairness in the distribution of social goods, and opportunities for citizen participation in the formation of public policy through membership in the voluntary associations of civil society and voting in electoral politics.

The warrant, or justification, for public moral claims does not lie in any independent pre-political ground, such as a common human nature, or shared comprehensive vision of the common good, but in the fact that they receive our mutual consent. Thus the institutions and processes by which the opinion of citizens is mobilized, and the deliberations of political and other elites through which public policy is determined, constitute the core of democratic governance. In the last analysis, popular sovereignty, the sovereignty of free and equal citizens, each following his or her private moral beliefs, and the collective will of the majority of such free and equal individuals as it is expressed through the election of their political representatives, rules all.

Thick interpretations of democracy, on the other hand, are much less prevalent yet may be found in the writings and speeches of literary, political and philosophical figures throughout the world. In this view, democracy is more than a political concept and more than a modern or European affair. The democratic tradition is a large tradition, the bearer of wisdom applicable to the governance of all the relationships of life, an idea evident, for example, in the usage of the term "democratic faith." Although democracy has flowered in the modern era, it has deep roots in the axial civilizations, Greek, Hebrew, Buddhist, Confucian, Hindu, Christian and Muslim, as well as many indigenous and native traditions.

Like all great traditions, democracy in its thick interpretation bears a multiplicity of goods, including the virtues of what is often called the "democratic way of life" – respect for others and respect for self, non-violence and mutual persuasion, critical reason and empirical inquiry, trust and hospitality for the stranger, and compassion for the suffering of others; as well as the distinctive political goods associated with the rule of law based on equal justice, reason and consent; universal political, civil, economic, social and cultural rights, conceived as essential preconditions for the cultivation of our common human capacities for physical, intellectual and moral development, and grounded in the right to life and security of one's person, property and community; a rich heritage of analogies between the democratic principles we employ to live together and the principles we employ in our treatment of the rest of nature, including a long debate over the rights of nature; and overall, a celebration of active citizenship that serves the common good.

Democracy, it is so held, bears a narrative which serves as the meta-narrative in terms of which all other narratives are deemed to be ethically judged; a narrative that recognizes human individuality as well as human interdependence, our capacities for moral greatness as well as unspeakable cruelty and evil; our inherent need to be free, yet also our vulnerability, frailty, and need for one another; a narrative that consequently finds in the struggle to assure the conditions for a life of free and equal dignity for all persons living in community with one another and the Earth as the ultimate end and meaning of human history.

Far from being free-standing, or historically contingent, this central driving purpose or ideal of democratic governance is true, authentic, real, because it is a universal good, a universal potentiality and – to a greater or lesser degree – a universal experience, grounded in human nature and in the creativity that drives the cosmos, however fallible and incomplete any particular claim to its truth or expression may be. Surely only some such confidence – hope or faith – can justify the covenantal commitments and sacrifices of those who have devoted their lives to democracy, such as those who signed Charter08 in China, or Aung Sung Suu Kyi on trial in Burma.

The long tradition of natural law and natural rights, and their evolving metaphysical foundations, extending in the West from Aristotle and the Stoics to Thomas Aquinas, Calvin, Grotius, and in substantial revision, Kant, and (in American thought) through Jefferson, Lincoln and Martin Luther King, Jr., has characteristically provided the philosophical and ethical grounding for thick interpretations of democracy. The moral ends of human action, including our most fundamental rights and obligations, are grounded in the constitution of nature itself, so that contrary to the so-called naturalistic fallacy, it is possible for a rigorous rational analysis of human experience to derive what "ought" to be from what "is," or as Holmes Rolston III argues, discover them together (Rolston 1986: 12–29).

Alfred North Whitehead well described the essence of natural law when he wrote:

> "… there is a large consensus, on the part of those who have rationalized their outlook, in favour of the concept of a rightness in things, partially conformed to and partially disregarded. So far as there is conscious determination of actions, the attainment of this conformity is an ultimate premise by reference to which our choice of immediate ends is criticised and swayed." (Whitehead 1926: 65)

He then went on to provide a natural law account of democracy in the following terms:

> "The basis of democracy is the common fact of value-experience, as constituting the essential nature of each pulsation of actuality. Everything has some value for itself, for others, and for the whole. This characterizes the meaning of actuality. By reason of this character, constituting reality, the conception of morals arises. We have no right to deface the value-experience which is the very essence of the universe." (Whitehead 1938: 151)

Franklin I. Gamwell at the University of Chicago, building on Whitehead's metaphysics and the social philosophy of John Dewey, has argued for a teleological interpretation of the comprehensive democratic moral ideal as "maximal individuality-in-association (maximal happiness) for all" (Gamwell 1984: 139). He is by no means alone in setting forth a naturalistic metaphysical foundation for a thick interpretation of the democratic tradition (Sturm 1988; Rockefeller 1991).

John Rawls's 1994 book, *Political Liberalism*, has become the focal point for debates in North America between advocates of thin democracy, such as Benjamin Barber, Ronald Dworkin, and Richard Rorty, and refutations of this position by communitarian and civic-republican philosophers such as Charles Taylor and Michael Sandel who seek to revive the Aristotelian notion that human beings are by nature social and political animals and therefore share a common moral nature whose fulfillment is the good of human life (Rawls 1993). From the standpoint of those of us who advocate a thick democratic perspective, many of the flaws inherent in the contemporary philosophical defense of thin democracy can be traced to the contractarian philosophies of Thomas Hobbes and John Locke. These flaws include a flawed ontology of an imaginary pre-political "state of nature," and the natural right to satisfy egoistic impulses this state allegedly bestowed.

Nonetheless, we need to be clear that government by the consent of the governed, government of, for and by the people, government through uncensured political debate with a peaceful transfer of power through free and fair elections, is no small achievement in the course of human affairs, as the current situation in Iran and elsewhere shows. Indeed, I would want to argue that, if anything, the "liberal" principle in thin democracy – cultivation of the human personality through freedom of thought and expression – deserves recognition in international law as a universal non-derogable human right! The problem with thin democracy is not with its principle of free and equal consent and its powerful narratives of resistance to tyranny, but that it does not provide adequate grounds for its principle, or a story large enough to do justice to its narrative; furthermore, its principle and narrative, while essential components of the democratic inheritance, are not a complete account of the moral and spiritual requirements of human self-governance.

In Tribute to Laura Westra

We must pay tribute to Laura Westra for her distinguished contributions to this discussion. Her contest with thin democracy is long standing and well-recognized in the literature. The arguments she presents for Aristotelian and Thomistic natural law are basic to the arguments she advances throughout her seminal work on the Principle of Integrity and its implications for environmental rights and strong ecological sustainability (Westra 2004: 29–50). She takes her place in a long conversation, beginning among Greek philosophers themselves, regarding the meaning of the Greek democratic experience for political theory, a conversation that has become quite vigorous in recent years, and includes Martha Nussbaum's hybrid liberal version of Aristotelian social democracy, and the spirited debate between Peter Euben and Benjamin Barber on the meaning of the Socratic dialectic for democracy in the recent volume *Demokratia: A Conversation on Democracies, Ancient and Modern* (Nussbaum 1990: 203–253; Ober and Hedrick 1996).

Although Westra has drawn substantially from Greek philosophy for the field of environmental ethics, she has not, to my knowledge, engaged the scholarship concerned for resources within Greek and Thomistic natural law philosophy for a thick interpretation of democracy (Westra 1997: 83–98). Instead, Westra's primary interlocutors have been North American environmental ethicists working in the pragmatic tradition, such as Bryan Norton, who cannot abide her insistence that democratic "choice" be limited by moral constraints – that there is a principle of ethical evaluation that judges the choices we make, and that establishes certain principles of rights and duties, such as those pertaining to the integrity of person and biosphere, that are non-derogable (Norton 2002: 11–32).

I share Westra's criticisms of thin democracy, especially her charge that by rejecting the notion that democratic societies can or should exercise collective self-government for an explicit and shared idea of the universal common good it is ultimately a form of ethical relativism. I agree that thin democracy provides only very indirect or weakgrounds for ethical critique of what in fact does function as the collective purpose of modern democratic societies – maximizing the economic means whereby individuals may freely and fairly pursue their private interests.

But I disagree with her decision to accept the thin interpretation of democracy as synonymous with the democratic ideal itself! I would like to argue that this is a mistake, that she belongs to the democratic tradition in its largest and thickest sense; and that moreover, her most distinctive contribution to environmental and legal philosophy, the Principle of Integrity, and her creative use of Aristotelian thought to justify its ethical priority in international law and governance, including our understanding of human rights, are essential components in the transformation of the democratic tradition and ideal that many of us here are trying to advance.

In order to sketch the skeleton of such an argument in the space available I want to first suggest how the classical natural law and rights tradition, in which Westra broadly stands, has been split in the post-WWII period between arguments for thick democratic principles of human dignity and arguments for strong ecological principles of nature's dignity; and then, in conclusion, propose that by affirming the common grounding and historical roots of these two streams of thought, and thinking through their implications for the unprecedented social and ecological crises we face today, we will be able to displace the prevailing thin interpretations of democracy and articulate a vision of the democratic ideal worthy of our loyalty.

Human Dignity and the Dignity of Nature

Over the centuries, in a variety of cultural contexts throughout the world, by way of philosophy, and by way of religion, natural law has inspired thick interpretations of the democratic ideal. The central theme has been the affirmation of the ontological reality and moral ultimacy of human dignity in the face of crises in human governance precipitated by violent manifestations of human injustice and domination. Of special relevance for our purposes was the interplay between history and natural law ethics which occurred with the founding of the United Nations and our contemporary regime of international law in the years immediately following World War II.

In 1946, while the judges of the Nuremburg Military Tribunal were pondering charges of crimes against humanity, half way around the world a group of political and legal philosophers were pursuing what was likely the most searching investigation in modern times of the Aristotelian and Thomistic natural law basis for human dignity, democracy and universal human rights. The prevailing sentiment among the group of European political philosophers who immigrated to the United States during the war years was alarm at the prevailing ethical relativism of American democratic thought, evident in the philosophies of scientific naturalism and legal positivism sweeping the United States in those years, and the neglect by American intellectuals of the classical rational and ethical standards of their democratic regime (Purcell: 1973). A number of these philosophers, among them Leo Strauss and the Catholic Thomistic scholars, Jacques Maritain and Yves Simon, were invited by Robert Maynard Hutchins, president of the University of Chicago, to join a stellar group of American colleagues at Chicago who shared their Aristotelian outlook and concerns, among them, Mortimer Adler, Scott Buchanan, Richard McKeon, John Courtney Murray, and Hutchins himself.

Although this circle had limited influence on American political philosophy as a whole, it played a critical role in philosophic debates on the meaning of democracy and human rights in the international discussions of the time. One reason they were able to have this influence is that the two formative UNESCO inquiries on human rights and democracy mentioned earlier were coordinated by Richard McKeon from his base at the University of Chicago. Another reason is that Jacques Maritain was a figure of international stature who advised the drafting committee of the International Declaration of Human Rights, and who effectively advocated, on natural law grounds, in political and intellectual circles throughout the world as well as within the Catholic Church, what he called the "Democratic Charter" – a proposal regarding the fundamental ethical principles which he believed should be publicly recognized as the comprehensive framework of the "law of nations."

In a series of lectures at the University of Chicago in 1951, Maritain shared his comprehensive democratic vision in terms very close to my earlier description of the central tenets of "thick democracy". Prefacing his remarks with the assertion that

> "it is supremely important to the common good that the practical assertions which make up the [democratic charter] be true in themselves. But the democratic State does not judge of that truth; it is born out of that truth …",

he concluded by underscoring the

> "obligations of each person toward the common good of the body politic and obligations of each nation toward the common good of civilized society, and the necessity of becoming aware of the unity of the world and the existence of a community of peoples." (Maritain 1951: 112–113)

Unfortunately, human rights and duties in relationship to the rest of nature were not part of Maritain's "Democratic Charter".

Yet not only man's domination of man, but man's domination of nature has spawned the response of natural law ethics over the course of human history. During exactly the same years Maritain and his colleagues at the University of Chicago were seeking to lay the foundations for an international regime that would be a bulwark for human dignity, only a couple hundred miles away, on an old farm on the Wisconsin River, an obscure wildlife game manager, Aldo Leopold, appalled at humanity's abuse of its natural environment, was hard at work on crafting what has since become recognized as the classic formulation of natural law environmental ethics, what he called the "land ethic". Leopold's 1948 book, *A Sand County Almanac*, marshals arguments from personal experience, scientific observation and theory, and good hard inductive and deductive reasoning for the ontological reality and moral ultimacy of natural integrity, or what Leonard Boff calls the "dignity of Earth", *dignatas terrae* (Boff 1995: 87). Consider this syllogism in which Leopold derives ought from is:

> "1. 'All ethics rest upon a single premise: that the individual is a member of a community of interdependent parts.'
> 2. We are members of the land community.
> 3. Therefore, we need to exercise the same constraints on our relations to the other members of the land community – soils, waters, plants and animals – as we do in our relations to people.
> 4. Thus the land ethic: 'A thing is right when it tends to preserve the integrity, stability, and beauty of the biotic community. It is wrong when it tends otherwise.'"(Leopold 1999: 311–313)

In 1973 Arne Naess published "The Shallow and the Deep, Long-Range Ecology Movement" in the journal *Inquiry*, setting forth an argument for "intrinsic value" as the axiological root for our duties to respect the dignity of nature (Naess 1973). In 1974, Holmes Rolston, III, published "Is There an Ecological Ethic?" and identified Leopold's ethic for what it was, a natural law argument for environmental ethics (Rolston 1986). In 1979, Hans Jonas published *The Imperative of Responsibility*, arguing on natural law grounds that only an ethic grounded in the inherent intentionality of each organism's "yes" to life could be strong enough to convince humanity to take the actions necessary to assure the planet's survival (Jonas 1985).

Leopold famously argued for "land citizenship" and the responsibility of private land owners to husband their land for the common good of human and natural communities. According to Susan Flader, he was "profoundly conscious of the American democratic tradition within which he was working and thought hard throughout his career about the meanings and implications of environmental stewardship (Flader 1999). He was also prescient in his perception that a tragic gulf separated 20th century conservation biology based on the evolutionary and ecological sciences and the philosophy, science and art of democratic citizenship and politics. While on a trip to Germany in 1935 he jotted down the following fragment:

> "One of the anomalies of modern ecology is that it is the creation of two groups each of which seems barely aware of the existence of the other. The one studies the human community as if it were a separate entity, and calls its finding sociology, economics, and history. The other studies the plant and animal community, [and] comfortably relegates the hodge-podge of politics to the 'liberal arts'. The inevitable fusion of these two lines of

thought will, perhaps, constitute the outstanding advance of the present century" (Meine 1988).

Yet he was not personally prepared, or sufficiently interested, to engage in a principled discussion of democratic theory. It is difficult to find much that is positive about democracy among post-Leopold environmental philosophers and their successors in the natural law tradition. Jonas, like Robert Heilbroner and William Ophuls, despairs of the potential of democratic societies to assume long range responsibilities for nature (Heilbroner 1980; Ophuls 1992). Naess, who took a leave from the University of Oslo to work with Richard McKeon on the Unesco study of democracy, and then went on to conduct a further study that found over 315 meanings of democracy in Western history from Athens to the 1950s, eventually throws up his hands and judges democracy to mean whatever we want it to mean (Naess et.al:1956; Rothenberg 1993:117-9). Most of the environmental philosophers who in recent years have positively addressed the relationships of democracy and environmental ethics, such as Bryan Norton, have interpreted democracy through the lens of one or more of its thin interpretations reflecting the widespread post-modern incredulity toward meta-narratives and metaphysics of any kind.

Earth Democracy

Laura Westra's greatest contribution, in my estimation, is her steady insistence that human dignity or integrity needs to be understood in relation to the dignity of nature or ecological integrity. For several decades she has been reweaving the natural law and rights of human nature and the natural law of nature as a whole, the *ius naturale*, that was severed with Grotius in the early modern period, and in the process setting new universal ethical and legal standards for human rights and the preservation of ecological integrity in the law of nations. Set against the background I have just sketched, this accomplishment marks a turning point in modern thought.

The Earth Charter is powerful testimony to the fact that thousands of people throughout the world are ready to endorse this turning point (Vilela and Roerink 2005). Its comprehensive principle of "respect and care for the community of life" is based on the ontological presupposition that "all beings are interdependent and every form of life has value regardless of its worth to human beings." Moreover, the Earth Charter not only explicitly affirms "ecological integrity" but sets forth a thick version of democracy through its principles of social and economic justice, non-violence and peace. Indeed, it was the conscious intent of the chair of the Earth Charter drafting committee, Steven Rockefeller, and myself to express through the Charter a thick vision of the democratic ideal (Rockefeller, Engel 1992: 139–172, 159–182). Toward the end of the drafting process, he and I worked hard to assure this intent was communicated by Principle 3. "Build democratic societies that are just, participatory, sustainable, and peaceful." Evidence that precisely such a vision is taking hold in the nations of the world may be found in the symbolism that informs the architecture of the new Constitutional Court of South Africa (The Constitutional Court Trust 2008). Nelson Mandela's famous pledge to sacrifice his life for democracy if circumstances required is remembered as central to the narrative that has now led to the rule of law under the protective limbs of the tree of life:

> "During my lifetime I have dedicated myself to this struggle of the African people. I have fought against white domination, and I have fought against black domination. I have cherished the ideal of a democratic and free society in which all persons live together in harmony and with equal opportunities. It is an ideal which I hope to live for and to achieve. But if needs be, it is an ideal for which I am prepared to die" (Mandela 2003).

With the foundations Westra and others have laid, we now have the unprecedented opportunity – and I believe, obligation – to make explicit a natural law argument for a truly universal ecological understanding of democracy – "Earth Democracy," let us call it, which will be at the same time a retrieval of some of the deepest historical and normative roots of the thick interpretation of democracy.

We have our work cut out for us. We have, for example, as Maritain stressed in his introduction to the 1949 UNESCO volume on "Human Rights", and as Westra reminds us today, to determine the "scale of values governing the exercise and concrete integration of ... various rights", derogable and non-derogable (Maritain 1949: 17). But most of all we have the task of demonstrating to the world that true human freedom lies not in the project of endlessly expanding the human domain by tragic attempts to manage evolution but in finding our way home to our true nature and end – our life together as democratic citizens responsible for sustaining the Earthly commonwealth.

That faith will conquer, Whitehead prophecized in *Adventure of Ideas*, "which can render clear to popular understanding some eternal greatness incarnate in the passage of temporary fact" (Whitehead 1933: 41).

References

Boff, Leonardo, 1995, *Ecology and Liberation: A New Paradigm,* Maryknoll, New York: Orbis.

The Constitutional Court Trust, 2008, *Art and Justice: the Art of the Constitutional Court of South Africa,* Parkwood, South Africa: David Krut Publishing.

Deneen, Patrick J., 2005, *Democratic Faith*, Princeton: Princeton University Press.

Dunn, John (ed.), 1993, *Democracy: the Unfinished Journey*, Oxford: Oxford University Press, 1993.

Gamwell, Franklin I., 1984, *Beyond Preference* (Chicago: University of Chicago Press.

Engel, J. Ronald, 1992, "Liberal Democracy and the Fate of the Earth" in Steven C. Rockefeller and John C. Elder, eds. *Spirit and Nature: Why the Environment Is a Religious Issue*, Boston: Beacon Press.

Flader, Susan, 1999, "Aldo Leopold and Environmental Citizenship," *Transactions of the Wisconsin Academy of Sciences, Arts and Letters* Vol. 87, 1999, 23.

Geertz, Clifford, 1973, *The Interpretation of Cultures*, New York: Basic Books.

Heilbroner, Robert, 1980, *The Human Prospect*, New York: W. W. Norton.

Jonas, Hans, 1985, *The Imperative of Responsibility: in Search of an Ethics for the Technological Age*, Chicago: University of Chicago Press.

Leopold, Aldo, 1999, *The Essential Aldo Leopold*, Curt Meine, Richard Knight, eds., Madison, Wisconsin: University of Wisconsin Press.

Mandela, Nelson, 2003, *Nelson Mandela In His Own Words*, New York: Little, Brown and Company.

Maritain, Jacques, 1949, "Introduction" to Unesco*, Human Rights: Comments and Interpretations*, New York; Columbia University Press.

—. 1951, *Man and the State*, London: Hollis and Carter.

McIntyre, Alasdair, *After Virtue: A Study in Moral Theory*, Notre Dame, Indiana: University of Notre Dame Press.

McKeon, Richard (ed.), 1951, *Democracy in a World of Tensions: A Symposium prepared by UNESCO*, Chicago: University of Chicago Press.

Meine, Curt, 1988, *Aldo Leopold: His Life and Work* (Madison, Wisconsin: University of Wisconsin Press.

Naess, Arne, 1973, "The Shallow and the Deep, Long-Range Ecology Movement," *Inquiry* Vol. 16.

Naess, Arne, Jens A. Christophersen, Kjell Kvalo, 1956, *Democracy, Ideology, and Objectivity*, Oxford: Basil Blackwell.

Norton, Bryan, 2002, "Democracy and Environmentalism: Foundations and Justifications in Environmental Policy" in Ben A. Minteer and Bob Pepperman Taylor, eds. *Democracy and the Claims of Nature*, Lanham, Maryland: Rowman and Littlefield.

Nussbaum, Martha, 1990, "Aristotelian Social Democracy" in *Liberalism and the Good,* eds. R. Bruce Douglass, Gerald M. Mara, and Henry S. Richardson, New York: Routledge.

Ober, Josiah and Charles Hedrick, eds., 1996, *Demokratia: A Conversation on Democracies, Ancient and Modern*, Princeton: Princeton University Press.

Ophuls, William, 1992, *Ecology and the Politics of Scarcity Revisited*, New York: W. H. Freeman and Co.

Purcell, Edward A., Jr., 1973, *The Crisis of Democratic Theory: Scientific Naturalism and the Problem of Value*, Lexington: University Press of Kentucky.

Rawls, John, 1993, *Political Liberalism*, New York: Columbia University Press.

Rockefeller, Steven C., 1992, "Faith and Community in an Ecological Age," in Steven C. Rockefeller and John C. Elder, eds. *Spirit and Nature: Why the Environment Is a Religious Issue*, Boston: Beacon Press.

—. 1991, *John Dewey: Religious Faith and Democratic Humanism*, New York: Columbia University Press.

Rolston, Holmes III, 1986, "Is There an Ecological Ethic?" in *Philosophy Gone Wild*, Buffalo, New York: Prometheus Books.

—. "Is There an Ecological Ethic?" op. cit.

Rothenberg, David, 1993, *Is It Painful to Think? Conversation with Arne Naess,* Minneapolis, Minnesota: University of Minnesota Press.

Sturm, Douglas, 1988, *Community and Alienation: Essays on Process Thought and Public Life*, Notre Dame: University of Notre Dame Press.

UNESCO, 1949, *Human Rights: Comments and Interpretations*, introduced by Jacques Maritain (New York: Columbia University Press.

Vilela, Mirian and Alide Roerink (eds.), 2005, *The Earth Charter in Action: Toward a Sustainable World* (Amsterdam: KIT Publishers.

Walzer, Michael, 1994, *Thick and Thin: Moral Argument at Home and Abroad* (Notre Dame: University of Notre Dame Press.
Westra, Laura, 1997, “Aristotelian Roots of Ecology; Causality, Complex Systems Theory, and Integrity,” in Laura Westra and Thomas M. Robinson, eds. *The Greeks and the Environment*, Lanham, Maryland: Rowman and Littlefield.
—. 2004, “Human Rights and Natural Law: Commonality and Divergences” in *Ecoviolence and the Law: Supranational Normative Foundations of Ecocrime*, Ardsley, New York: Transnational Publishers.
Whitehead, Alfred North, 1933, *Adventures of Ideas*, New York: Macmillan.
—. 1938, *Modes of Thought*, New York: Macmillan.
—. 1926, *Religion in the Making*, New York: Macmillan.

CHAPTER THREE

A MODEL OF VALUE-BASED DEMOCRACY AS CONDITION OF ECOLOGICAL SUSTAINABILITY

JOHANNES HEINRICHS

1. We speak of *economism* in order to label an intellectual attitude, which affirms the power and the pretended right of economics to determine the whole of society. The neo-liberal view was essentially such an economism (but so were some streams of Marxism, as well as some of the alternative monetary systems, the believers of which think: "if the economy is in order, the whole of society is automatically in order").

2. But there exists also an *ecologism*, which I would define as the intellectual attitude of those who have a highly developed ecological conscience and also a lot of knowledge about how to act in a better way with respect to nature, but not the essential insight that most of our ecological problems are not problems of nature, but of society, and consequently must be dealt with as problems of society and its institutions. (The Environmental Law Studies can be regarded as an *exception from this ecological mainstream*, under the condition that the juridical questions are not regarded as specialities, but as basic questions of the entire socio-political structure!)

3. Not only the pre-democratic nations at the limit of industrialization, but to a higher degree the "democratic" and fully industrialized nations have a fundamental lack of democratic institutions able to cope effectively with the ecological problems – and are lacking even consciousness of this fact. But this double lack, again, is not a special one, but concerns deeply the very understanding and development of democracy itself.

4. I see two main and principal deficiencies of our democracies (which we are not normally used to regarding as developing forms of government, but as a once-and-for-all achievement of social evolution). First, the whole of society depends structurally on economics. This is not *any* level of the entire society, but *the decisive level* – just as it is seen by the aforementioned economism. All our democracies, in their daily life, are not really based on human rights, human dignity and ultimate values, as they pretend to be, but on the ruling money system and its stock markets; the money system would be a discussion of current interest, but this is not my main topic here.

5. The second main deficiency is the functioning – or, better, the *not* functioning – of the system of political parties. These parties do not, and structurally cannot, really organize the political will of the people; they are not solution finders, but themselves an essential part of the democratic problem. Not to speak here of the Anglo-Saxon system of two teams that pretend to arrange the common wealth by simple alternation – a view of democracy that seems to me out of date – I take the example of the green parties in the European democracies. They have a highly developed ecological conscience, but little new perspective on the evolution of democracy, in spite of their self-proclaimed "basic democracy". In addition, they have an underdeveloped sense of the "cultural biotopes" called nations or cultural communities (at least in Germany). This is only one example. The general situation of the voters is that of a dilemma: they can vote for a political party with regard to one certain value, but, at the same time, in voting for this party they implicitly vote for values that they don't favour at all. This structural dilemma of the actual democracies concerning *all* political parties cannot be emphasized enough. It is suppressed by the "willing" majority of our political scientists – because they don't, and won't, see an alternative solution. As long as you see no alternative solution, you can hardly admit a non-functioning.

6. The rest of this chapter is dedicated to the question: is there a realistic alternative to this dilemma? My answer is that there is a *necessary* alternative; necessary because it is not merely an emergency measure but goes back to the very anthropological roots of human community- and society-building. At the same time, by adopting this alternative, the other main problem of present democracies – the illegitimate predominance of economics, which is so hostile to ecological standards – can be solved.

7. When I speak of "anthropological roots of society", the association of "anthropocentrism" must be kept well away. The so-called ecological anthropocentrism would better be called anthropo-egocentrism. Its view on nature is that of a predator, so not a very "human" or anthropological one. When I speak of the anthropological roots of society, I mean the link between individual action and the community or social system.

8. The link between individual action and the social system has been a matter of long discussion in the 20th century (especially the debate between Jürgen Habermas and Niklas Luhmann, after Talcott Parsons had failed to give a logical principle of his sequence of subsystems). In the view of a "reflection theory of society" (developed by me since 1975), the evident principle of a hierarchy of social subsystems is that of subsequent *reflection levels*. Reciprocal interpersonal reflection, or *practical* reflection, constitutes the decisive, socializing link between individuals, and thereby also between each individual and the social system. This is what neither the theorists of social action nor those of social system resolved.

9. The levels of direct interpersonal reflection are finite:

I. *simple objective relation:* I see the other one as an object.
II. *Subjective–reflexive relation:* I look at the other one as to another looking being, but I reduce him and his activity to my interests (and reciprocally; a reciprocity of *strategic attitudes* on each side).
III. *Double and responsive reflection:* I look at the other as someone who is capable of "our" reciprocity of looking (i.e. of a *communicative* attitude), in the same way as I am capable of it. (Whether we realize this ability of a communicative attitude is a further question. But the structural potency is given by this double and responsive reflection.)
IV. From that communicative potential each of us can take the position of a *metacommunicative relation*: I can reject the responsive reciprocity of the third level and devalue the other one, or I can continue a communicative life –establishing a *system of inner reciprocity* or *practical reflection*.

In this sense the interpersonal reflection circle (Figure 3-1) constitutes a social system, the four levels of which are reflection levels.

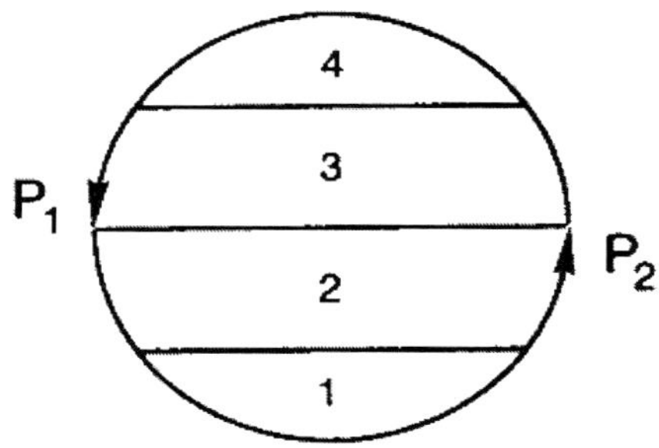

Figure 3-1. The unity of horizontal interpersonal reflection and vertical systemic reflection levels

The reflection circle can begin anew, but structurally it is finished by these 4 levels. The expressions "strategic" and "communicative" I take from Habermas, but he never developed a reflection theory; otherwise, the gap to the system theory of Luhmann (who also didn't develop it, because he didn't distinguish between theoretical and practical reflection) would have been overcome.

10. If now we change our perspective from that of the individual actors to one of collectivity (called "system reference" by Luhmann), we get not only *action levels* but *system levels*. This is a very important step of thought. The systemic levels are now:

I. The interaction system of exchange of goods: the level of *economic community* (unifying medium: money).
II. Interaction in terms of goal attainment and power: from which results a *community of law* (for law is nothing else than the regulation of power; a more or less just regulation, by the way). The common goal of the community of law is the attainment of common interests, which are not only economical ones. (unifying medium: law).
III. The interaction of real reciprocity or mutuality – a community in the proper sense of the word: a *community of communication*, which has its value in itself, not in the goal attainment of something outside the community (unifying medium: language).
IV. The interaction can, finally, direct itself upon the implicit value conditions of the former forms of community. The metacommunicative type of community is an *ethical and spiritual*

community (unifying medium: value axioms like dogmas and their expressive rites).

All these systemic levels can be regarded as implicit levels of every community, but also as explicit types of community or society.

11. These intentional levels or action levels defined above develop themselves as systemic levels, in a modern constitutional nation state, into more or less clearly differentiated subsystems (Figure 3-2).

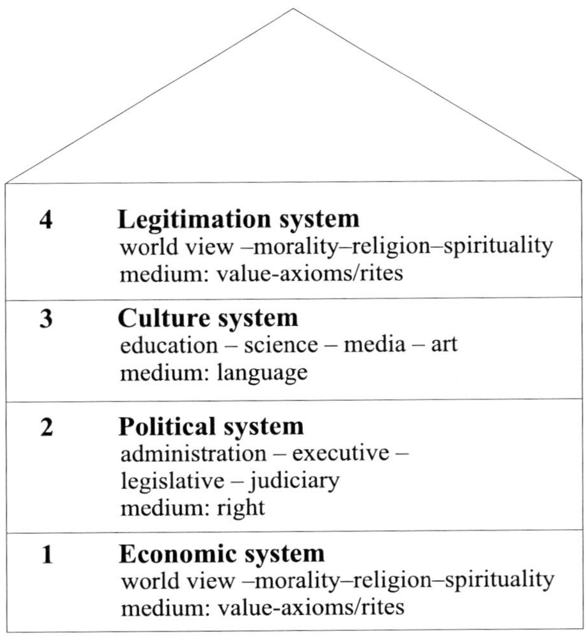

Figure 3-2. Hierarchy and inner division of the subsystems of a society

12. The historical evolution of a constitutional state was essentially characterized by the (at least theoretical) differentiation of religion (*pars pro toto* for the legitimation system) and politics, as well as by the differentiation of religion and an autonomous culture. The differentiation of politics (in the narrower sense) and economics/ecologics remains the task of our time! The pretended neo-liberal autonomy of economics is not the solution.

13. The effective regulation of a democratic society by democratic means becomes possible by the practical and institutional differentiation of all of these subsystems; namely, by the differentiation of the legislative "power" – by independent or interdependent partial parliaments (parliamentary "chambers" or "houses"):

- a basic-value parliament;
- a cultural parliament;
- a political parliament (in the narrower sense of "politics"); and
- an economical parliament.

Parliamentarism is the heart of democracy (at the state level, at least). A comparison: we have known since the time of William Harvey, the English physicist of the 17th century, that the physical heart of man has four chambers, which are essential for its functioning and which must be differentiated before the birth of a human being. Analogously, we should say that our democracies are not yet even completely born, because their heart chambers are not yet (theoretically and practically) differentiated.

More concretely: we need four expert-parliaments – or, better, trustee-parliaments (and respective executive organs) – elected independently from each other for each system level. Direct election and responsibility of the affected representatives for their specific fields is the remedy. Elections should be held, for example, each year for one of the specific parliaments. The character of the political parties would change by the force of these differentiated elections (and a little bit of juridical aid) from power-oriented to *matter-oriented* parties ("*Sachparteien*").

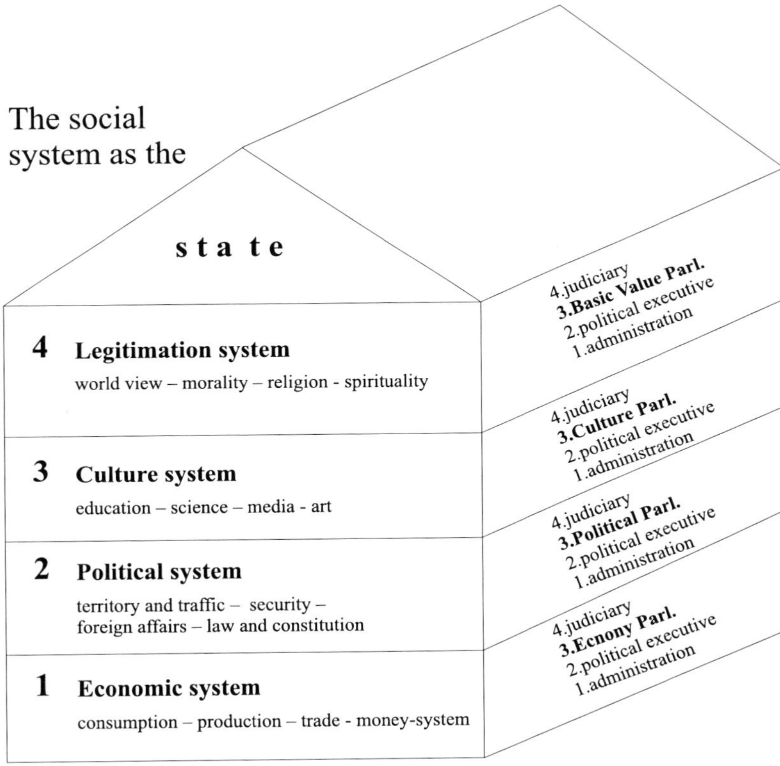

Figure 3-3. The social system as the state with extended division of "powers"

14. In this matter orientation of the new type of parties and elections lies an *inner* synthesis of direct and representational democracy. (Direct democracy alone without representative structures, applied to modern states, is a pure and even insane illusion!)

15. There is a hierarchical and circular interrelation of the subsystems. Without the correct hierarchy of values (economical, political, cultural and basic) – which means without an inversion of the present "order" (i.e. disorder, a real reversal of not only the theoretical but the practical order of values) – our democracies cannot become credible. And without a thoroughly new credibility they could not be saved in the long run!

But, on the other hand, it is not sufficient to reverse the materialistic dominance of economics to an idealistic dominance of the basic value representatives. Let us take the example of genetic engineering for plants, animals and humans, with all the ethical and economic questions arising from it. The general answer to this kind of interference of ethics and economics (or, more generally, interferences of all levels of the social system), is that there must be a feedback between these levels and their respective parliaments. Most democratic parliaments of the world already have the institution of a first, second and third reading of a legislatory draft. Thus, a circular inter-relation between the partial parliaments (or chambers) can easily be instituted.

It must be very seldom that there is a dispute over respective areas of responsibility between the parliaments, all the more since these "areas" are distinguished not always materially but formally – by the point of view of respective responsibility or value level. In the rare cases of competence dispute, the Supreme Court would have to decide about the competence.

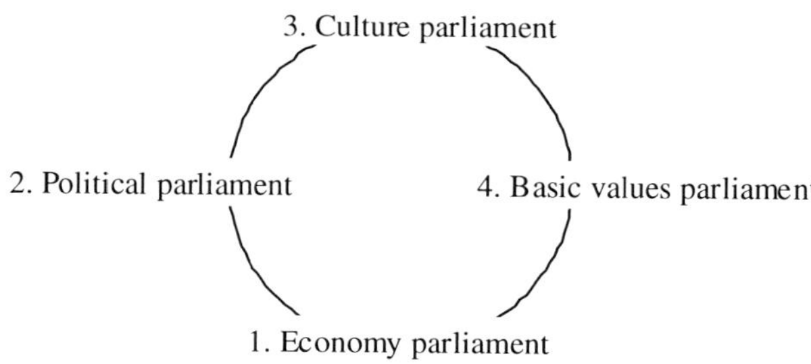

Figure 3-4. The systemic circle of the social as a parliamentary system

16. Ecological values are basic values in a double sense:

I. they are ultimate values of dignity of nature and life; even if ultimate values (as the religious ones) are not negotiable *per se*, there must and can be found a consensus about their "translation" into the political medium of a pluralistic society, which is the law; and

II. they are fundamental for human survival. Consequently, a common commission of the basic value-parliament and the economic parliament (i.e. the juridical institutionalization of ecological economics) would be the most realistic form of effectively achieving a sustainable form of economics. But this special arrangement seems to be realistic and effective only in the broader frame of the outlined "four-level democracy".

17. To summon up the essence of this chapter: most of our propositions about ecological ethics are in vain, and the rest are mere theory and wishful thinking, if there is no way of really implementing them in our societies, without further delay. We have no lack of value conscience, as is often stated, but rather a huge lack of value *realization*. This lack can only be overcome – democratically – by a compulsory system of legislative priority-rules: a hierarchical "framework" of legislation in a deeper sense of four frames, the most inclusive of which is the basic value frame.

After the courageous steps of our forefathers in the last two centuries (and much shedding of the blood of those who didn't succeed, as in 1848), we have to take a step today that could seem relatively tiny. In reality, it is a step as big and important as that of the founding fathers of the United States – which today are no longer a lighthouse of democracy (in spite of Obama). I propose that the "old" and the "new" Europe must take the initiative to prepare an indispensable further step – together with all democratic movements worldwide where institutional thinking is taken into account, not emotions alone. The indispensable further evolution of the structures of a democratic society are nothing less than the *natural laws*, in the modern sense of laws of action and of liberty, of a *communicative society*.

Notes

For further information see:
www.johannesheinrichs.de
www.viergliederung.de

Johannes Heinrichs, *Revolution der Demokratie. Eine Realutopie*, 444pp., Maas Verlag, Berlin, 2003

Johannes Heinrichs, *Demokratiemanifest für die schweigende Mehrheit*, 120pp., München, 2005

CHAPTER FOUR

BEHIND THE RED CURTAIN: ENVIRONMENTAL CONCERNS AND THE FALL OF COMMUNISM

JOSEPH W. DELLAPENNA

Introduction

In May 1992, I led a delegation of seven American environmental lawyers to the People's Republic of China to consult on the Three Gorges Dam. Ground had not yet been broken for the dam; today it towers 175 meters (540 feet) above the bed of the Yangtze River and creates a reservoir that is nearly filled – some 15 meters (47 feet) deep, some 500 kilometers (312 miles) upstream of the dam. The dam currently is the largest in the world. We (the delegation) were going both to learn about the dam and to provide insights about the project drawn from our experience as environmental lawyers. About a month before we arrived, the delegation lost significance when the National People's Congress voted to authorized the dam and by the time we arrived at the dam site (after meetings in Beijing), preliminary work had begun that would lead to the breaking of ground a few weeks later. It was apparent from the beginning that our function was to help justify the project by reporting that it would be compatible with a healthy environment. We were a disappointment to our hosts because we came away deeply skeptical of the project and said so in our report. We were not alone. Unmentioned by our hosts was that in authorizing the dam, the National People's Congress had encountered major public resistance locally and nationally and more than 600 members of the congress (out of more than 3,000 voting members) voted against the project. This was the first time since the events in Tiananmen Square in 1989 that anyone in the congress had voted against a government proposal – and only government proposals came before the congress. We found that ordinary people were willing to speak publicly of their opposition to the dam, at least in the region where the project was to be built. Yet nothing short of the collapse of the Communist system could have stopped the project and, of course, that did not happen in China. But the willingness of people to take risks to express their concerns about the environment in such a climate of repression led me to wonder what role, if any, such concerns played in the recent collapse of Communism in Eastern Europe and the former USSR. The answer, it turned out, was quite a bit.

Communism, the Environment, and the Collapse of the Communist Edifice

Communism has had a dismal record relative to the environment (Fishbach and Friendly 1992; Jancar 1987; Peterson 1993; Ziegler 1987). By 1989, sulfurous skies were killing people across the Soviet bloc. Single Russian factories were producing more pollution than all of Scandinavia (Specter 1994). Immediately after the end of the Soviet Union, the Russian Federation classified one-sixth of its territory as uninhabitable because of pollution – yet the inhabitants had nowhere to go (Zalygin 1992). Rivers were poisoned beyond anything found in western countries (Feschbach and Friendly 1992: 122), while Lake Baikal had become a paradigm of how not to manage a precious natural resource (Ziegler 1987: 55–58; *but see* Myers 2006b). The Aral Sea, in Central Asia, had become the prime example of "ecocide" (Bosnjakovic 2003: 16–18, 24–25, 41–42; Dukovny and Sokolov 2003; McCaffrey 2001: 260–265).

Communism performed so conspicuously poorly regarding the environment for five reasons. First, Marxism carried forward the western tradition of treating nature solely as providing resources for human consumption. As Vaclav Havel explained, Marxism saw humans as the "productive force" and nature as a "production tool," destroying the necessarily intimate relationship between the two (Crockett and Schultz 1990). This concept was succinctly captured in the "labor theory of value" that denied all value to natural resources consumed in productive processes (Kotov 2009; Maggs 1985). This was reinforced by a second feature of Marxism – its denial of individual responsibility. Because of its collectivist ethos, no one (or at least no one with any real power) felt responsibility for the natural environment, leading to reckless disregard of the environmental consequences of society's activities (Ziegler 1987: 46). Thirdly, the socialist goal of "transforming the world" led easily to "gigantomania" – a desire to achieve the largest and most grandiose technological feats (Zalygin 1992; Ziegler 1987: 24). Gigantomania is also seen in western countries, but structural features of Communism preventing effective counter-pressure that, at least sometimes, prevent such excesses in the west. This introduces a fourth factor – structural features rooted in Marxist ideology and the conspiratorial nature of Communism's rise to power – that are perhaps the most important. The "dictatorship of the proletariat" brooked no countervailing power centers (Stephen 1995; Ziegler 1987: 69–75). The Communist obsession with secrecy often kept problems hidden from both the public and the central authorities until catastrophe made the problem obvious to all (Galambos 1993: 217–218; Harman-Stokes 1995; Stephen 1995: 46–47). Finally, there was the importance of "fulfilling the plan." Success and promotion came only from fulfilling the plan, which generally was measured through quantitative achievements, resulting in pervasive poor quality production. New construction, moreover, is what the plan called for, not maintenance, while cost, in any rational sense, simply was not a factor (Galambos 1993: 209, 217; Robinson 1988: 359–362). The result, as a friend in China commented to me while I was living there before the market reforms, is that "They build old buildings here." That comment could also be applied across Eastern Europe and the former Soviet Union.

Communism itself may have looked like an old building in 1986, but it looked like a sturdily built building that would stand a long time. Yet that building crumbled in most of the nations included within its walls in the next five years. Attention has focused on the more immediately dramatic features of the collapse – the rise of *Solidarność* in Poland, the personality of Mikhail Gorbachev leading him to launch *perestroika* and *glasnost,* the opening of the Hungarian border to allow thousands of East Germans to pass through Austria to West Germany, open resistance to Soviet rule in the Baltic Soviet republics, Gorbachev's refusal to use the Red Army to maintain eastern European governments, the rise of Boris Yeltsin as a product of *glasnost,* and the attempted coup in Moscow in 1991. Furthermore, the experience of Communism varied significantly from place to place. For example, in Russia, China, Cuba, Vietnam, and Ethiopia, Communism was an indigenous force bound up with nationalism as much as it was in working out a universal ideal. On the other hand, in Afghanistan, the non-Russian parts of the USSR, and the eight "captive" nations of Eastern Europe, Communism was an alien force imposed by foreign bayonets on an unwilling people. Despite these central differences between Communist regimes, however, certain common themes appear in the final years of the formerly Communist nations (except perhaps Ethiopia). The most common theme, in both scholarly and popular accounts, is the growing yearning for democracy and human rights and for higher standards of living (Chirot 1991; Clark 1992; Meyerson 1991; Nagorski 1993; Stephen 1995).

While these features are intertwined in somewhat different ways in the various nations, they do not themselves form an adequate explanation even in a single nation. Thus while nationalism as well as the yearning for democracy produced an uprising in Hungary (1956), reform in Czechoslovakia (1968), and repeated unrest in Poland (1956, 1970, 1976, 1980), none of these succeeded and other east European Communist governments confronted no such unrest (Ash 1990). During this time, there is remarkably little evidence of wide-scale popular yearning for either democracy or human rights. Black markets throughout Eastern Europe and the USSR focusing on blue jeans and rock'n'roll hardly supply such evidence. Such dissatisfaction as there was came from the growing realization among workers that their efforts were accomplishing little and earning them less. Only after the process of dissolution of Communism was well underway did popular unrest begin to threaten the system. Even in Poland, the persistent attempts to create space for open opposition to the Communist government largely failed until near the end of the system. Attempts to create "civil society," primarily through the creation of "Helsinki Watch Groups," are simply not enough to explain why Communism so suddenly collapsed (Bernhard 1993). At least one aspect of the collapse is missing. The Communist governments had functioned for decades without serious internal or external challenge, and there was no obvious reason why they couldn't continue along the same path for decades more. Something was needed to crystallize the sentiment that the regimes not only had to go but could be overthrown.

In country after country, the something more turned out to be the environment. With the environment in such a problematic state, and structural and ideological problems precluding effective responses within a Communist system, it would be remarkable if the environmental problems did not play a major role in bringing down the system. And in fact, a major environmental disaster on such a scale that it could not be hidden precipitated crises when it became clear that the government could not cope. These crises discredited the government and emboldened its critics so much that the governments could not survive. Moderately large "green" movements with a modest degree of independence had been tolerated as a sort of window dressing for the official power structure (*e.g.*, Jehlička and Smith 2007). These groups, rather than the "forums" created to monitor compliance with the *Helsinki Accords* or to agitate for democracy or human rights, formed the organizational core of the toppling of the Communist governments. The toppling of some governments after an environmental crisis in turn weakened neighboring Communist governments, again with environmental groups leading the way. Yet the role of the environment (and more pointedly, of environmentalism) in bringing Communism to an end in Europe has largely escaped notice (*e.g.*, Feldbrugge 199[illegible]; Osyatinski 1990).

Chernobyl Opens a Crack in the Edifice

Dissidence had only barely survived in the USSR before the Chernobyl nuclear accident (Bernhard 1993: 311–312). The officially sponsored "peace movement" was kept under tight party control despite growing public unhappiness over the Afghanistan venture. The Soviet and party authorities allowed a modest degree of independence only for the environmentalist movement (Yanitsky 1993: 130–132). There were several reasons for this (Cole 1993: 54–58). First, most "Greens" were not ideological enemies of Marxism; many considered capitalism as necessarily worse for the environment than socialism or communism. Second, the Greens provided important information that would not otherwise reach the "Center" given the increasingly pervasive misinformation being passed up by the *apparatchiki.* Third, the "Greens" provided proof of "Soviet democracy" in action. And perhaps most importantly, "Greens" tended to be disaffected individuals who were isolated in temperament and thought from the mass of the people and thus could not present a real challenge to the regime.

The USSR had several nuclear accidents before Chernobyl, but they were in remote areas during periods of tight control of information and remained nothing more than unverified rumor (Kaptiza 1993: 7–8). Chernobyl, however, occurred in April 1986 in the middle of a large population and near the power centers of the USSR. And state control was loosening, President Gorbachev having launched *perestroika* (the restructuring of the economy). In this setting, the more perceptive members of the *nomenklatura* realized immediately that Chernobyl would be the end of the regime. For example, Marshall Sergei Akhromeev (who committed suicide after the failure of the 1991 military coup) commented, "The two most memorable days of my life were when I learned of the German invasion in 1941 and the day of the Chernobyl explosion. It was already clear to me how dreadful the consequences were." (*Economist* 1991). He foresaw that the Chernobyl disaster was on such a scale that ordinary Soviet citizens could not ignore the utter failure of the regime (Medvedev 1992; Medvedev 1993).

Millions lived with the areas subject to irradiation, which included the largest cities in Belarus (Minsk) and Ukraine (Kiyev); Kiyev was 115 km., 71 miles, away, Minsk was even closer. Costs of food within the region rose sharply because of the radiation rendered food grown nearby unsafe. The government actually kept agriculture in those regions in production, but put the resulting food into long-term storage because it was too radioactive to be consumed. Eventually, the government would mix the stored meat into sausages at a ratio of 1:10 to safe meat and then would disperse the sausages around the country to prevent "undesirable accumulations" of radioactivity in particular populations. The Soviet Minister of Health even advised people to breathe less in order to live longer (Fishbach and Friendly 1992). The Soviet government even proved incapable of furnishing adequate protection to workers sent to repair and restore the site, or to provide adequate medical care for those sickened by radiation (McAfferty 1996; Myers 2006a; Novokshchenov 2002). Secret records, released after the end of the USSR, would disclose that the government reclassified the amount of "acceptable radiation exposure" to a level 50 times higher than its pre-Chernobyl standard (*Economist* 1991).

One can easily imagine how politically corrosive public realization of such failing by any government would be. Consider the panic and suspicion generated by the comparatively mild Three Mile Island incident in the United States (Gray and Rosen 2003; Osif *et al.* 2004). In the USSR, the corrosive effect of the accident was exacerbated by the deliberate, but ultimately futile, government attempt to cover up the facts, coupled with the already pervasive distrust of the news media such that when the government began to tell the truth, nearly everyone assumed that the reality must be worse (*Economist* 1991). *Glasnost* ("openness," referring to the open flow of information) was little more than talk in early 1986 (Feldbrugge 1993: 51–54). Perhaps this talk was motivated by a sense that the USSR was falling behind the West in technological innovation and that state planning was breaking down due to the misinformation being passed by *appartchiki,* reinforced by the emergence of a competitive model for Communism in China's nascent reforms. Yet Gorbachev's first real break with the Soviet tradition of strict control of information was his reversal of the initial policies of restricting public knowledge of Chernobyl (Chirot 1991: 13). This change, 18 days after the accident, signaled to the Soviet press that "old habits were no longer good enough" (*Economist* 1991). Yet continuing efforts to cover up the extent of the disaster only served to further discredit the regime as Western broadcasts and refugees from the area spread the word (Kolbasov 1992).

In the wake of these failings, large-scale public criticism of the Soviet state and its policies emerged for the first time, within the Supreme Soviet and on the street. The increasing flow of information finally forced the government to admit that its "first-generation" of nuclear power reactors did not meet "modern safety standards," causing the Ministry of Atomic Power to order such reactors to be run at 70 per cent of capacity (Maloney-Dunn 1993: 398). Eventually, the Russian Academy of Sciences recommended that the reactors be shut down. The government's evident failures empowered its critics to speak out about other ecological disasters in the USSR. This movement spread rapidly in the non-Russian republics where it mingled with re-emergent nationalisms (Carere d'Encausse 1993; Deák 1990; Goldman 1992). The relatively tolerated environmental groups provided cover for those who might actually have been more interested in nationalism or other matters. It grew more slowly in Russia itself, perhaps because ethnic Russians were responsible for the errors of industrialization in the USSR.

The reaction was particularly intense in Belarus and Ukraine, two Soviet republics in which the sense of separate nationhood had been most thoroughly repressed before Chernobyl. An independence group, Rukh, was formed in Ukraine within weeks of Chernobyl. As Sergei Odarich, the founding Secretary-General of Rukh, commented, "Chernobyl helped us to understand that we were a colony" (*Economist* 1991). The first act of resurrected Ukrainian nationalism (in 1991, while Ukraine was still in the USSR) was enactment of a law on "The Legal Administration of Territories Contaminated with Radionucleides as a Result of the Chernobyl Disaster" (Sak 1993: 212–213). Even local Communists joined in the criticism in order to ensure their political survival when the "Center" fell. In Belarus, the reaction was slower, yet the Belarusian delegates to the Supreme Soviet of the USSR (almost all old-line Communists) took the lead in seeking an investigation of the handling of Chernobyl. When a wave of coal-mine strikes in 1991 reached Belarus, the strikers used their time on television to discuss Chernobyl and to demand the resignation of the President and Supreme Soviet of the USSR rather than the economic issues featured elsewhere in the Union (Shimansky 1991). The discrediting of the Soviet regime continued after the end of the Communist regime when its records, on Chernobyl and other ecological disasters, became public (Yaroshinskaya 1992; Yaroshinskaya 1993).

The Gabčikovo-Nagymoros Dam: Walls Crumble from the Edifice

Neither the Soviet government nor the long-troubled Polish government was the first Communist regime to lose power. That honor goes to the Communist government in Hungary, perhaps best symbolized by the decision to tear down the barbed wire along the Austrian border on 2 May 1989 (Judt 2005: 609–612; Ludwikowski 1991: 160). At about the same time, the Hungarian Socialist Workers Party (the Communists) ousted Janos Kadar, the party's Secretary-General since the failed 1956 revolution, and embarked on a broad program of reform that led to the regime's collapse (Bernhard 1993: 319; Revesz 1990: 125–127). Hungary technically was ruled by a "coalition" of allied parties under the "leadership" of the Socialist Workers Party, but the "allied" parties never took an independent line. Just before these events, the government allowed the emergence of *Fidesz* (the Alliance of Young Democrats) as an open opposition party, to be followed by several other opposition parties (Urban 1989). The significance of these events was obscured because long-drawn-out "discussions" about electoral reform delayed the first free election until 1990, after *Solidarność* had taken power in Poland (Bernhard 1993: 319–321; Pogany 1993: 336–339). These Hungarian developments, furthermore, were overshadowed by the dramatic events then playing out in Tiananmen Square in Beijing (Zhao 2009). With the scant attention given the events in Hungary, one is not surprised that few noticed that the Communist government in Hungary fell because of some dams.

The Danube River is the second longest in Europe (after the Volga). After dropping significantly across southern Germany and Austria, it flattens out as it forms the border between Hungary and Slovakia and spreads into three main channels that flow eastward across a broad flood plain, creating a network of islands and wetlands. These channels reunite at the Great Bend of the Danube after which it turns south to bisect Hungary. After falling 50 meters (154 feet) per 100 kilometers (62 miles) in Austria, the river falls only 50 meters across its entire 417 km. (260 mi.) length in Hungary (Galambos 1993: 177). This is inadequate to support hydroelectric generation unless a large dam were to be built to create an artificial fall. The Communist governments in Czechoslovakia and Hungary discussed such a project for 25 years before signing a treaty in 1977 to construct a series of dams, with the costs and the resulting electricity to be shared equally (*Treaty* 1977). Other planned benefits included flood control, improved navigation, storage of fresh water for use during dry periods, enhanced recreational opportunities, and the ability to extract sand and gravel from the siltation of the reservoirs (Williams 1994: 7–9). The plan called for construction of a dam at Dunakiliti in Hungary to "capture" the Danube. Virtually the entire flow of the Danube and several of its tributaries would then be diverted through a 17.5 km. (11 mi.) long elevated canal to a second dam and a 720 megawatt power plant at Gabčikovo in Slovakia. The canal, to lie atop mounded earth rising as much as 18 m. (60 ft.) over the landscape, was to be lined with asphalt and plastic to prevent seepage. As a result, the flow of water in the natural channel would drop from 2,000 m^3/sec. (67,000 cfs) to less than 1 m^3/sec (3 cfs). A third dam was to be built about 100 km. (62 mi.) downstream from Gabčikovo at Nagymaros in Hungary, to generate another 158 megawatts while directed the flow back to the natural channel of the river. The plan was for most of the flow of the river to be stored in a large reservoir (covering 60 km^2 (23 mi^2)) at Dunakiliti, to be released to support peak power generation at Gabčikovo. Water was to be released twice a day to flush the canal, allow the turbines at Gabčikovo to generate electricity when it was most needed, and to return to the river at Nagymoros in "12-ft. (4 m.) tidal waves" (Galambos 1993: 177–178; Williams 1994: 9–11).

This project was never popular in Hungary, largely because of serious ecological and social ills that the project would generate in Hungary (Galambos 1993: 198–202). Unpopularity rose when the Hungarian government proved unable to finance its share of the project and turned to Austria to provide the financing in exchange for the entire Hungarian share of the electricity for 20 years. Among the problems (all of which would largely affect only Hungary) were the risks of earthquake induced flooding, a drop of the water table by about 7 m. (22 ft.) that would dry up wells upon which farmers and villages had depended for centuries, and the drying of croplands and forests. Add to these fears of pollution and disease from leaching from the asphalt and an overall 25 per cent decline of the fish population (and perhaps a 90 per cent drop in the commercial fish harvest). Such problems were projected to affect about 2,000,000 people in Hungary (out of a population of 10,000,000). These problems would also affect about 1,000,000 people in Slovakia (out of a population of 4,000,000) (Galambos 1993: 178–181). While Hungary and Slovakia had together formed the core the Kingdom of Hungary before 1918, they had separate, troubled histories thereafter. After 1957, Hungary gradually developed "Goulash Communism" that was a relatively mild single-party state, but Slovakia, in the wake of the "Prague Spring" of 1968 was, in 1989, under control of what arguably was the most repressive regime in Europe. Yet the repression so obvious in the Czech lands was considerably less oppressive in Slovakia (Galambos 185–186). What perhaps accounts for the different reaction to the project in Czechoslovakia and Hungary was that most people put at risk in Slovakia by the project were ethnic Hungarians who received scant sympathy from the majority Slovaks.

With construction well underway, Hungarians took to the streets and fields in 1985 to challenge the project on ecological grounds (Diehl 1985; Szirmai 1993). Civil disobedience at the construction sites quickly evolved into an increasingly independent ecological movement (*Duna Kor:* Danube Circle), followed eventually by a peace group (*Dialogus:* Dialogue) (Bernhard 1993: 318–321; Galambos 1993: 180–182, 203–205; Pollack 1989: 212–215). By late 1988, as many as 60,000 people regularly rallied in the streets of Budapest at the call of *Duna Kor*, the first open opposition to the Kadar regime since the shooting stopped in 1956 (Galambos 1993: 182–183). Civil disobedience swept the construction site, with sit-downs before bulldozers and the occupation of the canal and dam sites by demonstrators. Petitions against the project collected 140,000 signatures, including signatures of 80 village mayors in the region to be affected. The protests quickly expanded to include demands for free elections, a revision of the "verdict" on 1956, support for the Hungarian minority in Romania, and, eventually, for the end of Communist rule. (Ash 1989: 226–227; Bernhard 1993: 318). In April 1989, the Communist regime cracked. The government suspended the Hungarian part of the project (the Dunakiliti and Nagymaros complexes) despite threats by the Austrian and Czechoslovak governments to demand hundreds of millions of dollars in compensation for breach of contract (Galambos 1993: 183–185; Williams 1994: 11–12). The suspension emboldened the demonstrators. The "allied" parties abandoned the government, causing it to lose its parliamentary majority and collapse (Urban 1989: 116). This collapse immediately inspired the negotiations in Poland by a now demoralized government and an increasingly confident *Solidarność* (Grabowski 1996; Judt 2005: 606–607). In Poland, with its well organized and active political opposition, no specific role can be assigned to environmental groups or concerns, although the increasingly evident environmental failures of the government in Poland did tend to undermine its support (Kabala 1993).

Evidence of any impact of the Gabčikovo-Nagymoros project on the "Velvet Revolution" in Czechoslovakia is far less definite than in Hungary. Yet despite the more intense repression in Czechoslovakia after the "Prague Spring" of 1968, an underground petition against the project netted 3,600 signatures in Prague in May 1989 (Galambos 1993: 206). Despite the well-established "woodcraft" groups in the Czech areas of the republic (Jehlička and Smith 2007), signers risked the sort of beatings, imprisonment, or worse, that had largely silenced "Charter 77" – an original "Helsinki Watch" group. Open criticism of the dam was tolerated in Slovakia (Galambos 1993: 206). More generalized environmental concerns seem to have played a role even if the dam project did not (Andrews 1993: 14). Still, linking the "Velvet Revolution" directly to environmental concerns is more tenuous than for Hungary or even Poland. Charter 77, led by Vaclav Havel, continued to speak out despite the repression and the government's failure to manage East German refugees flooding the West German embassy in Prague signaled a weakening of the regime. If the Gabčikovo-Nagymoros project was not central in ending Communism in Czechoslovakia, more generalized environmental concerns do seem to have played a role. Almost immediately after the "Velvet Revolution," Havel described the environmental destruction as one of the most serious problems facing the new government (Havel 1990). The Czech Ministry of the Environment attributed the loss of confidence in the regime that led to the "Velvet Revolution" to "the ecological crisis" (Czech Ministry of the Environment 1990).

The new government did not cancel the Gabčikovo-Nagymoros project (Galambos 1993: 206). This was perhaps because the Gabčikovo Dam was already 90 per cent complete when the Hungarians sought to cancel the project in October 1989 (Galambos 1993: 183–185, 209–213; Williams 1994: 11–13). Yet if the project was not central in ending Communism in Czechoslovakia, it did become central in the split of Czecho-Slovakia in two. Despite the belated emergence of an environmental group that opposed the project in Slovakia, Slovak leaders exploited fears of the Hungarians to build support for the dam as a nationalistic gesture, and then exploited that sentiment to break from Prague (Galambos 1993: 186–197, 206–207, 219–220). Meanwhile, Austria extracted US$ 240,000,000 from Hungary as compensation for the breach of the electricity contract (about three times as much as Austria had expended), money that then financed Slovakia's unilateral completion of the Gabčikovo Dam (Galambos 1993: 187–188, 191–192, 202, 218; Williams 1994: 13–14). The Slovaks re-designed the project to lie entirely within Slovakia. In November 1992, the Slovaks "closed the river" and began to fill their reservoir, leading to on-going litigation before the International Court of Justice (International Court of Justice 1997).

More Walls Fall: The German Democratic Republic, Bulgaria, and Albania

The German Socialist Unity (Communist) regime had always appeared particularly vulnerable because of the existence of the larger, non-Communist German Federal Republic that provided an alternative model that was visibly more prosperous and that also, simply because of its size, could claim to be "the real Germany." As a result, the German Democratic Republic always practiced a more severe repression that prevented the emergence of even the mild dissidence that characterized many of the Communist states in the 1970s and early 1980s (Bernhard 1993: 321; Sodaro 1983). No particular environmental crisis played a role in the collapse of the German Democratic Republic, but, as in Hungary, the Greens dominated the opposition, founded the *Neues Forum* (New Forum), and exploited popular discontent to topple the government. The Greens were unable to build a new regime, so their successes against the Communists yielded a united and decidedly un-Green Germany. Yet as part of their struggle against West Germany and NATO, the German Communists did allow the emergence of ostensibly independent pacifist and environmentalist movements after 1981 that were intended to cooperate with their counterparts in western Germany (Allen 1991; Woods 1986). The Communists seemed to have overlooked that influence could run both ways.

The East German pacifist and environmental groups were made up mostly of clergy and intellectuals. Because the Communists had worked hard to make such people marginal in society, they apparently did not fear that they could mobilize wide support should they escape control. When the government failed to prevent the flight of large numbers of people through the Hungarian border and the West German embassy in Prague (Judt 2005: 612–613), it lost control of the pacifist and green movements at home. Having had years of organizing experience, they now turned those skills against the government, leading demonstrations against the government (particularly in Leipzig). They founded a new organization (*Neues Forum*) that succeeded in crippling the government, but was reluctant to replace the Communist regime; *Neues Forum* even opposed the opening of the Berlin Wall (Rossman 1990). *Neues Forum*'s hesitancy at the critical moment did not prevent the collapse of the Communist regime once Soviet President Gorbachev indicated that he would not allow the Soviet Army to sustain the regime against its own people (Judge 2005: 610–616). Yet opponents of the Communists – whether in fleeing through Hungary or Prague, or streaming through the wall – seemed more interested in obtaining more goods and a better standard of living than in environmental issues. The first thing most people did when the wall opened was to go window shopping in West Berlin. Early projections that *Neues Forum* would sweep the interim elections in early 1990 proved false because of its organizational weaknesses and unpopular policy stands, particularly its choice to oppose German reunification in the hope of creating an alternative to both Communism and capitalism (Bernhard 1993: 321–323; Darnton 1990). The *Neues Forum* leaders should have known which way the wind was blowing when their crowds changed their chants from "*Wir sind die Volk*" ("We are the people") to "*Wir sind ein Volk*" ("We are one people") (Ash 1990: 72). The result of the collapse of *Neues Forum* in East Germany temporarily rendered the Greens politically insignificant in both parts of Germany. The East German Greens won only two seats in the Bundestag in 1990, while the West German Greens lost all of their seats (Markovits 1991; Morris 1990; Rossman 1990).

The story in Bulgaria, however, is even more strongly centered on their "Greens" than is the story of Hungary – but in a curious way. Before the middle of 1989, despite (or perhaps because of) the active persecution of ethnic Turks and Bogomils living in the country, Bulgaria seems to have lacked any active dissident movement, let alone significant opposition to the regime. Bulgaria was so quietly obedient to Moscow that it was sometimes called the "Sixteenth Republic" of the USSR (Tismaneanu 1992: 13). Then, quite suddenly, the government collapsed, entirely because of the government's minimal toleration of an apparently tame environmental movement known as *EkoGlasnost* (Ecological Openness) (Buamgartl 1993). This came about when the Bulgarian government hosted a three-week-long meeting of the Conference on Security and Cooperation in Europe (CSCE) to discuss increased cooperation on the environment, particularly in light of Chernobyl (Beattie 1989a). *EkoGlasnost* was in fact technically an illegal organization, but the Bulgarian government did not pursue its members. But when the CSCE meeting, ostensibly open to the public, began on 15 October 1989, the police barred members of *EkoGlasnost* from entering the building. The ensuing scuffles were witnessed by Western delegates to the meeting. On 27 October 1989, the Bulgarian government apologized to the delegates and to *EkoGlasnost* (Beattie 1989b). This apology touched off demonstrations against the government that led it to announce only two days later its intent to reform itself (Reuters 1989). As in other Communist states when they made similar announcements, this simply touched off greater protests until Sofia (the capital) was paralyzed (Beattie 1989c). Unofficial trade unions, human rights groups, and political reform groups sprang up overnight. Within a few days, Todor Zhivkov, long-time Secretary-General of the party, resigned; a few days later and the government recognized the legitimacy of *EkoGlasnost* (Antonov 1989). Bulgaria's apparently stable Communist regime had collapsed like hollow shell.

Bringing down Communism in Albania took more time and did not immediately flow from the leadership of Green groups, but those emerged very early in the process. Albania's Communists always employed one of the most repressive regimes in Europe and held on for a full year after most of the East European Communist governments had fallen. Finally, on 28 November 1990 anti-Communist riots broke out in Elbasan, in southern Albania. Elbasan was not fortuitous. It was the most heavily polluted industrial center in Albania, with heavy air and water pollution. One estimate is that as much of 10 per cent of the cement produced in the town was released into the air (BNA 1992). Other pollution came from plants producing chrome, coke, nickel, and steel in the city. The Elbasan riots led, two weeks later, to the formation of two independent parties – the Democratic Party (dedicated initially to protecting peasants' interests, but quickly broadening out to represent intellectuals and industrial workers) and the Albanian Ecological Party (founded in Elbasan and dedicated to addressing a growing ecological crisis) (Damon 1990). This led to free elections in March of 1991, which brought a coalition of the Communists and the Democrats to power (Williams 1991). The Albanian Ecology Party remained in opposition, charging that the Communists had "ruined the environment here more than in any other East European country" (Ikonomi 1992). The coalition proved fragile, collapsing in December 1991. This, however, led only to rampant lawlessness and further environmental degradation (de Boerr 1994; Ikonomi 1992). The inability of the Albanian Ecological Party to make an electoral impact led to the creation of an environmental activist group, *Perla* (de Boerr 1994). *Perla,* however, made only a small impact as the nation's attention became diverted to the fights of compatriots for autonomy or independence in Serbia (Kosovo) and Macedonia.

Walls Tumble for Other Reasons

In Europe, only in Romania and Yugoslavia was blood shed in the overthrow of the Communist regime. And only there did environmental concerns seem wholly irrelevant to the course of events. Instead, the transition from Communism seems to have come about largely through an intraparty coup that once started could not be stopped short of a toppling of the Communists, in form if not in fact (Judt 2005: 616–626; Osiatynski 1990: 840–844) The National Salvation Front, which was set up after the Ceauşescus were summarily executed was, in fact, dominated by *apparatchiki* of the old regime (Rekosh 1995). Only gradually and fitfully did the government evolve towards democracy (Light and Phinnemore 2001). Even Romania, however, was deeply influenced by the events in neighboring states, all of which exhibited, to varying degrees, changes brought about because of environmental concerns. Similarly, the Communist Party in the former Yugoslavia dissolved in power struggles over the succession to long-time dictator Josip Tito in which the leaders of the constituent republics sought to exploit a resurgent nationalism in an effort to strengthen and secure their positions. The collapse of the Yugoslav Socialist Federation thus led to a violent explosion of nationalism that seemed to have little to do with environmental concerns (Thomas 2003).

The situation in Ethiopia was much like Romania and Yugoslavia – a violent fall, this time driven by more than 20 years of civil and foreign war that seemed to have little to do with environmental concerns (Zewde 2001). One might also add Afghanistan to this list. The failure of the Communist regime in Afghanistan resulted from pure nationalism, expressed religiously (Borovik 2001). These seemingly remote and disparate failures of Communism are not without significance to our main story. In particular, the Soviet failure in the Afghan war, coming only 3 years after Chernobyl, contributed significantly to the corrosion of public support for the Soviet government, particularly among the military, contributing to the collapse of that government.

Why Did So Much of the Edifice Collapse?

Environmental concerns played a pivotal role in corroding and collapsing Communism, but they hardly explain everything about the collapse. I do not suggest a simple, deterministic theory of environmental crises bringing down Communist (or perhaps other) authoritarian governments. There was a highly generalized disaffection with the regime, whether derived from frustrated consumerism, resurgent nationalism, or a real yearning for democracy and human rights (Olson 1995). The willingness of the Red Army to turn against the regime in a failed coup after the disastrous war in Afghanistan also contributed to the fall. And there is the completely opposite roadmaps for reform followed in China and in the USSR and Eastern Europe. None of this provides a fully satisfactory explanation, however. Inept governments have survived for decades or centuries until some precipitating event causes their demise. In the end much depends on how people in general interpret the crisis – an interpretation that the government itself can manipulate so long as it retains effective control over the flow of information. Much also depends on the personality of leadership. Mikhail Gorbachev, no less than Louis XVI or Nikolai II, bears a heavy responsibility for the collapse of his system of government. Gorbachev was the first person to hold supreme power in the USSR since Lenin who had a legal education, and he surrounded himself with others with legal educations (most of whom, like Gorbachev, had not actually worked as lawyers). So little attention has been given to this feature that entire books were written about Gorbachev without mention of his legal training and his legalistic approach to crises (*e.g.*, Hanson 1989). The impact of the legalistic mindset (the lawyer's way of constructing reality, as it were) on the unraveling of these regimes would make an interesting case study (Feldbrugge 1993: 65–69; Sharlet 1992), if only because the most conspicuous exception to the dissolution of Communist control (China) has the least legalistic culture of the lot.

Moreover, as bad as environmental concerns have been (and continue to be) in Communist countries, people seem able to tolerate an enormous amount of environment degradation even to the point of rising death rates so long as the degradation seems routine and high generalized, dismissing it as just part of the "natural background" within which humans work out their lives (Dempsey 2009; Eberstadt 2008; Feschbach 2008; Specter 1995; Wines 2000). Yet as the collapse of Communist regimes has shown, when a dramatic event (a nuclear accident, a potentially disastrous dam) attributable to the government arises within this pattern of generalized degradation, that event can galvanize hitherto diffused discontent against the government, so long as there is some minimal flow of information outside the control of the government and some groups are able to operate outside of effective government control. The government can survive these challenges if it can credibly blame the event on external forces (for example, a force of nature) or if it is willing and able to repress opposition ruthlessly. Compare in this regard Gorbachev's legalistic response to Deng Xiao Ping's response to the demonstrations in Tiananmen Square (Feldbrugge 1993; Zhao 2009). When these conditions fail, governments fall – or, at least, they lose elections (Pan 2009).

When there is a broad range of information flowing and organizations operating within a society, people will find many reasons and avenues for opposing the government than just environmental concerns. When, as in the former Communist governments, environmental groups are the only even minimally tolerated forms of dissent or criticism, all (or nearly all) opponents of the regime will gravitate to those groups. These groups stand ready to challenge the government when an environmental catalyst occurs. It was not an accident, then, that in so-many formerly Communist states, the impetus for change and the source of organized opposition came from the environment and environmental groups. In such a context, one should also not be surprised that after the government falls, many who joined or supported the environmental groups prove to have little real commitment to ecological values and that environmental concerns ended up far down on the list of priorities of the new government (Bowman and Hunter 1992; Jehlička and Smith 2007). And as it turned out, the non-Communist regimes in the former Communist states have often not committed much resources to cleaning up the mess inherited from the Communists or even to preventing continuing or new sources of pollution (Cole 1995; Cummings 199[illegible]; Earhart 1997; Friedberg and Zaimov 1994; Harman-Stokes 1995; K[illegible]bel 1994; Knickerbocker 1994; Levin 1992; Libertiny 1993; Simons 1994; Specter 1995; Williams 1995: 450–455). Even the highly dangerous Soviet-era nuclear power plants, like that a Chernobyl, remained in operation for a considerable time (Marples 1993; Myers 2005; *New York Times* 1993). Chernobyl itself was not shut down until 2000 (LaFraniere 2000). In fact, whatever environmental improvements did occur, whether immediately after the collapse of the Communist regimes or at later points over the last 20 years, seem to have resulted from declines in industrial production and other economic activity rather than because of environment-friendly government policies (Henry and Douhovnikoff 2008). These disruptions in turn created pressure to prioritize restoring economic activity ahead of environmental concerns. Thus, in the long run, the overturning of Communism may have been motivated initially by environmental concerns in many instances, but that overturning did far less to address the environmental problems Communism left behind than one might have expected.

What Remains of the Edifice Today?

Communism in China did not fall, despite the near miss in Tiananmen Square in 1989 (Zhao 2009). Yet China is hardly a testament to the durability of Communism – China is only nominally Communist. It is ruled by a Communist party, but is fully committed to a capitalist economy, without the checks and balances, ineffective as they sometimes are, that are built into democratic capitalist systems. The rampant corruption is suggestive of the “robber barons” who dominated American capitalism a century earlier. The Three Gorges Dam was built, and caused even more severe problems than were predicted in 1992 (Shi 2009b). Reports are circulating that other, smaller reservoirs may have triggered severe earthquakes (LaFraniere 2009; Shi 2009d), which, if true, create truly major worries about the Three Gorges project. The Three Gorges Dam and its reservoir are, after all, the largest in the world. And China continues to destroy its environment to feed the ravenous appetite of its industries, particularly in the face of the global economic crisis (Chen 2009a; Chen 2009b; He 2008; Shi 2009a; Shi 2009e). The air in many cities is simply poisonous (Shi 2009c). Attractive historic buildings are torn down with little or no compensation to the owners in order to clear land for highly polluting factories or expensive modern shops and houses (Chan 2009a; Jacobs 2009; Wines 2009c; Zhang 2009). The north of China faces severe water shortages caused by pollution and overuse, with little prospect of resolving those problems despite investing vast sums in attempting to import water from a thousand kilometers or more away in central China (He 2009; Shi 2009a). These projects have been delayed in part from strong public resistance to the water importation project prompted by ecological concerns – unlike the Three Gorges project where resistance had no impact (Shi 2009f). China has become the major contributor of greenhouse gases to climate disruption and will reap its reward from impending "mega-disasters" (Chen 2009c).

That such reports are circulating despite efforts of the government to suppress them perhaps suggests that the government is losing control of the flow of information. Whether other conditions will combine with that spreading information to undermine the government remains to be seen. The fact is that environmental concerns (whether conceived narrowly in terms of pollution and ecological disruption or broadly in terms of the destruction of the built environment as well as the natural environment) are the driving force before the steadily rising incidence of "mass incidents" – anything from a modest protest to strikes to anti-government riots – that has the government so alarmed (Guo 2009; Huang 2009; Ma 2009; Moore 2009; Tam 2009; Wines 2009a). The government, however, is sufficiently concerned about the possibility of chaos and disorder that it has issued warnings to local governments to be especially vigilant in this year of multiple anniversaries (90 years after the "May 4 Movement"; 60 years after the founding of the People's Republic of China; 50 years after the "liberation" of Tibet; 20 years after the Tiananmen demonstrations) (Chan 2009b). It has also undertaken to harass, disbar, beat, or arrest lawyers who seek to use the courts to challenge these wrongs (Ng 2009; Wines 2009b).

Looking around the world, we find several other nominally Communist systems (Cuba, Laos, and Vietnam) that have largely followed the Chinese path, and several proto-Marxist regimes (Bolivia, Nepal, Nicaragua, and Venezuela) that may go down the classic Leninist-Stalinist-Maoist path or take the Chinese path. Only in North Korea do we find a classic Communist state, and that state is an environmental basket case (Cho and Zang 1999). It may be that all political-economist systems, whether classically Communist, capitalist, or following some "third way," will end up like North Korea if we do not constrain our excesses, but there is still something to be said for getting there more slowly than classic Communism seems to. And that is perhaps the ultimate reason that classic Communism collapsed.

References

Allen, Bruce 1991. *Germany East: Dissent and Opposition.* Montreal: Black Rose Books.

Andrews, Richard 1993. "Environmental Policy in the Czech and Slovak Republic," in Vari and Tamas 1993, at 5–48.

Antonov, Nikola 1989. "Bulgaria Lifts Ban on Environmentalist Group," *Reuters Library Report,* 11 December, Monday BC Cycle.

Ash, Timothy Garton 1989. *The Uses of Adversity: Essays on the Future of Cetnral Europe.* New York: Random House.

—. 1990. *The Magic Lantern.* New York: Random House.

Baumgartl, Bernd 1993. "Environmental Protest as a Vehicle for Transition: The Case of *EkoGlasnost* in Bulgaria," in Vari and Tamas 1993, at 157–175.

Beattie, Meriel 1989a. "35 Nations Gather for Environmental Forum," *Reuters Library Report,* 15 October, Sunday BC Cycle.

—. 1989b. "Bulgaria Says Police Who Beat Ecologists Went Too Far," *Reuters Library Report,* 27 October, Friday BC Cycle.

—. 1989c. "Bulgarian Activists Find 20-Day Platform to Voice Dissent," *Reuters Library Report,* 5 November, Sunday BC Cycle.

Bernhard, Michael 1993. "Civil Society and Democratic Transition in East-Central Europe," *Political Science Quarterly* 108: 307–326.

BNA 1992. "Regional Center Finds Economic Woes Prevent Albania from Addressing Pollution" *BNA International Environmental Daily,* 17 January.

Borovik, Artyom 2001. *The Hidden War: A Russian Journalist's Account of the Soviet War in Afghanistan.* New York: Grove Press.

Bosnjakovic, Branko 2003. *Negotiations in the Context of International Water-Related Agreements.* Geneva: UNESCO SC-2003/WS/45.
Bowman, Margaret and David Hunter, "Environmental Reforms in Post-Communist Central Europe," *Michigan Journal of International Law,* 13: 921–980.
Carrere d'Encausse, Hélène 1992. *The End of the Soviet Empire: The Triumph of Nations.* New York: Basic Books.
Chan, Kelly 2009a. "Demolition Gang Drags Residents from Homes," *South China Morning Post,* 20 January May, at 6.
—. 2009b. "Beijing Advises Local Officials to Lend an Ear to Peititioners," *South China Morning Post,* 2 May, at 5.
Chen, Stephen 2009a. "Birth Defects Caused by Environmental Pollution," *South China Morning Post,* 9 January, at 7.
—. 2009b. "Lowest Marks over Environment," *South China Morning Post,* 9 January, at 6.
—. 2009c. "More Mega-Disasters Ahead as Environment Deteriorates," *South China Morning Post,* 12 May, at 6.
Chirot, Daniel (editor) 1991. *The Crisis of Leninsim and the Decline of the Left.* Seattle: University of Washington Press.
Cho, Myung Chul and Hyoungsoo Zang 1999. *The Present and Future Prospects of the North Korean Economy.* Seoul: Korea Institute for International Economic Policy.
Clark, Barry S. 1992. "Political Economy, Democracy and Eastern Europe," *International Journal of Social Economics* 19: 259–272.
Cole, Daniel H. 1993. "Marxism and the Failure of Environmental Protection in Eastern Europe and the USSR," *Legal Studies Forum* 17: 35–72.
Cole, Daniel H. 1995. "Poland's Progress: Environmental Protection in a Period of Transition," *Parker School Journal of East European Law* 2: 279–319.
Crockett, Tamara Raye and Cynthia B. Schultz 1990. "Environmental Protection Issues in Eastern Europe," *Bureau of National Affairs International Environmental Report,* 13 June, p. 258.
Cummings, Susan S. 1993. "Polish Environmental Regulations: The State of Poland's Environment, Governmental Authorities and Policy," *Suffolk Transnational Law Review* 16: 379–404.
Czech Ministry of the Environment 1990. *Environment of the Czech Republic: Evolution, Situations, and Trends to the End of 1989.* Prague: Czech Ministry of the Environment.
Damon, Dan 1990. "Albanians Form Green Party," *Daily Telegraph,* 17 December, at 7.

Darnton, Robert 1990. "Campaign Trails: West German Politics Go East," *The New Republic,* 26 March, at 21.

Deák, István 1990. "Uncovering Eastern Europe's Dark Secret," *Orbis,* 34: 51–65.

de Doerr, Hilary 1994. "Business and the Environment: Freedom to Own and to Pollute – The Effects of Albania's New Economic Activity," *Financial Times,* 29 June, at 20.

Dempster, Judy 2009. "Study Looks at Mortality in Post-Soviet Era," *New York Times,* 16 January, at A8.

Diehl, Jackson 1985. "Danube Plans Rile Hungarians: 3 Environmental Groups Bring Pamphleteering Art to East Bloc," *Washington Post,* 12 December, at E1.

Dukhovny, Viktor and Vadim Sokolov 2003. *Lessons on Cooperation Building to Manage Water Conflicts in the Aral Sea Basin.* Geneva: UNESC0 SC-2003/WS/44.

Earnhart, Dietrich 1997. "Enforcement of Environmental Protection Laws under Communism and Democracy," *Journal of Law and Economics* 40: 377–402.

Eberstadt, Nicholas 2008. "Rising Ambitions, Sinking Population," *New York Times,* 25 October, at A23.

Economist 1991. "Not Just a Nuclear Explosion," *The Economist,* 27 April, at 19 (international edition).

Feldbrugge, F.J.M. 1993. *Russian Law: The End of the Soviet System and the Role of Law.* Dordrecht: Martinus Nijhoff.

Feschbach, Murray 2008. "Behind the Bluster, Russia Is Collapsing," *Washington Post,* 5 Oct., at B3.

Feschbach, Murray and Alfred Friendly, jr. 1992. *Ecocide in the USSR: Health and Nature under Siege.* New York: Basic Books.

Friedberg, James and Branimir Zaimov 1994. "Politics, Environment, and the Rule of Law in Bulgaria," *Duke Journal of Comparative and International Law* 4: 225–263.

Galambos, Judit 1993. "An International Environmental Conflict on the Danube: The Gabčikovo-Nagymoros Dams," in Vari and Tamas 1993, 176–226.

Goldman, Marshall I. 1992. "Environmentalism and Ethnic Awaking," *Boston College Environmental Affairs Law Review* 19: 511–514.

Guo, Al 2009. "Further Study on Beijing's Waste Plant Ordered after Public Protest," *South China Morning Post,* 12 March, at 12.

Grabowski, Tomek 1996. "The Party That Never Was: The Rise and Fall of the Solidarity Citizens' Committes in Poland," *East European Politics and Societies* 10: 214–254.

Gray, Mike and Ira Rosen 2003.*The Warning: Three Mile Island.* New York: Norton and Co.

Harman-Stokes, Katherine M. (1995). "Note: Community Right-to-Know in the Newly Independent States of the Former Soviet Union: Ending the Culture of Secrecy Surrounding the Environmental Crisis," *Virginia Environmental Law Journal,* 15: 77–138.

Hanson, Stephen 1989. "Gorbachev: The Last True Leninist Believer?," in Chirot 1989, at 33–59.

Huang, Cary 2009. "Taskforces Set up to Keep Lid on Protests," *South China Morning Post,* 28 February, at 1.

Havel, Vaclav 1990. "Our Freedom," *Washington Post,* 3 January, at A15.

He Huifeng 2008. "Guangdong Admits It Is Unlikely to Meet 2010 Pollution Targets," *South China Morning Post,* 12 May, at 8.

—. 2009. "Beijing Faces Major Water Shortage until at Least 2014," *South China Morning Post,* 12 May, at 6.

Henry, Laura A. and Vladimir Douhovnikoff 2008. "Environmental Issues in Russia," *Annual Review of Environment and Resources* 33: 437–460.

Ikonomi, Ilir 1992. "Stump Lines the Avenues in Wood-Burning Albania," *Reuters Library Report,* 22 March, Sunday BC Cycle.

International Court of Justice, *The Gabčikovo-Nagymoros Case (Hungary v. Slovakia),* 1997 ICJ No. 92.

Jacobs, Andrew 2009. "Finding Treasures in a City's Disappearing Past," *New York Times,* 19 January, at A8.

Jancar, Barbara 1987. *Environmental Management in the Soviet Union and Yugoslavia: Structure and Regulation in Federal Communist States.* Durham, NC: Duke University Press.

Jehlička, Petr and Joe Smith 2007. "Out of the Woods and into the Lab: Exploring the Strange Marriage of American Woodcraft and Soviet Ecology in Czech Environmentalism," *Environment and History* 13: 187–200.

Judt, Tony 2005. *Postwar: A History of Europe Since 1945.* New York: Penguin Press.

Kabala, Stanley J. 1993. "Environmental Affairs and the Emergence of Pluralism in Poland: A Case of Political Symbiosis," in Vari and Tamas 1993, at 49–66.

Kapitza, Sergei P. 1993. "Lessons of Chernobyl: The Cultural Causes of the Meltdown," *Foreign Affairs* 72(3): 7–11.

Kibel, Paul Stanton 1994. "Russia's Wild East: Ecological Deterioration and the Rule of Law in Siberia," *Georgetown International Environmental Law Review* 7: 59–76.

Knickerbocker, Brad 1994. "Hopes of Ecological Bliss Elude the Former Soviet Bloc," *Christian Science Monitor,* 16 March, at 6.

Kolbasov, Oleg S. 1992. "Ecological Disaster Area: The Chernobyl Case Study," *Boston College Environmental Affairs Law Review* 19: 637–640.

Kotov, Vladimir 2009. "Russia: The Historical Dimensions of Water Management," in Joseph W. Dellapenna and Joyeeta Gupta (editors), *The Evolution of the Law and Politics of Water* 139–155. Dordrecht: Springer Netherlands.

LaFraniere, Sharon 2000. "Chernobyl Closing: A Relief for Many," *Washington Post,* 15 December, at A30.

—. 2009. "Possible Link between Dam and China Quake," *New York Times,* 6 February, at A1.

Levin, Julia 1992. "Russian Forest Laws – Scant Protection during Troubled Times," *Ecology Law Quarterly* 19: 685–725.

Libertiny, Karen S. 1993. "Cause for Cautious Celebration: Hungarian Post-Communist Environmental Reform," *Michigan Journal of Environmental Law* 14: 518–552.

Light, Duncan and David Phinnemore (editors) 2001. *Post-Communist Romania.* New York: Palgrave.

Ludwikowski, Rett R. 1991. "Searching for a New Constitutional Model of East-Central Europe," *Syracuse Journal of International Law and Commerce* 17: 91–170.

Ma, Josephine 2009. "Security Panel Chief Warns of More Unrest and Social Tension," *South China Morning Post,* 13 January, at 6.

Maggs, Peter B. 1985. "Marxism and Soviet Environmental Law," *Columbia Journal of Transnational Law* 23: 353–372.

Maloney-Dunn, Kathleen 1993. "Russia's Nuclear Waste Law: A Response to the Legacy of Envrionmental Abuse in the Former Soviet Union," *Arizona Journal of International and Comparative Law* 10: 365–430.

Markovits, Andrei S. 1991. "Germany: Power and the Left," *Dissent* 38: 354–359.

Marple, David R. 1993. "The Post-Soviet Nuclear Power Program," *Post-Soviet Geography and Economics* 34(3): 172–184.

McAfferty, Nell 1996. "Life after Chernobyl," *Audubon Magazine,* May–June, at 86.

McCaffrey, Stephen C. 2001. *The Law of International Watercourses: Non-Navigational Uses.* Oxford: Oxford University Press.

Medvedev, Grigori 1992. *The Truth about Chernobyl* (translated by Evelyn Rossiter). New York: Basic Books.

—. 1993. *No Breathing Room: The Aftermath of Chernobyl* (translated by Evelyn Rossiter). New York: Basic Books.
Meyerson, Adam 1991. "Ash Heap of History: Why Communism Failed," *Policy Review* 58: 4–5.
Moore, Michael 2009. "China Tells Police to Plan for Mass Unrest in Downturn," *Daily Telegraph* (London), 28 February, at 18.
Myers, Steven Lee 2005. "Belarus Resumes Farming in Chernobl Radiation Zone," *New York Times,* 22 October, at A3.
—. 2006a. "First at Chernobyl, Burning Still," *New York Times,* 26 April, at A8.
—. 2006b. "Putin Reroutes Oil Line to Avoid Landmark Lake," *New York Times,* 27 April, at A8.
Morris, Nomi 1990. "How Greens Withered in Germany: Mainstream Adopted Environmental Causes," *San Francisco Chronicle,* 7 December, at A25.
Nagorski, Andrew (editor) 1993. *The Birth of Freedom: Shaping Lives and Societies in Eastern Europe.* New York: Simon and Schuster.
New York Times 1993. "Ukraine to Keep Chernobyl in Operation" *New York Times,* 22 October, at A8.
Ng Tze-wei 2009. "Rights Lawyers Worry about Licence Renewal," *South China Morning Post,* 28 May, at 6.
Novokshchenov, Validimir 2002. "The Chernobyl Problem," *Civil Engineering,* May, at 74–83.
Olson, Manur jr. 1995. "Why the Transition from Communism Is So Difficult," *Eastern Economic Journal* 21: 437–461.
Osiatynski, Wiktor 1991. "Revolutions in Eastern Europe," *University of Chicago Law Review* 58: 823–858.
Osif, Bonnie A., Anthony J. Baratta, and Thomas W. Conkling 2004. *TMI 25 Years Later: The Three Mile Island Nuclear Power Plant Accident and Its Impact.* University Park, PA: University of Pennsylvania Press.
Pan, Philip P. 2009. "Russian Voting Tinged with Green," *Washington Post,* 10 April, at A1.
Peterson, D.J. 1993. *Troubled Lands: The Legacy of Soviet Environmental Destruction.* Boulder, CO: Westview Press.
Pogany, Istvan 1993. "Constitutional Reform in Central and Eastern Europe: Hungary's Transition to Democracy," *International and Comparative Law Quarterly* 43: 332–355.
Pollack, Maxine 1989. *East European Fault Lines: Dissent, Opposition and Social Activism.* Boulder, CO; Westview Press.
Rekosh, Edwin 1995. "Romania: A Persistent Culture of Impunity," in Naomi Roht-Arriaza (editor), *Impunity and Human Rights in*

International Law and Practice 129–159. Oxford: Oxford University Press.
Reuters 1989. "US Welcomes Encouraging Signs of Reform from Bulgaria," *Reuters Library Report,* 30 October, Monday AM Cycle.
Revesz, Gabor 1990. *Perestroika in Eastern Europe: Hungary's Economic Transformation.* Boulder, CO: Westview Press.
Robinson, Nicholas A. 1988. "*Perestroika* and *Priroda*: Environmental Protection in the USSR, *Pace Environmental Law Review* 5: 351–424.
Rossman, Peter 1990. "Dashed Hopes for a New Socialism: East Germany," *The Nation,* 7 May, at 632.
Sak, Pamela Bickford 1993. "Environmental Law in Ukraine: From the Roots to the Bud," *UCLA Journal of Environmental Law and Policy* 11: 203–253.
Sharlet, Robert 1992. *Soviet Constitutional Crisis: From De-Stalinization to Disintegration.* Armonk, NY: M.E. Sharpe, Inc.
Shi Jiangtao 2009a. "Experts Blame Drought on Human Error," *South China Morning Post,* 7 February, at 4.
—. 2009b. "100b Yuan Outlay to Cope with Dam Impact," *South China Morning Post,* 23 April, at 5.
—. 2009c. "Rise in Number of Cars Causes More Bad Air, Environment Chief Says," *South China Morning Post,* 23 April, at 5.
—. 2009d. "Geologist Defies Ban to Discuss Possible Link between the Disaster and a Hydroelectric Dam," *South China Morning Post,* 12 May, at 9.
—. 2009e. "Environment Takes a Back Seat: Rushed, Politically Driven Reconstruction Is Causing Alarm," *South China Morning Post,* 12 May, at 9.
—. 2009f. "Public Concern and Costs Blamed for Water Scheme Delay," *South China Morning Post,* 29 May, at 6.
Shimansky, M. "The Minsk Strike Committee Sets Forth Its Demands," *Izvestia,* 11 April, at 2 (in Russian).
Simons, Marlise 1994. "East Europe Sniffs Freedom's Air, and Gasps," *New York Times,* 3 November, at A1.
Sodaro, Michael 1983. "Limits to Dissent in the GDR: Fragmentation, Cooptation, and Repression," in Jane Leftwich Curry (editor), *Dissent in Eastern Europe,* at 82–116. New York: Praeger.
Specter, Michael 1994. "Far North in Russia: The Mine's Fatal Blight," *New York Times,* Mar. 28, p. A1.
—. 1995. "Plunging Life Expectancy Puzzles Russia," *New York Times,* 2 August, at A1.

Stephen, Paul B. III 1995. "The Fall – Understanding the Collapse of the Soviet System," *Suffolk University Law Review* 29: 17–49.

Szirmai, Viktoria 1993. "The Structural Mechanisms of the Organization of Ecological–Social Movements in Hungary," in Vari and Tamas 1993, at 145–156.

Tam, Fiona 2009. "Villagers Riot after Row with Officials," *South China Morning Post,* 25 March, at 5.

Thomas, Raju G.C. (editor) 2003. *Yugoslavia Unraveled: Sovereignty, Self-Determination, Intervention.* Lanham, MD: Lexington Books.

Tismaneanu, Vladimir 1992. "Between Liberation and Freedom," in Vladimir Tismaneanu and Patrick Clawson (editors), *Uprooting Leninism, Cultivating Liberty* 1–46. Lanham, MD: University Press of America.

Treaty Concerning the Construction and Operation of the Gabčikovo-Nagymoros System of Locks, 1977, Czechoslovakia–Hungary, UNTS 1109: 235–248.

Urban, Laszlo 1989. "Hungary in Transition: The Emergence of Opposition Parties," *Telos* 79: 114–118.

Vari, Anna and Pal Tamas (editors) 1993. *Environment and Democractic Transition: Policy and Politics in Central and Eastern Europe.* Dordrecht: Kluwer Academic Publishers.

Williams, Carol J. 1991. "Albania – A Step to Democracy," *Los Angeles Times,* 26 March, at 6.

Williams, Paul R. 1994. "International Environmental Dispute Resolution: The Dispute between Slovakia and Hungary Concerning Construction of the Gabčikovo and Nagymoros Dams," *Columbia Journal of Environmental Law* 19: 1–58.

—. 1995. "Can International Legal Principles Play a Positive Role in Resolving Central and Eastern European Transboundary Environmental Disputes?," *Georgetown International Environmental Law Review* 7: 421–462.

Wines, Michael 2000. "For All Russia, Biological Clock Is Running out," *New York Times,* 28 December, at A1.

—. 2009a. "China's Leaders See a Calendar Full of Anniversaries, and Trouble," *New York Times,* 10 March, at A5.

—. 2009b. "China Said to Harras Rights Lawyers," *New York Times,* 28 May, at A1.

—. 2009c. "To Protect an Ancient City, China Moves to Raze It," *New York Times,* 28 May, at A1.

Woods, Roger 1986. *Opposition in the GDR under Honecker, 1971–1985: An Introduction and Documentation.* New York: St. Martin's Press.

Yanitsky, Oleg 1993. "Environmental Initiatives in Russia: East–West Comparisons," in Vari and Tamas 1993, 120–145.

Yaroshinskaya, Alla 1992. "40 Secret Protocols of the Kremlin Elders – The Lie about Chernobyl' Is Just as Terrible as the Catastrophe Itself," *Izvestia,* 24 April, at 3 (in Russian).

Yaroshinskaya, Alla 1993. "CPSU Central Committee's Secret Protocols on Chernobyl'," *Izvestia,* 18 April, at 1.

Zalygin, Sergei 1992. "A Brief Summary of the Report 'Ideology and Ecology,'" *Boston College Environmental Affairs Law Review* 19: 635–636.

Zewde, Bahru 2001. *A History of Modern Ethiopia, 1855–1991.* Athens, OH: Ohio University Press.

Zhang, Lilian 2009. "Couple Appeal over 'Illegal' Land Grab and Demolition Ordered by Officials – Not Court," *South China Morning Post,* 8 May, at 8.

Zhao Ziyang 2009. *Prisoner of the State.* New York: Simon and Schuster..

Ziegler, Charles 1987. *Environmental Policy in the USSR.* Amherst, MA: University of Massachusetts Press.

CHAPTER FIVE

FROM METAPHYSICS TO ENVIRONMENTAL ETHICS: ARISTOTLE OR ZHU XI?

MAY SIM

Arne Naess distinguishes the shallow from the deep ecology movement.[1] Whereas shallow ecology, according to Naess, prioritizes the "health and affluence of people in the developed countries," so that man is externally related as a "man-in-environment," deep ecology favors "the relational, total-field image," urging an intrinsic, constitutional relation between man *and* the environment.[2] Naess's deep ecological philosophy insists on grounding non-ultimate arguments on ultimate bases. He says, "the limitation of the shallow movement is not due to a weak or unethical philosophy, but due to a lack of explicit concern with ultimate aims, goals, and norms."[3] The philosophies of Aristotle and Zhu Xi (1130–1200 CE, Song Neo-Confucian) share the explicit concern with ultimate aims and norms about which Naess speaks. Comparing Aristotle and Zhu Xi, I aim to illustrate how each one's metaphysics, or concern with the ultimate, affects his ethics – and can offer better or worse resources for an environmental ethics. More specifically, my goal is to show how each author can or cannot justify something like the first two of Naess's eight-point platform principles for the deep ecology movement which stress intrinsic values. More specifically, Naess's first principle states: "The flourishing of human and nonhuman life on earth has inherent value. The value of nonhuman life-forms is independent of the usefulness of the nonhuman world for human purposes."[4] His second principle asserts, "richness and diversity of life-forms are also values in themselves and contribute to the flourishing of human and nonhuman life on earth."[5] If a philosophy (such as Aristotle's or Zhu Xi's) can account for these principles, then it can also provide the resources for defending the kind of intrinsic value Naess maintains in the deep ecology movement, thus illustrating my claim that one's metaphysics has direct implications for one's account of ethics and man's relation to the environment.

Both Aristotle and Zhu Xi agree that human beings are good when they possess the virtues. Whereas Aristotle sharply divides virtues into two types, intellectual and moral, Zhu Xi does not make such a division, though he too talks about knowing the reason why of things and right actions. Both maintain that there are first principles of virtue. Given Aristotle's distinction between two types of virtue, his principles governing the virtue of scientific knowledge (*epistêmê*) culminating in contemplative wisdom (*sophia*) would also differ radically from those governing the moral virtues culminating in practical wisdom (*phronêsis*).[6] Contrariwise, since Zhu Xi does not distinguish between theoretical and moral knowledge, his first principles of virtue are continuous between the nonhuman and human realm.[7] Comparing these thinkers' views of first principles and virtues, I explore the resources they might offer for an environmental ethics.

That Aristotle's and Zhu Xi's philosophies are compatible with, and can be used to support the kind of reasoning by more recent environmental thinkers, can be seen in a few quick examples before considering each of their views in detail. Consider Philip Cafaro's "Thoreau, Leopold, and Carson: Toward an Environmental Virtue Ethics," appeal to Henry David Thoreau's virtue of simplicity to argue for an environmental virtue ethics, where one's appreciation and respect for the environment can contribute to the development of one's virtues.[8] Cafaro's position can be supported by Aristotle's argument for *theôria* as the highest human virtue. In spite of Cafaro's criticism of Aristotle's anthropocentrism by according virtue to humans alone,[9] Aristotle's argument for the priority of *theôria*[10] is of a piece with Thoreau's account that simplicity enables stability and focus in our lives so that we can develop our virtues. Like Thoreau, Aristotle too values the kind of stability and focus coupled with the limited use of external goods Cafaro praises, as is evident in the value he attributes to contemplation and how contemplation requires fewer external goods. Similarly, take Cafaro's appeal to Aldo Leopold's attribution of virtues to the nonhuman world when he speaks of the chickadee's "valor" and the plover's "grace," and more generally, how "human and nonhuman beings may share some virtues because [they] are in some respects similar."[11] The attribution of intrinsic value to the nonhuman realm, especially to animals, can be supported by Zhu Xi's metaphysics which argues for the immanence of the same moral principles in all things – human and nonhuman. Zhu Xi explains how different animals can exhibit different moral virtues too, albeit in a more limited way when compared to human beings. Moreover, Zhu Xi's account of how our knowledge of things around us can contribute to our morality can be used too for justifying the argument by Thomas Hill, Jr., that our ability to feel gratitude toward nature is a precondition for possessing the virtue of gratitude toward human beings.[12]

Let us consider Aristotle's and Zhu Xi's metaphysics in greater detail. To understand the similarities and differences between Aristotle's and Zhu Xi's first principles of virtue, how one knows them and their necessity for practice, I explore the ultimate reality that is the source of these principles for each. These are the Great Ultimate (*taiji* 太極) and the Non-Ultimate (*wuji* 無極) for Zhu Xi, and the primary substance or God for Aristotle. Comparing the ultimate reality of each, seeing if it is one or many, immanent or transcendent, self-sufficient or dependent, I show how it is the reason for each one's view of human nature, goodness and our relation to the environment. Human nature determines the intelligibility of the first principles of virtue since the human soul for Aristotle, and correspondingly, the human mind (*xin* 心) for Zhu Xi, make possible knowledge. Knowing the way that one's mind grasps the first principles of virtue enables us to understand if the intelligibility of these principles is essential for practicing goodness. Apart from agreeing about the role that ultimate reality and its knowledge play in the norm of goodness, Aristotle and Zhu Xi would agree too that virtues are to be cultivated.

Aristotle's human nature can be understood from his conception of substance or essence. For Aristotle, essence causes the 'what it is to be' (*to ti en einai*) of something and satisfies the conditions of being definable, knowable, separable, prior in cause and time.[13] With respect to a concrete thing such as a man or a vase, Aristotle maintains that it is composed of a certain form and matter. Though both form and matter are needed, form is causally prior to matter since it satisfies the conditions of definability, knowability and separability better than matter (*Met.* 1036a1-2). Consider an artificial object such as a vase. It has a certain shape or form informing a certain matter, such as clay. The form is prior in cause and time since it has to exist in the craftsman's mind prior to his making the vase. This shape makes this clay into a vase rather than something else, such as a bowl. Because a vase is for holding fluid and flowers, it has to have a certain shape and matter. Form and function go hand in hand, determining the kind of matter something should have. Though a human being is natural and more complicated than vases, his nature is similarly determined by his form – namely, soul – rather than by his matter – namely, flesh and bones. A human soul, one capable of rationality, distinguishes man from plants and animals (*NE* I.7). The rational soul unites the form-matter whole by leading the non-rational animal desires, appetites and emotions. Man is capable of knowing what is true or false, and undertaking actions that are good or bad.[14]

Zhu Xi too has an account of human nature that is like Aristotle's. He too distinguishes human from physical nature. He says, "Man's nature is nothing but principle … It is not a concrete entity by itself which is to be destined as nature and which neither comes into nor goes out of existence."[15] In contrast, "when *physical* nature is spoken of, it refers to principle [*li* 理] and material force [*qi* 氣] combined."[16] By material force or *qi*, Zhu Xi means the expansive energy that is common to matter and mind. Human nature for Zhu Xi is identified with principle (*li*) alone, which is eternal[17] immaterial,[18] prior to physical form,[19] "harmonious, … undifferentiated … invariably good"[20] – that which causes man to be.[21]

The two also agree that the concrete man cannot be separated from matter. Aristotle maintains that the human being cannot be defined apart from matter (*Met.* 1036b23–24) and Zhu Xi maintains that "wherever there is the nature as endowed by Heaven, there is also the physical nature."[22] Because of the inseparability of human from physical nature, plus the inferiority of physical to the non-physical, both agree that human function is to act according to the non-physical, human nature. Thus Aristotle defines human function as reasoning and Zhu Xi speaks of the proper feelings as the function of one's nature.[23] For Zhu Xi, this means that one ought to feel commiseration, shame, deference and compliance. Further, one is to distinguish right from wrong because one is endowed with the moral principles of humaneness (*ren* 仁), appropriateness (*yi* 義), propriety (*li* 禮), and wisdom (*zhi* 知).[24] Put otherwise, one's human nature dictates the proper functioning of one's feelings. Likewise, for Aristotle, excellent human functioning consists in the virtues, which necessarily involve appropriate feelings regardless of one's situation. Apart from subordinating the physical to the non-physical, both concur that human nature is realized in knowing the truth, mastering one's feelings and doing what is right. These activities make one virtuous for both.[25]

Despite their agreements about: (i) the distinction between human and physical nature; (ii) the two natures' inseparability; and (iii) proper function as the subordination of the physical to human nature, disagreements are rife because of their respective metaphysics of form and matter and *li* 理 and *qi* 氣. Aristotle's form is not exactly like Zhu Xi's *li*, nor is his matter like Zhu Xi's *qi*. Comparing their metaphysics will reveal the role that metaphysics and knowledge of the first principles of virtue play in their views about the environment.

It is no accident that Aristotle prioritizes form over matter in causality, knowledge, and ultimately, in defining human nature and function. This results from his view of God as the first cause of everything, and His characteristics. God cannot be movable or He would have to be moved by something else, which would then be prior to Him in time and cause. To be unmovable, God cannot be material or He would be moved in turn when He moves another. Being immaterial also makes God indestructible; material things are combinations of form and matter, which weaker unities are always destructible. Since God is immaterial, indestructible and most united, He is also most stable. Thus, God is most definable and knowable, unlike concrete composites which change, lacking the stability to be defined and known. To remain stable and prior in cause, time and knowledge, God cannot engage in activities that rely on other things. Such reliance would prevent God from being self-sufficient and His activity from being continuous. As such, God's most appropriate activity is thinking of Himself.

From these criteria of God, Aristotle derives the criteria for defining man and his virtues. It is the form of man, namely, rationality that defines man and his function. Since there are immaterial and unchanging things like mathematical objects and God, and material and changing composite things like natural and artificial entities such as men and houses, Aristotle divides the rational soul to accommodate them. The speculative part enables distinction of the true from the false whilst the deliberative part enables distinction of right from wrong. Our controllable, non-rational appetites, desires and emotions, become united to rationality by adhering to decisions from deliberation. Both correctness of deliberation and rightness of our appetites, desires and emotions are dependent on years of experience and habituation. When deliberation is perfected, one has the virtue of *phronêsis* or practical wisdom, where one always does the right thing, at the right time, in the right way, to the right person(s), etc. When appetites, desires and emotions are perfected, one has the virtues of character such as generosity, justice, courage, temperance, etc. A *phronimos* possesses all the moral virtues for these tell him what ends are to be pursued, e.g., justice, courage, and temperance, etc. *Sophia*, on the other hand, perfects speculation, providing wisdom of unchangeable entities.

God is primary substance for Aristotle because of His oneness, stability, self-sufficiency and knowability, which criteria are also the standards of goodness. Consequently, contemplative wisdom or *theôria* is ranked higher than *phronêsis* and the moral virtues because it better satisfies these criteria. Aristotle tells us that the activity of *theôria* is higher than *phronêsis* because it is more self-sufficient. *Theôria* does not need external goods, such as wealth or friends, unlike practical wisdom which cannot be exercised without such goods, since one needs wealth in order to be generous and friends to help one accomplish virtuous actions. The discontinuity of *phronêsis*, because it ends with the completion of each virtuous action, also makes it less pleasant than *theôria* which can be continued and enjoyed for longer, further causing it to be more united and stable, like God.

Like Aristotle's form which priority is based on his first cause, the priority of Zhu Xi's *li* 理 is also based on his first principle. Zhu Xi's *li* is prior to everything as their cause. Zhu Xi says, "Principle refers to the Way, which exists before physical form [and is without it] and is the root from which all things are produced."[26] Apart from being a first source, *li* is also immaterial since it is prior to physical form. Thus, Zhu Xi's *li* is similar to Aristotle's substance in being immaterial and prior. *Li*, like Aristotle's substance, is the standard of goodness for it too, is the highest good (Chan, 640, 49: 11b). Zhu Xi says, "What is right is the Principle of Nature [*tian li*, 天理 Heaven's principle], and what is wrong is in violation of the Principle of Nature."[27] Thus Zhu Xi urges people to cultivate themselves by overcoming and eliminating selfish desires which inhibit their knowledge of *li* 理.[28] Since both authors maintain that first principle is the measure of ultimate goodness, they agree too that knowledge of this is essential to being virtuous.

In spite of these similarities between their first principles and virtue, there are numerous differences. One major difference is the presence of a distinction in Zhu Xi's first principle absent in Aristotle's. Whereas Aristotle's principle is simply God, Zhu Xi's first principle is divided into the Non-Ultimate [*wuji*] and the Great Ultimate [*taiji*]. Whilst Aristotle's first principle is always active in thinking itself, Zhu Xi's *wuji* is without activity, whilst his *taiji* is active. By the *wuji*, Zhu Xi does not mean a state of total nonexistence. Rather he means a state where only principles exist, without any activity whatsoever – only tranquility.[29] As such, *wuji* is also removed from the *yin* 阴 and *yang* 阳 *qi*, and is thus removed from the physical realm.

Taiji, on the other hand, is closer to *qi* for Zhu Xi: "The Great Ultimate exists only in the *yin* and *yang*, and cannot be separated from them."[30] The invisible principle becomes visible through *yin* and *yang*: "Principle attaches itself to *yin* and *yang* as a man sits astride a horse. As soon as *yin* and *yang* produce the Five Agents 五行 (*wuxing*), they are … fixed by physical nature and are … differentiated into individual things each with its nature. But the Great Ultimate is in all of them."[31] *Taiji* then, is a kind of mediator between the totally immaterial, inactive and transcendent *wuji* and the physical world. It shares some characteristics of the *wuji* and is immanent in everything.

Given that Aristotle's God is always active, He differs from Zhu Xi's *wuji* which is prior to activity. God is also unlike *taiji* which is said to "begin" activity. The apparent direct contact between *taiji* and *qi*, its becoming visible as a result of *yin* and *yang qi*, and its immanence in them, are all characteristics alien to Aristotle's God who is not in contact with anything, let alone immanent in them. Unlike *taiji*, Aristotle's God also does not become visible nor is He ever dependent on anything else. Whilst causation for Zhu Xi seems to be inseparable from *qi*, first causation for Aristotle is completely separate from matter since God is the final cause who moves everything as if by love. The fact that both of Zhu Xi's Ultimates contain the concrete principles of all things also distinguishes them from Aristotle's God who is one and self-sufficient. Other things are only related to God in a *pros hen* manner, i.e., insofar as they more or less satisfy the same criteria as He (such as being definable, knowable, individual, one and being prior in time and cause), they are more or less one in being.

What implications for environmental ethics follow from these thinkers' different conceptions of first principle(s), which in turn determine their views of goodness? Given Aristotle's God's self-sufficient activity, which is totally separate from all physical and changeable things, the norm for ultimate human goodness is the self-sufficient activity of contemplation rather than the other-related activity of *phronêsis*, which is the norm for secondary goodness. Notice that even *phronêsis*, for Aristotle, is the virtue of human relations instead of relations between human beings and the environment in general.[32]

On the other hand, neither Zhu Xi's *wuji* nor *taiji* is self-sufficient. Each seems to be essentially related to all things, both human and non-human, by containing their principles. Whilst the *wuji* is not active at all but tranquil, the *taiji* relies for its activity on *qi*, or the material energy that is common to mind and matter. Accordingly, Zhu Xi's metaphysics offers an interdependent view of first principles and things. Since his first principle, like Aristotle's, is also the norm of goodness, Zhu Xi's interdependent view of human and nonhuman things means that their goodness are interrelated too. More importantly, *taiji* not only causes all things, but is immanent in them. This means that *taiji* is not only the norm of *human* goodness, but also the norm of the goodness of all things. In brief, the human or moral good for Zhu Xi, is identical with the goodness of everything. This is the kind of resource that is useful to environmental thinkers keen on defending the intrinsic value of the non-human world. If the same goodness exists in everything, then it is inconceivable for what is good for human beings to contradict the good of the non-human world. Moreover, if the source of everything for Zhu Xi is the non-ultimate which is tranquil, then proper human desires and actions would preclude the exploitation of their environment for such exploitation would be a kind of excess. The norm of tranquility and harmony would also mean a balance of growth and decay/destruction in the natural world, protecting healthy ecospheres from excessive manipulations by humans for the sake of humans.[33]

That our relation to the first principle is dependent on our relation to the natural world for Zhu Xi is clear when he discusses the *wuji* as the Way. He says, "Principle refers to the Way, which exists before physical form [and is without it] and is the root from which all things are produced."[34] Just as *taiji* mediates the transcendent *wuji* and the myriad things, it is also the key to knowing *wuji*. He says, "Way … means something indefinite, boundless, and different in infinite variety, where concreteness is not to be seen. Only when we look into nature – do we find the Way in its concreteness."[35] To know the Way, we must look into nature, or *li*, in the sense of *taiji*. To know *taiji* is to know the concrete principles in it. Such knowledge enables one to know the proper ways of things and events, removing perplexity, leading to knowledge of proper actions.[36] Moreover, Zhu Xi believes that because *li* is one for all things[37] the investigation of things (*gewu* 格物) will not only lead to knowledge of these things, but also to right actions. Since things and man share the same *li* 理, which *li* is always good, nothing in the world is devoid of moral principles. Thus, *gewu* and knowledge of things can lead to moral goodness for Zhu Xi. Zhu's united conception of the knowledge of things and morality differs radically from Aristotle's separation of the two.

Despite Zhu Xi's unity of theoretical knowledge and moral actions, to the extent that Zhu Xi elevates human over physical nature, he too has an explanation for how human beings are higher and better than animals, plants and nonhuman things. This lies in his conception of *qi*.[38]

Zhu Xi insists that principle must exist in *qi* (634) to become manifest and active. The kind of *qi* in which principles must inhere is mind (*xin* 心). He says, "The mind embraces all principles and all principles are complete in this single entity, the mind,"[39] and "There must be mind before nature can be gotten hold of and put into operation."[40] So although one and the same *tian li* (天理 Heaven's principle) is always completely endowed in all things, not everything can access it completely. Something's limited access to *li* 理 is due to its *qi*. Depending on the clarity or turbidity, purity or impurity of *qi*, something will be more or less obstructed from *li*. In spite of how men, animals and things are born with the same *tian li*, the *qi* of things and animals obstruct their access to *li*; they can never penetrate this obstruction. Zhu Xi says, "The principle received by things is precisely the same degree as the material force received by them. … the physical constitution of dogs and horses being what it is, they know how to do only certain things."[41]

Zhu Xi holds that what makes man superior to things and animals is his ability to possess the Great Ultimate (*taiji*) completely, or his ability to realize his nature containing the principles of *ren*, *yi*, *li* 禮 and *zhi*. These principles enable man to "love people gently and ... benefit things;" activities that harmonize with the mind of Heaven and Earth which is to "produce things infinitely"[42] so that he can cooperate with Heaven and Earth in working for the benefit of the universe.[43] Man's ability to exercise these virtues depends on his ability to master *qi* and control selfish desires. Zhu Xi says, "The good that emanates of course proceeds from the mind. But the evil that emanates is all due to selfish material desires endowed by material force."[44]

These thinkers' views of knowledge of the first principles again have implications for environmental ethics. Because of Aristotle's separation of *theôria* from *phronêsis* and his elevation of *theôria's* God-like self-sufficiency over the troublesome nature of *phronêsis*, our relations to others and our environment are at best secondary goods to our well-being. Our primary happiness of *theôria* offers knowledge impractical for both relations to humans and to the environment. In radical opposition, Zhu Xi insists on uniting ultimate reality, our knowledge of it and correct actions in both the human and nonhuman world. Because we share the same moral principles as everything else, and yet we are best able to know them, we are responsible for working with heaven and earth to benefit everything else.[45] Since the same moral principles are in us and the nonhuman world, theoretical study of nature for Zhu Xi can lead to an understanding of correct actions toward humans and our environment, and actions can lead to knowledge too.[46] Such a unity of theory and practice is not only different from Aristotle's separation, but also makes possible multiple accesses to Zhu's first principles. Thus Zhu's first principles are more accessible than Aristotle's which are limited only to those who contemplate.

The danger of Aristotle's account is that our environment is relegated to the realm of luck of which our flourishing seems completely independent. Contrast this with Zhu Xi's injunction to pursue knowledge of the Way which is intertwined with cultivation of seriousness (靜 *jing*).[47] Seriousness is a way of controlling our body and mind so that we can attain a kind of tranquility and calmness, exemplified by the *wuji*.[48] Tranquility in turn prepares us for extending knowledge – through both study and action – enabling a harmonious relation with humankind and the environment. No doubt, Aristotle's teleological account of nature ought have implications for what is natural for our environment. But because the environment plays a minimal role in his talk of human virtues, we also do not seem to have a role to preserve it. Contrariwise, Zhu Xi's interconnected view of nature, self and the first principles also means that we are essentially bound up with our environment. Ultimately, knowledge and action cannot preclude consideration of what is good for the environment, which good is dictated by *wuji's* tranquility. Contrary to Aristotle's celebration of independence from our environment, Zhu Xi urges interdependence between man and environment and offers a norm for extending our humaneness to the rest of the universe.[49] Both Aristotle and Zhu Xi offer first principles which are the norms for actions and our relations to the environment. I suggest that examining critically their metaphysics can illuminate how we are to properly relate to our environment, and on this issue, Zhu Xi can offer more resources for the kind of deep ecology Naess champions.

Glossary of Chinese Terms

dao 道	Way
gewu 格物	Investigation of things
jing 靜	Seriousness
li 理	Principle
li 禮	Ritual Propriety
qi 氣	Material energy
ren 仁	Humaneness
taiji 太極	Great Ultimate
tian li 天理	Heaven's principle
wuji 無極	Non-ultimate
wuxing 五行	Five agents
xin 心	Heart-mind
yang 阳	Positive principle of cosmological change: identical with light and activity; opposite to and dependent on *yin*
yi 義	Appropriateness
yin 阴	Negative principle of cosmological change: identical with darkness and passivity; opposite to and dependent on *yang*
zhi 知	Knowledge/wisdom

Notes

[1] *Ecology, Community and Lifestyle: Outline of an Ecosophy*, (Cambridge University Press, 1989) p.28.
[2] Ibid, p. 28.
[3] Ibid, p. 33.
[4] *Ecology of Wisdom: Writings by Arne Naess*, Alan Drengson and Bill Devall, eds. (Berkeley: Counterpoint, 2008) p. 111.
[5] Ibid, p. 111.
[6] *Nicomachean Ethics* (hereafter *NE*) Bk. VI.
[7] As Zhu Xi puts it, "Of the thousands and tens of thousands of human beings and all things, there is none independent of these moral principles." Wing-Tsit Chan, *A Sourcebook in Chinese Philosophy* (Princeton University Press, 1963), p. 617, 42:13a–b.
[8] Chapter 2 in *Environmental Virtue Ethics* (hereafter, *EVE*), eds. Ronald Sandler and Philip Cafaro (Rowman and Littlefield Publishers, Inc., 2005), pp.31–44, see especially pp. 32–33.
[9] Ibid, p. 34.
[10] See *NE*, Bk. X, op cit.
[11] *EVE*, p. 34 and p. 38.
[12] Thomas Hill, Jr., "Ideals of Human Excellence and Preserving Natural Environments," ch. 3 in *EVE*, especially pp. 53 and 57.
[13] *Metaphysics*, (hereafter *Met*) 1028a31 and 1029a28.
[14] For a detailed account of Aristotle's view of the human soul and function, see my *Remastering Morals with Aristotle and Confucius* (Cambridge University Press, 2007), chapters 1 and 5.
[15] *A Sourcebook in Chinese Philosophy*, op cit, p. 614, 42: 5a. See also Zhu Xi, *Reflections on Things at Hand*, (hereafter, *RTH*) trans. Wing-Tsit Chan, (Columbia Univ. Press, 1967), "The nature is the same as principle," ch. 1, no. 38, p. 28, and "One's nature is the one source of all things and is not one's own private possession," ch. 1, no. 48, p. 32. Compare these with Zhu Xi, *Further Reflections on Things at Hand*, (hereafter, *FRTH*) trans. Allen Wittenborn (Univ. Press of America, 1991), ch. 1, no. 15, p. 61, "Nature is what is not yet moved. Feelings are what is already moved."
[16] *A Sourcebook in Chinese Philosophy*, (hereafter, Chan) op cit, p. 613, 42: 4b. Chan translates '*qi*' as 'material force'. See also *RTH*: "What exists before physical form [and is therefore without it] constitutes the Way. What exists after physical form [and is therefore with it] constitutes concrete things," ch. 1, no. 19, p. 18.
[17] Chan, p. 624, 43: 3a–b, see 618, 42: 18b.
[18] Chan, p. 615, 42: 6b–7a.
[19] Chan, p. 634, 49: 1b.
[20] Chan, p. 597, Zhu Xi's commentary on Cheng Hao's view of nature, *I-shu*, 1: 7b.
[21] Chan p. 612, 42: 1a–b, p. 615, 42: 6b–7a, see also Chan's comment of 49: 1b on p. 634.

[22] Chan p. 616, 42: 6b–7a. See *RTH*, ch. 1, no. 21, "What is inborn is called nature. Nature is the same as material force and material force is the same as nature. They are both inborn," p. 19.
[23] Chan p. 612, 42: 1b.
[24] Chan p. 615, 42: 6b–7a. *RTH*, ch. 1, no. 15: "That which is inherent in things is principle. That by which things are managed is moral principles," p. 16.
[25] As I have already pointed out, such human virtues are used by environmental thinkers like Philip Cafaro and Thomas Hill, Jr. to argue for how the preservation of the integrity of the environment can contribute to the cultivation of human virtues. In addition, Cafaro and Holmes Rolston III also stress that a proper environmental ethics moves beyond a mere concern for the cultivation of human virtues to the realization that it is also essential for the well-functioning of the nonhuman realm like plants, animals and other organisms in an ecosystem (Holmes: p. 73, Cafaro: p. 38, op cit). See Holmes Rolston III, "Environmental Virtue Ethics: Half the Truth but Dangerous as a Whole," ch. 4 in *EVE*, pp. 61–78.
[26] Chan, p. 636, 49: 5b.
[27] Chan, p. 606, 2: 2a. See *RTH*, ch. 1, no. 38: "The nature is the same as principle. Traced to its source, none of the principles in the world is not good," p. 28. See also no. 48.
[28] Chan, p. 594, Zhu Xi's commentary on *Analects* 12: 1.
[29] He says:

> While the state before activity begins cannot be spoken of as the Great Ultimate, nevertheless the principles of pleasure, anger, sorrow, and joy are already inherent in it. … In the initial stage the four are not manifested, but their principles are already there. (Chan, p. 639, 49: 11a–b).

Tranquility, for Zhu Xi, "is not non-being as such. Because it has not assumed physical form, we call it non-being. It is not because of activity that there is being. Because (activity makes) it visible, we call it being" (Chan, p. 629, 44: 6b–7a). See also *RTH*, ch. 1, no. 1, p. 5–7.
[30] Chan, p. 630, 45: 1a.
[31] Chan, p. 641, 49: 14a. See Chan, p. 635, 49: 3a–b, "As there is this principle, therefore there is this material force operating, nourishing, and developing." I think that *taiji* either causes *qi* to start creating things according to its principles so that *qi* pre-exists it, or it causes *qi* itself and its subsequent creations. See also *RTH*, ch. 1, no. 15, p. 16 and *FRTH*, ch. 1, no. 49, p. 66: "Nature is the principle of heaven. All things innately receive it."
[32] Nevertheless, that there are resources in Aristotle's ethical system for the environment if we were to pursue his account of external goods, is a topic which cannot be explored within the scope of this essay. I am sympathetic with Laura Westra's argument for "a constant interface between environment and human life and flourishing that renders environmental rights as foundational to basic human rights," (81). See Laura Westra, "Virtue Ethics as Foundational for a Global Ethic," ch. 5 in *EVE*, pp. 79–91.
[33] For the practical effectiveness of norms for conservation biology, see Arne Naess, *Ecology, Community and Lifestyle: Outline of an Ecosophy*, op cit. Naess

follows Michael Soulé's four norms which form the basis of conservation biology. They are the goodness of: diversity in organisms, ecological complexity, evolution and biotic diversity (p. 46). Naess says, "The normative character of conservation biology results in recommendations and decisions which can be made as hastily as our current political structures require. These have … influenced the political decisions in Western New Guinea … to establish a series of national parks" (p. 46).

[34] Chan, p. 636, 49: 5b.

[35] Chan, p. 619, 42: 21b–22a.

[36] Chan, p. 611, 3: 34a.

[37] "Principle is one," (Chan, p. 612, 42: 2b); "From the point of view of principle, all things have one source, and of course man and things cannot be distinguished as higher and lower creatures," (Chan, p. 622, 42: 27b–29a); and "there is only one Great Ultimate … each of the myriad things has been endowed with it and each in itself possesses the Great Ultimate in its entirety," (Chan, p. 638, 49: 10b–11a).

[38] Chan, p. 623, 42: 31b–32a.

[39] Chan, p. 606, 2: 4b. See also Chan, p. 615, 42: 6b–7a, "At bottom nature has neither physical form nor shadow. It is merely the moral principle possessed by the mind."

[40] Chan, p. 616, 42: 6b–7a. See also Chan, p. 630, 44: 14a, "The mind is the principle of production."

[41] Chan, p. 620, 42: 26b–27a. See *RTH*, ch. 1, no. 51, p. 34 and *FRTH*, ch. 1, no. 26, p. 63: "Nature is the same, but energy is different."

[42] Chan, p. 595, 67: 20a–21b. See also Chan, p. 613, 42: 3a–b, "if I investigate principle to the utmost and fully develop my nature, then what I have received is wholly Heaven's moral character, and what Heaven has endowed in me is wholly Heaven's principle," and Zhu Xi's quotation from Zhang Zai: "If one investigates principle to the utmost and fully develops his nature, then his nature will be in accord with the character of Heaven and his destiny will be in accord with the Principle of Heaven" (Chan, p. 613, 42: 3b). See *RTH*, ch. 1, nos. 1, 20, 48 and 51.

[43] In interpreting Zhu Xi's account as one which makes human beings superior to the nonhuman world, I am subscribing to a kind of anthropocentrism and the promises which the Confucian and Neo-Confucian highest virtue of humaneness (*ren* 仁, the highest virtue where familial love is extended to the rest of the universe) has to offer to the environment. Just as Philip Cafaro and Holmes Rolston III urge the extension of virtue ethics to the nonhuman realm so that we are attuned to the intrinsic value of our environment, I believe that Neo-Confucianism offers us the resources for such an extension. As such, I differ from Toshio Kuwako who urges that the most important contribution of Neo-Confucianism is its "interpretive turn of *jen*" from humaneness to embrace both humans and their environment so that "Chu Hsi's philosophy can be seen not only as antianthropocentrism but also as antihumanism" (155). See Toshio Kuwako, "The Philosophy of Environmental Correlation in Chu His," in *Confucianism and Ecology: The Interrelation of Heaven, Earth and Humans*, eds. Mary E. Tucker and John Berthrong (Harvard University Press, 1998), 151–168.

[44] Chan, p. 628, 44: 2b–3a. See also, Chan, p. 618, 42: 14b–15a, "in its original state the Principle of Nature is free from human desire," and Chan, p. 608, 3: 3a, "If the Principle of Nature exists in the human mind, human selfish desires will not, but if human selfish desires win, the Principle of Nature will be destroyed." See also *RTH*, ch. 1, no. 14, p. 14–15 and no. 38, p. 28–29 and *FRTH*, ch. 1, no. 46, p. 66: "Mind means ruler."

[45] Zhu Xi, *FRTH*, ch. 1, no. 4, "Our heaven-endowed nature is everywhere. But when we look for it we must start with ourselves. This is why we say that one's nature is the form and essence of *Dao*," p. 59.

[46] Zhu Xi, *FRTH*, ch. 3, no. 15, p. 89: "Investigating the phenomena of things is the beginning of knowledge. Keeping our thoughts sincere is the beginning of practice."

[47] See *RTH*, ch. 1, no. 26, p. 23 and no. 1.

[48] As Zhu Xi puts it, "Tranquility nourishes the root of activity and activity is to put tranquility into action ... For example, when the feelings are aroused and all attain due measure and degree, that is tranquility in activity" (Chan, p. 607, 2: 38a). See Zhu Xi, *FRTH*, ch. 4, no. 37, p. 102: "The mind of the gentleman is like deep and clear water: even small waves will not form."

[49] It is Zhu Xi's metaphysical norm, founded upon the *wuji*, that makes his approach superior to the relational approach Marion Hourdequin and David B. Wong propose in "A Relational Approach to Environmental Ethics," *Journal of Chinese Philosophy* 32: 1 (2005): 19–33. Hourdequin and Wong try to justify an environmental ethics by arguing for a relational conception of the self that is constituted by the self's relation to the nonhuman world (p. 20). While I agree with them that the self is constituted by the environment by being shaped by it, nothing in their account offers what is normative for the environment. To gather from the Inuit and Western Apache traditions, as they do, that humans are not sharply separated from the environment, does not yet tell us what our right relation to the environment is (p. 26). We might have strayed so far from a correct relationship with the environment that looking at these different traditions' practices does not suffice for arriving at a right relation. On the other hand, I think that Zhu Xi's account of the *wuji's* state of tranquility, as well as the motions of *yin* and *yang qi* which determine the nature of things, can provide us with a more concrete content for relating to our environment regardless of how far we might have departed from the norm.

CHAPTER SIX

EARTH DEMOCRACY: INSTITUTIONALIZING SUSTAINABILITY AND ECOLOGICAL INTEGRITY

KLAUS BOSSELMANN

> "(E)cological wisdom does not consist in understanding how to live in accord with nature; it consists in understanding how to get humans to agree on how to live in accord with nature." (Wilber 2001: 268)

The Institutional Bias against Sustainability

The central problem faced by the Global Ecological Integrity Group, or any other initiative towards a sustainable future, is not so much ecological integrity or sustainability itself, but how to achieve *agreement*. The importance of this point cannot be overstated. For it is not that we generally have no solutions to the problems of carbon emissions, global poverty, wealth inequality, resource depletion and so on. We do and anyone doubting this should ask themselves what sort of information they rely on. The point is that we are failing, as a global community, to agree to *implement* those solutions.

Generally speaking, the chances for agreement depend on what participants feel most strongly about. If the economy is seen as paramount such view will dictate the extent to which environmental concerns can be accommodated. At present, the main actors of the global community are corporations, professional networks, states and their organizations at international level. They all see the world through economic lenses focused on finances and growth. This does allow for a view sympathetic to sustainability, but only as a balancing act of economic, social and environmental issues. From this perspective, there is no need for readjusting priorities (the "weak" sustainability model).

By contrast, most people do not see the world through economic lenses. Their perspective is wider and more receptive to fundamentals such as social security or environmental sustainability. To most, economic prosperity is important, but not at all cost. There is a widespread feeling that political leaders and corporate managers, in their money and power games, are too preoccupied to notice the real needs of people and the planet. As we are heading for social and ecological disaster, our leaders ought to take a broader perspective and be more responsible. Such view would consider sustainability as more important than economics (the "strong" sustainability model; Bosselmann 2009).

With such vastly different viewpoints chances for agreement are slim. So far, the agreement has only been made between states and only on the lowest common denominator, called "sustainable development" (Bosselmann 2008: 55–57).[1] Loosely defined, sustainable development calls for a balancing of environmental, social and economic needs. But even this weak sustainability model is only starting to inform the political economy. Governments aim for more integrated policies and some businesses promote triple-bottom-line accounting. Such measures are a first step towards integrated governance, but not more.

The real challenge is shifting the attention of governance from growth to sustainability. Only then is integration meaningful. As long as economic growth dictates direction and extent of "sustainable" development, we will not move forward. Like hamsters in their wheels we will be running in circles.

To overcome their growth fixations, institutions need to focus on prosperity as a central objective of macroeconomics. Prosperity is a much larger concept that involves material wellbeing, but not automatically growth. "Prosperity without growth" (Jackson 2009) is not only feasible, but the only way to ensure that the economy operates within ecological boundaries. To this end, institutions will have to become more ecologically literate, but also mindful of a simple fact: the more economic output there is, the greater the environmental impact. Today, the global economy is nearly five times the size it was fifty years ago, i.e. at the beginning of the environmental crisis. If it continues to grow at the same rate, the economy will be 80 times that size by the end of this century (Jackson 2009: 102).

That growth cannot go on forever is a triviality. To draw the right conclusions economically and politically, however, seems almost impossible. How can this be? One reason is probably the success rate of capitalism over the last 200 hundred years. Driven by growth, the economy has always been a promise for the future: more outputs, more jobs, better life for all. This promise has only ever been kept for some, of course, and certainly to the detriment of the environment, but for proponents of capitalism the success mythos has endured the changes of times almost unharmed.

Another reason may be more daunting. If the institutions of governance believe that there is no alternative to economic growth even in the face of ecological and human disaster, then we must doubt their ability to govern responsibly. It is not that states are incapable of governing at all. Recently, we have witnessed just how willing states have been to help the economy in times of crisis. Governments intervened by lending to banks and companies in return for more regulation and oversight. That a more regulative economic system has not emerged matters less than the very fact of intervention. Obviously, states saw it as their responsibility to act in a situation of economic crisis. What prevents them to extend this responsibility to the task of adjusting the economy to ecological realities, is their obsession with growth. Behind all sate interventions we have ever seen was the expectation of growth. To turn this around and promote a sustainable economy, quite simply, seems beyond the imagination and ability of governments (Jackson 2009: 94).[2]

Any progress here depends on engaging a wider community for advice. This has not yet happened, but may be critical for restoring already lost credibility of governments. In addition, there is the vital question of leadership. Consulting with the wider community, as important as it is, does not in itself lead to better decisions. It takes strong leadership to intervene with people's values and aspirations. If we are to achieve a sustainable society, we all must change the logic of consumption: from more is better to less is more. Such reversed logic flies into the face of consumerism and capitalism and all of us have responsibilities here. However, that governments could see it as their prime responsibility to wean us off from consumerism is almost unimaginable. Yet, only a common effort by those who govern and those who are governed could bring about the necessary behavioral changes.

From what has been said so far it should become clear that the institutional bias against sustainability is more than negligence and flawed thinking. Governments argue from the point of economic rationality, whereas common sense dictates the wider perspective of ecological rationality. This perspective resides largely outside the institutions of governance. The challenge is, therefore, to find new ways of institutionalizing it.

To bring the two perspectives closer together, economic rationality must, first of all, be reduced to size. It has its origins – and rightful space –

in the development of business and market strategies. To this end, it is also part of the public sphere and an important one at that. But economic rationality has little to offer to some of society's basics, i.e. motivations, values and ethical guidance for policy, law and governance. Economic rationality has dominated the public sphere so much that even these basics are seen through the lens of an imaginary "economic man". In the words of J. Emerson (Emerson 2009): "The problem with economic man is not that he is an imaginary theoretical fiction, but there are too many of them. Economic rationality, which is supposedly purely a formal assumption, has bled out into the community and has become a foundation of ethics." The authority economic man or *homo economicus* is totally undeserved and unacceptable considering the disastrous ecological effects of the economic system now ruling globally.

Ecological rationality (Bosselmann 2008: 319, 322), on the other hand, perceives economic concerns in a wider conceptual framework comprising social, environmental and economic relationships. The environment is, however, not just a pillar next to the social and economic pillar, as economic rationality would define sustainable development. The environment is foundational to everything. Sustainable development is, therefore, only possible, within the ecological limits.

Seeing economic and ecological rationalities in their respective places is foremost a task for our educational institutions (including media). Environmental literacy and sustainability knowledge should have priority in school and university curricula (Bosselmann 2001: 167–186). But even an educational revolution of this kind will not in itself change the general political discourse. The powerful players would be challenged by people with new ideas, but would they change. They are in power after all and dominate the public sphere (again, including media). Basically, the entire public sphere is captured by economic rationality and needs to change.

At present, public institutions are not equipped and willing to embrace the sustainability agenda. But they are not blind to it either. Some political parties have been serious about the priority of ecology over the economy, so have ethically motivated business and responsible media. It is doubtful, however, to see them as early signs of a general greening of society. The economic wind is blowing into their face and probably never more so than now amidst the global economic crisis. Moreover, the green agenda that most Green parties, business or media promote, has been shallow and light, playing to the hopes of those who are not yet disillusioned with the status quo: you can eat your cake and save it too, just eat with greater care.

The big issues like economic growth, consumerism, population growth or Western world dominance are hardly mentioned in any parliamentary debate. And the number of voters who live simply so others can simply live, is arguably very low; most Green voters prefer some feel-good deal between prosperity and sustainability. No, the consequent implementation of sustainability will not be promoted by our public institutions, at least, for the time being.

Whether the economic bias that our institutions and indeed the whole world is plagued with can be overcome is, of course, everybody's guess. There may be good reasons to not believe in such a possibility and to rather "get on with life". Such a conclusion may be one of denial, a strong characteristic of the human psyche. Or it may be one of wisdom as can be seen in the many practical examples of living in accordance with nature. On the other hand, we should not discount the ability of human intelligence to change and transform the conditions that they have created for themselves. This is true for natural environments as much as for social environments and their institutions. In the end, the institutional bias against sustainability is historically determined and hence not unchangeable. The question can be asked, therefore, what it may take for public actors and institutions to transform their economic bias into a wider ecological perspective.

This question is one of the right form governance. How can we perceive governance for sustainability (Bosselmann, Engel and Taylor 2008)?

Basically, governance involves the formation and operation of social institutions. It consists of sets of rules, decision-making procedures, and programmatic activities that serve to define social practices and to guide the interactions of those participating in these practices (Young 1997; Young 2002). Governance does not necessarily require material entities or organizations of the sort normally associated with governments, therefore governance must be distinguished from government which implies a centralized institutional arrangement as the basis of authority and order. All forces that can influence human behaviour are potential tools of governance (Zaelke, Stilwell, and Young 2005: 38). Governance can encompass networks, informal regimes, and customary linkages, with a minimum degree of bureaucratic centralism and a maximum amount of political space for exploration and diversity. Governance is created as a result of individuals recognizing that they are interdependent; that the actions of one can affect the welfare of others. Such interdependence can lead to conflict, when members of society recognize that the efforts of some to achieve their goals will interfere with the efforts of others to achieve different goals. However, interdependency can also be a basis for cooperation when group members recognize opportunities to improve the welfare of the group by coordinating their behaviour. In a world of increasing interdependence among people, societies and with nature, new approaches to governance must be developed at the local, national, regional, and global levels.

It is within this context that the search for appropriate governance must occur. This includes the possibility of developing forms of governance that are new to the modern Western form of democracy. Nevertheless, there are pertinent reasons why any search must begin with an analysis of the status quo. So, what are the tensions between democracy and sustainability? And what would it take to reconcile the institutional form of democracy with the value-laden idea of sustainability?

Tensions between Democracy and Sustainability

A widely held assumption of Modernity is that there is no alternative to democracy. From the sustainability perspective that confidence is dubious as the track record of democratic societies has not been great. This is not the same as saying that other forms of governance may be more desirable. A dictatorship is probably more effective in implementing sustainability goals, yet less attractive overall considering the importance of human rights and individual freedom. The point is to look through all the various models of governance known to us to see how accountability of both, of governments and citizens can be maximized.

Fundamentally, we are facing a dilemma. If democratic societies have failed, so far, to meet the sustainability challenge, does democracy stand in the way of sustainability? Or is democracy a *conditio sine qua non* for achieving agreement on sustainability? Overlooking the political debate we can find evidence that can be used for either view. A "benevolent" authoritarian system may well be able to enforce more sustainable behaviour, but equally it can block any such moves. Likewise, democratic societies may be more conducive to sustainability initiatives of their citizens, but may also lack the necessary leadership to install change.

China offers an example in case. The ability of a centralized power to mobilize resources for a shift towards sustainable development can be contrasted to the populism and consumerism of democratic states. Yet the reality of environmental degradation in China belies the idea of political leadership, at least so far. Any attempt to find an example of successful governance for sustainability among existing states must fail. It simply does not exist.

The links between democracy and sustainability are far from being clear and seem obscured by the fact that both, proponents and critics of democracy are unified in their advocacy for more sustainability. The historical association of democracy with capitalism has obviously blurred the differences between the political and economic system. However, while it seems obvious that capitalism and sustainability are at odds with each other, the real challenge is to distil an idea of democracy that is conducive with, and supportive of, sustainability. How can democracy in a "pure" sense be perceived and then linked to sustainability? The answer to this question, important as it may be, cannot be given; at least, not at present. For the moment, we have to acknowledge the reality of a close nexus between liberal capitalism, economic growth and democratic societies (Speth 2008;Eckersley 2004).

Unraveling the nexus between economic growth and democracy

Yet despite this close nexus, democracy has the sovereign of the people at its core. Democratically elected governments may have consistently promoted economic growth at the expense of ecologically sustainable development, but that does not tell us the full story of democracy as a political system. Governments are elected by the *demos*, i.e. its citizens. It is only by virtue of citizens that governments are able to reaffirm the idea of economic growth, hence missing the point of sustainability. It would therefore be wrong to characterize democratic societies as inherently unsustainable societies. Such a view would overlook the crucial importance of civil society. As constituency and counterpart of governments, civil society can either be critical or supportive of economic growth. Likewise, civil society can either be indifferent or proactive with respect to sustainability. The issue may be more to what extent sustainability concerns feature in the democratic processes. Given the fact that both, the environmental debate and the sustainability agenda are closely associated with civil society, the prospects for democratic, sustainable societies are ultimately determined by the role that civil society is willing and able to play (Bosselmann 2009).

To assess the tensions between democracy and sustainability it is useful, therefore, to introduce the notion of civil society as a third category. It will help us to better understand why and to what extend democratic societies are self-reflexive in the sense of being able to reflect, learn and change.

Using the three categories of civil society, democracy and sustainability, we can make some interesting observations. Recently, a conference on "Democracy and Sustainability" was held in London

(Elkington and Lothrington 2008). At its centre stood the question whether democracy can be decoupled from the economic growth model" To answer this question the relationships between democracy, civil society and sustainability were examined. As a basis for this examination a paper had been commissioned that identified four very different propositions (Christie 2007). They are as follows:

1. *Democracy is crucial for humane and just sustainable development.* Democracy can be shown to be very closely associated with high standards of ecological protection and effective implementation of environmental law. We cannot begin to tackle big environmental challenges without democracy and all it is based on – rule of law, open society, free media, experimentation and low levels of corruption. The worst cases of unsustainable development at local and regional scales are being exacerbated above all by the misrule of authoritarian regimes. The many non-democracies can tackle unsustainable development only by adopting democratic processes and moving to open societies based on the accountable rule of law.

2. *Democracy poses huge problems for sustainable development.* In the advanced liberal capitalist states, democracy is tightly coupled to the promise of economic growth, ever-rising consumption and individual freedom. Representative democracies have become sclerotic and throughout OECD member states there is a widespread decline of public trust coupled with political apathy.. Politicians cannot challenge vested consumer and producer interests for fear of losing votes, media support, lobby and associated funding. This makes democracies incapable of mobilizing citizens to tackle collective action problems on a big scale – above all, climate disruption and the need for deep emission cuts.

3. *Sustainability NGOs are a massive success for civil society worldwide.* Without them, we would not have anything like the progress we have seen in the past half-century in protecting the environment, cutting pollution, raising resource efficiency, highlighting linked issues of environmental and social injustice, and saving wildlife and habitats from destruction. Without them, the discourse and practice of sustainable development would not have become established in governments worldwide, and huge issues such as climate disruption would not have been acknowledged or tackled sufficiently by governments and businesses. NGOs have been at the forefront of civil society's emergence in authoritarian states and have played a key role in fostering democratic trends and challenges to abuses

of power.

4. *Sustainability NGOs are a massive failure by their own standards.* For nearly 50 years now they have campaigned and educated citizens, governments and businesses worldwide ; yet ecological damage continues on a vast scale, environmental injustices abound, and dangerous climate disruption seems to be unavoidable. NGOs have achieved limited gains in specific areas of policy but have failed to mobilize and energize citizens on a large enough scale to put real pressure on politicians and businesses in the West and beyond.

Prima facie, the four propositions seem in conflict with each other. Why are democracies seen as crucial for achieving sustainability when they have shown so little interest in tackling the real issues? How can the claim of massively successful NGOs be reconciled with the verdict of massively failing NGOs? And is there any justification for democracy as a political model when both, civil society (NGOs) and governments have failed to turn around unsustainable development?

At a second look, the various propositions are not conflicting, but complementary. They form a certain pattern. When civil society was able to lift the environmental problem to public prominence, even against the will of those in power, then this is democracy in action. When, on the other hand, the democratic system continues to give more prominence to short-term economics over long-term sustainability, then this is a system failure assuming that the *demos* has an interest in long-term security. Perhaps such an assumption cannot be made – people may not practice what they preach – but that does not change the underlying pattern, namely the crucial importance of civil society. In the end, all institutions and forms of governance take their mandate from citizens acting individually as well as cumulatively. This would explain why civil society has been capable of promoting the sustainability agenda, yet unwilling to face its consequences. If governments stuck to their rhetoric of economic prosperity, social justice and environmental protection without actually changing things for the better, then they have only responded to the needs of the system, for example, the short-term cycles of governmental power. Obviously, the d*emos* wasn't sufficiently interested to demand a change of the system. In this regard, the ball is still in the court of civil society, not "the system". However, such a conclusion does not take into account the problem of large scale "brain-washing" caused by all-dominating economic rationality. We will look into this shortly.

The Critical Role of Civil Society

To further understand the role of civil society, on the one hand, and the democratic system, on the other, Jürgen Habermas's distinction between "system-world" (*Systemwelt)* and "life-world" (*Lebenswelt*) may be helpful (Habermas 1985). For Habermas, democracy is about creating some kind of communication between these two worlds. While the established democratic process, through elections and media, recreates abstract systems of money and power (the "system-world"), the wider public sphere is also influenced by perceptions, values and everyday experiences of people (the "life-world"). Human beings act in both worlds, but the legacy of Modernity is the colonisation of the life-world by the system. The challenge, therefore, is one of decolonisation and reclaiming "life" over, or within, the "system".

Leaving aside, for the moment, the pertinent question of how such a challenge could be mounted, the thesis of a colonized democracy is important. This would imply the, at least, theoretical perspective of decolonizising something that is inherently capable of addressing the real issues. In other words, democracy would appear as not *per se* hostile to agreeing on sustainability. According to Habermas no domination and colonization process can ever be complete since human motivations, values and ideas about the good life underpin all forms of communication. So even most dominating structures of money and power cannot alter the human condition. People will always defend some space they consider fundamental for their lives.

What happens then when, for example, the catastrophic impacts of climate change are strongly felt in the life-world, but continue to be marginalized by the system-world? If the gap between the two becomes too wide, the real experience of real people just might prevail and eventually create a new quality of debate. It is certainly for the members of civil society, not political parties or governments, to insist on the reality, not "theory", of climate change. Only then the denial character of the system-world becomes obvious. And only then it may be conceivable to bring the system-world back to addressing the concerns of the life-world.

But how realistic is Habermas's thesis of a remaining autonomous life-world considering the fundamentality of the growth paradigm? The entire project of Modernity rests on the optimistic idea of progress and growth. This is not to say that Modernity cannot transform to something else – whether incrementally or as a postmodern construct – but it does show that the stakes are high. How reliable are the values of the life-world, we must ask, when the Aristotelian idea of the "good life" today can so easily be confused with materialism? And how can we be sure that the life-world has lost its confidence in the system-world's obsession with economic growth? The answer is we don't know and perhaps cannot know until the assumed fundamental values are put to the test. For the time being, we can only hope that the "civil" in civil society continues to have meaning.

Ultimately, civil society will determine whether and to what degree public concerns enter the democratic process. The "system" can never be entirely detached from civil society.[3] What is important here, however, is how sustainability concerned citizens and groups are generally perceived. Are they seen as a mere voice in a chorus of manifold voices (so typical for "pluralistic" democracy)? Or are they, at least potentially, seen as vanguards of an emerging new form of democracy?

The difference between being a mere voice and a vanguard may be gradual in practice, but conceptually we are talking about another paradigm shift. The usefulness of democracy is in doubt, if it has no mechanisms to prevent people from human and ecological disaster. Parts of modern democracy are the rule of law and the guarantee of human rights. Both are important mechanisms to prevent the ruling majority from being a tyranny to the lives of people. But do they also prevent from becoming a tyranny against life itself? If the rule of law and human rights do not account for nature and responsibilities of humans towards nature, then democracy could convert into a tyranny to nature and future generations. It is of great importance, therefore, to regard radical critique as vital to the functioning of democracy. And going back to the root *(radix)* is necessary if we are serious about trying to establish whether, and to what extent, democracy is suited to embrace sustainability.

Formulating the Problem

The fact is that sustainability has been kept at the sideline of democratic decision-making for too long. As an essentially ethical and utopian idea, sustainability was never integrated into social, cultural, scientific, economic, political and legal systems. As Niklas Luhmann maintained, systems are self-referential with each operating under its own type of rationality (Luhmann 1995). This makes it impossible to impose the idea of ecological sustainability from the "outside" (Luhmann 1989). Only if and insofar the idea finds "resonances" within a system, it can be integrated. With respect to law, for example, constitutional debates surrounding ecological responsibilities created resonances with integrating effects for sustainability (Bosselmann 1995). On the whole, however, the sustainability dialogue has not changed the respective rationalities of society's systems. Wolfgang Sachs has argued that the promotion of "sustainable development" discourse (since the Brundtland Report in 1987) completely diverted the sustainability discourse. In his view, the absorption of sustainability by the composite term "sustainable development" has rendered its foundational character. Ever since the Rio Summit 1992 the "system world" has prolonged economic growth in mind when talking about sustainable development.

For such ecological deafness we could blame the voice of sustainability as not being audible enough. The call came from the "outside" and could not be heard inside the system. But there is another way to look at it. The voice of sustainability really comes from deep within and perhaps people have only differed in their ability to listen to their inner voice. The fact that virtually everyone – including managers of big corporations – agrees that sustainability does matter is an indication of "resonance" from within. What seems to be missing is the ability to relate such inner resonance to the system-world that – to varying degrees – we are all responsible for. This is where the vanguard comes in. Sustainability concerned citizens have no reason to restrict themselves as merely one voice among others. Their concern for the future positions them "ahead" not left or right of the mainstream. "Neither left nor right, but ahead" has been the slogan of the Green movement, but also captures what more and more people realize: political leaders do not really lead. Caught in the "dilemma of growth" (Jackson 2009: 98) politicians want to serve both, the needs of growth and the needs of sustainability making it impossible for them to provide direction. This leaves "ordinary" citizens with the task to exercise leadership.

The worldwide sustainability movement has championed the idea to integrate the future into the presence. And this vanguard role is political in its most fundamental and radical sense. It adds a future dimension to politics that it currently does not have. For this reason sustainability can be seen "as a democratic term" and the strategy for sustainability as a development towards "democratic sustainability" (Hansen, Tind and Clausen 2005).

Sustainability could fill the empty space of democracy. The money-and-power system may have colonized the democratic process, however, not completely leaving a vacuum of silence that can no longer be ignored. If, as many argue, sustainability is the most profound challenge that humanity ever faced, then now is the time to break the silence and allow the voice of sustainability to speak loudly and clearly for democracy to listen.

Images of sustainability as filling an empty space, providing a foundation or giving direction are expressions of what is missing in current democratic governance. Its institutions lack a fundamental sense of responsibility for the future. But this may only be true for the dominant *form* and not necessarily for the *concept* of democratic governance. The dominant form is representative democracy with institutions and timeframes that favour short-term gains over long-term responsibility. Trying to overcome this failure may well require a fundamental rethinking of democracy (Heinrichs 2003). A culture of respect for the planet does not currently exist and needs to be nurtured. We must address this defect and ask how respect and care for people and respect and care for the planet can go in tandem.

We also must address the defect of a state-centered concept of international law that gives national demands of sovereign states priority over global responsibilities.

Institutionalizing Sustainability: The Project of Earth Democracy

Earth is the home of all living things. This home or "house" (Greek *oîkos*) is the subject of ecology and taking care of the entire house, not just individual rooms (states, economies, companies, shareholders etc.) must be the priority of all human governance.

The idea of Earth democracy[4] essentially requires a shift from economics to ecology realizing their common ground, i.e. Earth our home. This new focus can be understood as "Earth" governance. The notion of "democracy" encapsulates citizenship and self-governance of citizens and promotes participatory, transparent and accountable governance.

Present forms of governance – whether democratic or not – do not follow the ecological approach as mentioned. They are fragmented and reductionist and in most instances purely concerned with unsustainable forms of economic growth. In the current setting, environmental governance remains a minor concern, an add-on or a minimalist, shallow program. It is the poor cousin of economic governance. Until recently, many have perceived economic governance as vitally important or even reasonably successful. This is no longer the case. The recent breakdown of the global financial and economic system has left most people stunned and searching for new answers. While governments continue to bail out and restore what is essentially broken, more people than ever before realize that the breakdown is systemic and beyond repair.

Humanity is confronted with three crises and they are all linked. The economic crisis affects all people in material terms, but most of all those who are already poor and powerless. Based on false assumptions about human nature and the limits of the biosphere, neo-liberal economics have destroyed any security of individuals, communities or nations. The environmental crisis is characterized by the destruction of the very living conditions and natural resources that economic wealth is based on. However, as economic degradation is increasingly being realized as a consequence of ecological degradation, we are beginning to understand not only the systemic nature of both, but also their common cause. This third crisis is the crisis of the human spirit, as reflected in a culture of alienation, disorientation, and a loss of meaning and purpose. The crisis of human consciousness underlies all material economic and environmental manifestations and is the ultimate cause of both, self-destruction and its opposite, i.e. sustainability. Consequently, the current path of human self-destruction is not inevitable.

A new model of governance that integrates environmental, economic and social dimensions is not only needed, but a real prospect if we consider the importance of basic values and consciousness.

To remedy the flaws of existing governance models, a number of issues need to be addressed and all have to do with overcoming current reductionist awareness. Overcoming this reductionism is foremost a matter of growing ecological consciousness, and arguably this is happening today at a noticeable scale. There are various currents at work in today's world. And while existing power structures may yet be unaffected by currents of ecological (re-)awakening, they do exist in all cultures and are likely to grow faster the more obvious the failure of dominant economic, social and political forces becomes. The key is to understand the inherent unsustainable nature of governing against, or in ignorance of, natural processes. The good news is that the inescapable reality of human dependence from nature can only be ignored to a point.

This point may well have been reached now. We live in a situation where climate change, biodiversity loss, toxification of food chains, water shortages and unhealthy consumerism are felt on a daily basis, while all public money is being pumped into economic systems that created these problems in the first place. The response of governments to the current crisis reveals not just helplessness, short-sightedness and denial, but genuine signs of schizophrenia. In one reality, governments continue their modest environmental agenda of energy efficiency and climate measures, in another reality, they do everything to undermine even the most modest policies of environmental protection. Caught in their economic fixations, governments are literally losing a sense of reality.

The project of "Earth democracy" should not be seen as an effort to bail out governments. Too much is at stake and the record of governments has been dismal. Rather, Earth democracy should foremost be understood as governance by, for and through people. In a democratic political context all people are citizens. They ought to act accordingly. Citizens are ultimately in charge and can – and always will! – determine the forms, functions and effectiveness of governance. If we lament bad governance we ultimately lament our own inability or unwillingness to take charge. And while it is true that millions of people live in undeserved, hopeless powerlessness, it is also true that millions of people do not. They must use the opportunities they actually have. For Western democracies, in particular, the notion of active citizenship or positive freedom continues to have meaning. Citizens can make a real difference, but only if and only in so far they act as responsible citizens. In essence, this means a rejection of the consumer status that economies and increasingly governments have bestowed upon us in lieu of citizenship.

So, what ideally does democratic citizenship mean?

In the political and legal literature, the discourse of citizenship has been identified as a useful vehicle for generating a more sustainable relationship between individuals, society and the environment. The concept of democratic *ecological citizenship* (Bosselmann 2008: 196–205; Bosselmann et. al.: 40–45) epitomizes this endeavour. In its re-configuration of the relationship between the human and non-human, ecological citizenship raises several key issues:

- the relationship of the concept of citizenship to sustainability (that is, the theoretical and practical utility of a concept of ecological citizenship in effecting sustainability);
- the re-formulation of responsibilities entailed in idea of citizenship; and
- the interaction between ethical responsibility and political activism in the meaning of citizenship.

Most importantly, ecological citizenship poses a new relationship between humans and the natural world that stresses non-reciprocal obligations and responsibilities. The adoption of an ethical standpoint that embraces ecocentrism involves a shift in social and political thought to a new politics of responsibility. The new politics of responsibility will shape relations between people living today and their common relationship to the future.

Implicit in much of the discussion on ecological citizenship is the need for a vibrant and active civil society. The concern for the active engagement of civil society in designing and practicing governance for sustainability is mirrored more generally in the discourse of sustainable development and recent literature on governance and sustainability.

The core condition of such governance is the idea of ecological citizenship coupled with active civil society. Citizenship and civil society are also at the core of democracy. For this reason, the desired new form of governance must be democratic. With its focus on ecological interdependences between humans and nature, the new form of democracy is aptly named "Earth democracy".

Dimensions of Earth democracy

Earth democracy is not a one-size-fits-all concept for governance. Historically, sustainable forms of governance have existed in a variety of cultural, social and local settings.

The notion of "Earth democracy" is more conducive to today's global realities where governance needs to be multidimensional, i.e. global and local, ecological and economic, cultural and political, rational and spiritual. These dimensions of Earth democracy are all related to each other, however, they can also be differentiated as each reflects a different concern.

Examples include the following:

Covenantal democracy describes a political system that fundamentally rests on consent and shared norms and responsibilities across generations and all forms of life.

Political democracy provides for public processes and institutions to facilitate active citizens, the rule of law and human rights.
Deliberative democracy aims for processes and institutions that include all citizens at every level of public discourse.

Representative democracy refers to the election of representatives who operate within the appropriate body (e.g. parliament or council).

Participatory democracy happens when members of the public are more directly involved in decision-making.

Cultural democracy promotes cultural diversity and multicultural approaches to governance and decision-making.

Global democracy aims for more effective participation of citizens in existing or future international organizations.

Local democracy enables people to directly participate in decision-making that affects their community.

Ecological democracy promotes decision-making that is reflective of the relationships between human and non-human spheres and their ecological balancing.

Economic democracy ensures participation of all people affected by the economic system and the just production and distribution of material wealth.

Health democracy reflects a political system that aims for meeting the health needs of people, communities and ecosystems.

Gender democracy strives for gender balance in decision-making and equal access of women to (natural, educational, social etc.) resources.

Spiritual democracy holds that spirit is the source of all life and aims for decision-making that is inspired in this way.

These (and possibly other) dimensions can be further classified in terms of their functions. Some are normative, others institutional in character. We can, for example, identify attributes such as covenantal, cultural, ecological, gender or spiritual as normative categories and the attributes political, representative, participatory, global and local as institutional categories. Within institutional categories a further distinction could be made between methods (political, representative, participatory) and levels (global, supranational, national, regional, local, corporate).

Normative foundations

All of the mentioned normative aspects should be understood as being captured by the notion "Earth democracy". They may not all or not to the same degree be present in the actual concept of Earth democracy, but should inform and guide it as much as possible. The essence of Earth democracy is a value-based concept of democracy. The values relevant here are "basic" and reflective of ecological realities as mentioned above (under 1). They are addressed in the composite terms of covenantal democracy, cultural democracy, ecological democracy, gender democracy and spiritual democracy. One summarizing term could be "ecological wisdom" to mean that the various forms of interrelatedness of people and the planet should guide democratic decision-making (at all levels).

Associated values can be identified in a number of ways. First, a reasonable starting point would be to further explore the findings of the preceding *"Governance of Sustainability"* (GfS) project (Bosselmann et. al. 2008). In a way, Earth democracy is nothing else than giving democratic, institutional and legal cognizance to sustainability. The GfS project explored the relationship between democracy and sustainability in a conceptual and in a practical way.

Conceptually, some tensions between (existing) democracy and sustainability were articulated (Bosselmann et al. 2008: 10–16). For example, democracy, in its present form, poses huge problems for sustainable development. In advanced capitalist states, democracy is tightly coupled with economic growth, consumerism and individual freedom (replacing the citizen with the consumer). The question before us today is how to join the precious legacy of authentic democracy, with its affirmations of human dignity and individual moral responsibility, with true sustainability, i.e. the ability to live within the ecological boundaries of the planet. To this end, it seems more helpful to explore the potential of covenants (as the GfS report does) rather than relying on modern democracy based on liberalism and pluralism.

Practically, the GfS report compiles 19 case studies from around the world that demonstrate challenges, obstacles and successes of a sustainability driven democratic process (Bosselmann et. al. 2008: 69–251). There are successful examples and those of failure where the obstacles of power-driven vested interests were too high to overcome. Patterns of success or failure can be detected. For example, the existence of a covenantal bond – typically underpinning traditional communities – is likely to generate more sustainable processes and decisions. Similarly, effective democratic institutions are more likely to result in good quality decisions. Another important feature is a sense of "place" and citizenship. People are, in most instances, more motivated within neighborhoods or communities than in anonymous cities with less defined local governance structures. Overall, practical experiences with linking sustainability values to public decision-making seem to indicate a need to define democracy as locally based and participatory rather than national and representative ("act locally, think globally").

Second, there are many other sources that can be explored to define relevant normative values. Among these are the rich literature on social and *environmental ethics* and the even richer literature on *religious and spiritual traditions*.

A further source of relevant normative values are international agreements, in particular, soft law agreements such as Agenda 21 or the *Earth Charter*. The Earth Charter with its origins in global civil society and subsequent endorsements, for example, by UNESCO and IUCN is of particular importance. It represents the most comprehensive, truly global, i.e. cross-cultural rather than international, ethical statement to date. The Earth Charter's principles with respect to ecological integrity, human rights, democracy and governance are clearly relevant to Earth democracy. To a large extent, the Earth Charter describes the normative dimensions of Earth democracy (Bosselmann 2008: 108–109, 204–208).

Institutional aspects

Among further institutional aspects of Earth democracy are those of democratic method (e.g. representative or direct or participatory) and those of governance level (e.g. global or national or local).

In terms of method, democratic models vary between highly representative forms, more participatory forms and direct forms of decision-making. Theories of democracies need to be examined in terms of their suitability for our age. It seems obvious, however, that direct involvement of citizens and citizens groups is vital for the build-up of democratic processes and alterations to representative democracy models.

For the examination of various governance levels, the GfS project provided for a suitable structure. In the report a distinction was made between various levels and each included a discussion of current trends (Bosselmann et.al. 2008: 17–40):

1. Global governance (beyond UN reform)
2. Regional governance (EU a model?)
3. National governance (the state as mediator)
4. Local governance (incl. women and indigenous peoples)
5. Corporate governance (economics as if people matter)

While each of the levels, discussed in the report, deserve more analysis, they can also be used to explore new ways how Earth democracy could be institutionalized. For example, there is a lot discussion on trusteeship models for governing the global commons. The trusteeship model is not only suitable for the areas outside national jurisdiction, but also within national jurisdiction. Here it can help to redefine the concept of territorial sovereignty (a key pillar of traditional international law). There is also a wealth of literature surrounding "ecological councils" or sustainability councils to bring ecological wisdom to democratic decision-making.

The overall aim should be to develop a multilevel governance model of Earth democracy that serves the needs of cultural diversity as much as sustainability unity. In other words, institutions need to serve individual needs of people, communities and societies while also providing for global orientation and ecological responsibility.

Objectives and possible structure of the Earth democracy project

The overall aim of the project is the development of a concept of Earth democracy based on sustainability. The concept should be developed in form of workshops, conference and publications, but also by way of active involvement in the political process. (e.g. dialogue with NGOs, international organizations and governments).

The project should not be completed at a particular time, but rather be understood as an ongoing concern. To this end, it may be possible to work towards particular objectives that involve a specific timeline (e.g. book publication) and also pursue long-term objectives (e.g. co-operations with new partners) that can be open-ended. All this is, of course, subject to actual resources available. For the time being, ESG acts as the coordinating body of the project.

As an initial step, a working group of interested colleagues has been formed to start research on key areas. The following structure, while not prescriptive, identifies some of the areas of research.

I. The idea of Earth democracy

Basic concept; similarities with and differences to related ideas

II. Normative dimensions

1. Covenantal governance
Covenantal foundations of Earth democracy

2. Value-based democracy
Common ground of democracy and sustainability

3. Framing Earth democracy
Earth Charter and democracy

III. Structural dimensions

1. Ecological citizenship and democracy
Citizenship and ecology as ideological core of Earth democracy

2. Global civil society
Civility and citizens groups beyond national boundaries

3. Ecological democracy
Ecological realities as the basis of governance

4. Economic democracy
Ecological realities and ethics as the basis of economics

IV. Institutional dimensions

1. Global democracy
Institutions of global governance (UN)

2. Regional democracy
Supranational governance (EU)

3. National democracy
Role of the state

4. Local democracy
Communities as smallest units of Earth democracy

5. Corporate democracy
Corporate governance for sustainability

V. The Way Forward

Actions to promote Earth democracy
(this could include case studies or current government activities or any form of action towards sustainability governance)

Again, the above structure is not intended to be prescriptive or complete, but to illustrate what the concept of Earth democracy may entail. The overall aim is for initiating a wider discourse on how to incorporate sustainability values in public decision-making.

Conclusion

This chapter aimed for introducing to the project of Earth Democracy. This project has only just begun and while it is structured with project partners, meetings, timelines and publications, it is open ended. Its subject, the institutionalization of ecological integrity, has to be an ongoing concern with many levels of action. Academic writings would not in themselves be sufficient as they are not usually accessed by politicians and law-makers. To them, the public perception of politics matters more. It is necessary, therefore, to influence the public sphere through the media, public meetings, action days and such like.

The sustainability movement must reach out to all sectors of the public sphere. Only then we will stand a chance of penetrating the "system-world". Our "life-world", that may well include the coziness of academic life, can no longer remain outside the political arena. Too much is at stake.

References

Bosselmann, K., 1995, *When Two Worlds Collide: Society and Ecology* (Auckland: RSVP).

—. 2001, "Sustainability and University: Compatible Agendas?", 33 *Educational Philosophy and Theory* 2.

—. 2008, "The Way Forward: Governance for Ecological Integrity", in: L. Westra, K. Bosselmann and R. Westra (eds.), *Reconciling Human Existence and Ecological Integrity*, Earthscan, London, UK

—. 2008, *The Principle of Sustainability: transforming law and governance* (Aldershot: Ashgate).

—. 2009, "Sustainable Aotearoa New Zealand Incorporated (SANZ)", *Strong Sustainability for New Zealand: principles and scenarios* (Auckland: Nakedize).

Bosselmann, K., R. Engel and R. Westra, 2001, *Governance for Sustainability – Issues, Challenges and Successes* (Bonn: IUCN Environmental Law and Policy Series Vol.70).

Christie, I., 2007, *Democracy, Civil Society, Sustainability,* at http://democracy.sustainability.com/insight_and_analysis/christie.cfm

Eckersley, R., 2004, *The Green State: Rethinking Democracy and*

Sovereignty (Cambridge, Massachusetts: The MIT Press).
Elkington, J. and J. Lothrington, 2008, http://democracy.sustainability.com/conference/ A *Summary Report* by can be viewed at http://democracy.sustainability.com/downloads/Democracy_&_Sustainability_summary_report_11-04-08.pdf
Emerson, J., 2009, "What does Economic Reality Do" (available at http://www.idiocentrism.com/economic%20rationality.htm
Habermas, J., 1985, *The Theory of Communicative Action, Volume 2: Lifeworld and System: A Critique of Functionalist Reason* (Beacon Press).
Hansen, H.P., E Tind and L. Tolnov Clausen, 2005, *Democracy and Sustainability – a challenge to modern nature conservation*, Paper presented at the 11th International Symposium on Society and Resources Management, Östersund, Sweden June 2005, 4; available at www.ruc.dk/upload/application/pdf/b0d8da54/Democracysustainability.pdf
Jackson, T., 2009, *Prosperity without Growth? The transition towards a sustainable economy* (London: Sustainable Development Commission).
Luhmann, N., 1989, *Ecological Communication* (University of Chicago Press).
Luhmann, N., 1995, *Social Systems* (Stanford University Press).
Speth, G., 2008, *The Bridge at the Edge of the World: Capitalism, the Environment, and Crossing from Crisis to Sustainability* (New Haven: Yale University Press).
Wilber, K., 2001, *A Brief History of Everything*, (Gateway: Gill and Macmillan)
Young, O., 1997, 'Rights, Rules and Resources in World Affairs' in O. R. Young, (ed) *Global Governance: Drawing Insights from the Environmental Experience*, (Cambridge: MIT Press).
—. 2002, *The Institutional Dimensions of Environmental Change, Fit, Interplay and Scale* (Cambridge: MIT Press).
Zaelke, D., M. Stilwell, and O. Young, 2005, 'What reason demands; making law work for sustainable development' in *Making Law Work: Environmental Compliance and Sustainable Development* (eds) D. Zaelke, Kaniaru, D. Kruzikova, E. and May, C. (London: International Law and Policy).

Notes

[1] That the use of this term serves many and mostly dubious purposes should, however, not conceal the potential validity of the underlying concept. If understood correctly, i.e. as ecologically sustainable development, governments would be on the right track.

[2] Tim Jackson makes the point that governments intervene constantly with people's lives on social matters whether they like it or not (*ibid.* 94). This would suggest a sheer lack of imagination that intervention may also be indicated when it comes to environmental matters.

[3] Apart from NGOs, civil society consists of various other groups such as trade unions, business associations, religious bodies, academic institutions, student organizations, ethnic lobbies, community groups, and so forth. Similarly defined in Dower and Williams (eds.), *Global citizenship: A Critical reader* (Edinburgh: Edinburg University Press, 1999, xxi.

[4] The term is borrowed from V. Shiva, *Earth Democracy: Justice, Sustainability and Peace* (London: Zed Books, 2006).

PART II –

STATE SOVEREIGNTY, INDIGENOUS GOVERNANCE AND INTERNATIONAL LAW

PART II INTRODUCTION

KLAUS BOSSELMANN

One central obstacle against prospects for global democracy and justice is state sovereignty. States interpret their sovereignty as a right to develop on their own terms and to prosper as they see fit. Usually, this means exploitation of the planet's resources and, in many instances, people, especially indigenous peoples. Internationally, states serve financial markets and corporate interests at the expense of humanity's future. How can state sovereignty be legitimate?

Arguably, state sovereignty is legitimate. It is a crucial means for controlling power, including economic power. Internally, the state's monopoly of power is vital for peaceful settlements of social conflict. Externally, sovereignty is vital for peaceful conflict resolution between states. To this end, the idea of the sovereign state and international law is indispensable. Despite all its failures and despite all aspirations for a just global order (UN) or a global community of people (world republic), states have not lost their legitimacy. Who else could guarantee procedures of fair treatment of people and fair distribution of goods?

The political debate pro and contra state sovereignty often circulates around these kinds of issues. However, for a significant part of the world's population this debate appears strangely removed from reality. To them, the legitimacy of states is irrelevant or, at least, secondary to the most fundamental right that people have, i.e. the right to live in accordance to their tradition, culture and identity. While most peoples believe that their right to self-determination is best served through a state, this may not be true for all. The Kurdish people, or the peoples of Palestine or Tibet, may prefer the idea of an independent state, but other indigenous peoples may not, as they are accustomed with own traditions of governance and alien to the idea of a sovereign state. What matters in each case, is the right to self-determination, guaranteed, after all, under the Charter of the United Nations.

The four chapters of this part have the concern for self-determination at their heart. Whether the polar region (Morse and Zakrison), land rights (Coyle and Bulan), or governance (Manno), at stake is always the right of indigenous people to govern themselves. "Why should they?" the international lawyer may ask, as the model of international law is based on state-centered assumptions about self-determination. "Why shouldn't they?" the environmental lawyer asks, or ought to ask, as people and their environment are central to self-determination.

The key challenge of this Part of the book is to place indigenous rights amongst the established categories of international law and, at the same time, appreciate and promote autonomous indigenous governance. Despite the diversity of authors and topics, they all agree on one thing. Governance for sustainability desperately needs indigenous knowledge and wisdom. As, for example, Jack Manno comments: "(T)he power of governance rests on the internalization of social norms involving collective and participatory decision processes, demands on personal duties and the high regard placed on the powers of persuasion, a skill of Native people that numerous settler chroniclers remarked."

In their chapter on the destruction of the Canadian Arctic region Bradford Morse and Michelle Zakrison show how the integrity of this delicate ecosystem is being threatened. Global warming impacts on precipitation patterns, soils stability etc., ozone layer depletion affects plant growth, persistent organic pollutants and other air and water-borne contaminants taint the food chain, and all of these and additional factors such as species introduction and tourism contribute to changes in biodiversity. The indigenous peoples of this Polar region are particularly affected as their health and livelihood depend on this finely balanced ecosystem. The Canadian government and international environmental law are only beginning to address the issues of survival and self-determination that are involved here.

A further illustration of the conflict between the state and indigenous governance can be found in Michael Coyle's chapter. Using the examples of colonial history in Canada and New Zealand, Coyle gives insights into "power imbalances". *Prima facies*, the colonialist approach of treaty making and, more recently, the process of restorative justice through native land claims (Canada) and negotiated settlements (New Zealand) appears more constructive than anywhere else in the colonized world. It would seem unfair to dismiss any such attempts towards reconciliation as just a variation of Western imperialism. And yet, as Coyle demonstrates, all these attempts have followed the logic of the state rather than mutual agreement (not to mention remorse). The power imbalance is of structural nature: as long as the state determines the nature and extent of "dialogue", indigenous peoples miss out. Neutral dispute resolution mechanisms simply do not exist. A way out of this dilemma could be true indigenous self-governance and independence, but would states ever grant them?

An encouraging message comes from Jack Manno. Not the example of the Great Law of Peace *per se* is being presented as a model for global environmental governance, but its underlying ethics of respect, gratitude and care. The Confederacy of the Six Nations of the Haudenosaunee (Iroquois) comprises the Great Lakes district (across the boundaries between the USA and Canada) and represents one of the oldest examples of living democracy. In Western terminology, the foundational Great Law of Peace could be seen as treaty for transboundary ecosystem management. This is in itself remarkable, however, what makes it so relevant for the discourse on democracy and sustainability, is the strong voice that speaks from, and about, this treaty. Haudenosaunee leaders have for decades now warned about the global ecological crisis and advocated the strengthening of indigenous peoples as the greatest hope for humanity at large. Manno, himself deeply involved in this cross-cultural dialogue, discusses how the fundamental principles of the Great Law of Peace provide lessons for global governance. Like the previous chapter, this chapter describes empowerment of indigenous peoples not as a rights-granting exercise by states – as important as it may be – but as a challenge to the very definition of what constitutes the state and the law, perhaps conceivable through a new covenant between indigenous and non-indigenous peoples.

The final chapter by Rami Bulan leads us into a state where the indigenous population is not a minority, but with 65% forms a strong base for government. Malaysia has a constitutional mix of customary law of Malay and other indigenous peoples, Islamic law and British common law. In some of Malaysia's thirteen states indigenous law is more prominent than in others, and the coherence between the various customary laws is limited. Malay land rights are constitutionally entrenched, but not entirely spared from the possibility of "replacement" or "disappearance". Land rights of other indigenous peoples are even more at risk both, legally and practically. Discussing a number land rights cases, Bulan concludes that inconsistencies and discriminatory effects are due to Malay dominance and colonial history.

CHAPTER SEVEN

DESTROYING THE CANADIAN ARCTIC: A GLOBAL CRISIS WITH DEVASTATING LOCAL RAMIFICATIONS

BRADFORD W. MORSE
AND MICHELLE ZAKRISON

Introduction

We are currently living through the most dramatic and negative changes in the Polar Arctic in centuries, if not in human history. For the people who reside in the region, and especially for the Inuit whose ancestors have been living there for millennia, these changes pose an almost incomprehensible threat to what remains of their traditional way of life as well as their very health. Environmental degradation is occurring on an immense scale, for which almost all of the causes stem from sources 1000s of kilometres away and far beyond their direct reach to influence. As this chapter will describe, the many sources of pollution along with the emerging effects of global warming is already raising concerns about human health in northern Canada, is harming the local economy, reducing income opportunities, making transportation more dangerous and undermining the reliability of traditional environmental knowledge. The capacity of Inuit people collectively to alter this picture is extremely limited, leaving them reliant once again upon their historic ability to adapt to an extremely difficult environment but now with it warming and contaminated in previously never experienced ways. Their only other alternative is to look to the peoples of the world, as well as new binding and effective international treaties to change human conduct drastically to reduce the effects of transboundary pollutants and climate change.

The first part of this essay is devoted to summarizing the gravity of the existing and projected transformations with due regard paid to the incredible complexity of the impacts upon the Polar Region. One can only begin to consider what might qualify as providing possible solutions if the full panoply of negative circumstances is appreciated. We briefly describe the major environmental threats, including climate change, persistent organic pollutants (POPs), air and water borne contaminants, water pollution, changes in biodiversity, and the presence of radioactive nuclear remnants. The particular impact of these new realities upon the Indigenous peoples of the North, and particularly the Inuit, is identified. The authority of the Inuit and of other northerners to remedy these risks is then considered as part of self-governance. While various governments possess legal jurisdiction concerning aspects of environmental protection, the ability of any state or sub-state government to be effective in removing contamination from other than local point sources is minimal. Thus, this chapter will end with a consideration of the importance of international law and agreements in changing human and corporate behaviour to help cleanse the Arctic and ensure it remains fit for human habitation and able to nurture future generations of Inuit within their homeland.

I. Environmental Threats in the North

Despite numerous varying descriptions, the Arctic is usually defined as the geographic region north of the Arctic Circle at 66.6 degrees latitude, but also extends south in various areas to encompass similar geological space and habitats.[1] The Arctic is an extremely environmentally sensitive area and is, therefore, one of the most acutely affected regions globally by various forms of environmental degradation.[2] The presence of so many different factors and distinct sources of danger augment the complexity and heighten human fears. Each environmental concern will be described in turn in order to convey properly the breadth of the separate, yet interweaving, issues that they present.

Climate Change

The expression "climate change" is used to describe a fundamental shift in the usual weather patterns of a certain region or globally, which includes changes to temperature, precipitation, wind velocity and major storms that may arise or have already occurred. The Intergovernmental Panel on Climate Change (IPCC), the scientific body created by the World Meteorological Organization (WMO) and by the United Nations Environment Programme (UNEP), defines climate change narrowly as "a change in the state of the climate that can be identified (e.g. using statistical tests) by changes in the mean and/or the variability of its properties, and that persists for an extended period, typically decades or longer."[3] Interestingly, this definition varies from that of the United Nations Framework on Climate Change (UNFCCC), which refers to climate change as a change "attributed directly or indirectly to human activity that alters the composition of the global atmosphere and that is in addition to natural climate variability observed over comparable time periods".[4]

Climate change is immensely significant in that it is projected to generate highly unpredictable and potentially disastrous consequences. The IPCC published their Fourth Assessment Report in 2007 that declared there is concrete evidence that climate change is connected to anthropogenic activity and, therefore, its existence can no longer be disputed.[5] Furthermore, the Report stated that "there is a 90% chance that anthropogenic greenhouse gas (GHG) emissions have caused most of the warming since the mid-20th century".[6] The IPCC also concluded that most warming increases observed in the last 50 years are very likely due to the increase of anthropogenic activities that have raised GHG concentrations.[7] Since it is now definitively accepted that climate change in the form of global warming is affected by human activity it is imperative that the global community adjust lifestyles and create policies to dramatically reduce the magnitude of climate change. Failure to act will simply mean that the world is going to continue to experience its dramatic and erratic impacts with ever increasing and devastating consequences.

In 2004, The Conservation of Arctic Flora and Fauna (CAFF) reported that the Polar Regions are likely to be impacted "soonest and to the greatest extent".[8] The impacts of climate change in the two Polar Regions are especially profound due to "their large cryospheric components that also dominate their hydrological processes and water resources".[9]

Observed changes

While most believe that the worst of climate change has yet to come, significant effects have already been observed to the Arctic waters and sea ice.[10] Although there is uncertainty due to year-to-year variability, both scientists and Indigenous peoples have reported major signs in the Arctic.[11]

The waters have increased since the 1930s "in the combined flow from the six largest Eurasian Rivers", which is estimated at 7%. In the late 20th century, increased runoff to the Arctic Ocean has been noticed "from circumpolar glaciers, ice caps and the Greenland ice sheet".[12] This is thought to be "comparable to the increase in combined river inflow from the largest pan-Arctic rivers"[13]. A further future challenge is the possibility of increase in flooding caused by floating freshwater ice within Arctic rivers and lakes.[14] Although there is no apparent increase as of yet in flood frequency or magnitude "ice-cover duration has decreased in much of the sub-Arctic".[15]

One aspect of the northern ecosystem that has been undergoing a noticeable transformation in recent years relates to permafrost, which plays an important role in effecting surface water and snow melt. "At higher latitudes, initial thaw is thought to have increased surface ponding and lake abundance whereas, at lower latitudes, lake abundance has declined as more extensive and deeper thaw has permitted ponded water to drain away to the sub-surface flow systems".[16] In addition to affecting water levels and drainage, "the biological composition of lake and pond aquatic communities has been shown to respond to shifts in increasing mean annual and summer air temperatures and related changes in thermal stratification/stability and ice-cover duration".[17]

Projected changes

Despite difficulties in estimating effects, it is generally projected that the runoff from major Arctic catchments will increase by 10–30%.[18] While glaciers are currently retreating inland resulting in "increasing streamflow, a gradual disappearance or a new glacier balance at smaller extents will eventually result in lower flow conditions, particularly during the drier late-summer periods, critical periods for aquatic Arctic biota".[19] Arctic warming also implies that there will be a continuation of the recent trends observed of ice freezing up later in the year and breaking up earlier on rivers and lakes as well as reductions in ice thickness.[20] This will lead to changes in the quality and quantity of under-ice habitat, in lake thermal structures, as well as affecting river-ice jamming and its related flooding.[21] These transformations also threaten sustaining healthy riparian ecosystems, which depend on the spring inundation of water, sediment and nutrients.[22]

These changes to the hydrological system undoubtedly affect the plant and animal systems that they support.[23] Since climate change may interfere with water and air currents, this will likely alter migration routes for marine mammals, birds, fish, and insects.[24] Furthermore, the loss of up to half of the tundra due to the treeline moving northwards will leave many existing tundra species with a reduced ecosystem.[25]

As the climate warms, thereby thawing permafrost, an increase in soil instability and erosion will occur.[26] The impact upon both land and sea ice is already beginning to make "snow surface travel dangerous, traditional environmental knowledge (TEK) unreliable, hunting harder or impossible, igloo-building snow scarcer, erosion and landslides more common, permafrost unstable, and weather patterns more erratic".[27]

Global warming has been changing the temperature of the North in recent decades, which in turn affects the permafrost and sea ice, despite some evidence of cooling in the eastern regions of Canada's North.[28] For example, "the average Arctic temperature has increased at twice the global rate over the past century. In the Northwest Territories, the Yukon and Alaska, warming has been seen at an unprecedented rate of 2 degrees Celsius per decade since the 1960s. Snow cover extent has decreased in the Arctic by 5–10% and the extent of sea ice cover has shrunk by 7% since 1972. Furthermore, ice thickness in the Arctic basin has decreased by 40% since the 1950s".[29] CAFF states that due to sea ice retreat, some subsidiary Arctic seas might completely disappear.[30] The Council has stated that reductions in sea ice, particularly in the Hudson Bay area, threaten polar bears that depend on regular seasonal ice cycles.[31]

As one can imagine, this temperature increase is also affecting the Arctic Ocean. Although the 70% of the Arctic Ocean is currently permeated by ice,[32] the IPCC "reports that in recent decades the ice coverage during the summer months has thinned by approximately 40%"[33]. The United States Arctic Research Commission has also estimated that the Northwest Passage will be ice free and easily navigable throughout the summer by 2050.[34]

Increased temperatures are already affecting the environment in many ways, the first of which is a shift in precipitation patterns and a risk of summer droughts.[35] Soil in some regions will grow drier and soil temperatures are expected to rise by at least 2 degrees Celsius in the 21st Century.[36] This will result in a longer growing season and a deeper thaw of the active layer in the summer months[37] influencing both the natural environment and the culture of the region. The northward migration of birds, animals and insects previously only in lower latitudes is already occurring[38] with reported sightings of southern songbirds as far north as Iqaluit, grizzly bears roaming in traditionally polar bear-only habitat, and increasing incidence of winged insects.[39] The migration of moose, mosquitoes, and predators into caribou habitat is threatening calf survival rates through competition for habitat and food resources. The presence of longer growing seasons with warmer temperatures has reduced the extent and duration of sea-ice cover, thereby affecting seal and polar bear populations.[40]

Effect of Arctic climate change on the globe

Shifts in the Arctic in turn exacerbate global affects of climate change in three critical ways. It will result in variations to ocean circulation, the decreasing reflectivity of the Earth (albedo), and the massive release of carbon due to permafrost thawing.[41]

The Earth's water systems are connected on a global level transferring water and energy throughout the oceans.[42] The existence of dense Arctic waters significantly controls oceanic circulation.[43] A considerable influx of freshwater, which is warmer and less dense than the colder salty Arctic waters, could alter patterns of ocean circulation, thereby causing an increase in precipitation and further melting of glaciers.[44] This is thought to be of particular grave concern because if the Greenland Ice Sheet melts, "which holds 10% of the world's fresh water [it] would raise sea levels by 7m".[45]

The melting of ice structures, such as glaciers, will reduce the overall reflective land mass of the Earth and leave larger areas of darker open water.[46] This can trigger a positive feedback loop in that the dark water areas will increase the temperature of the oceans and further melt the glaciers so as to increase sea level.[47]

Permafrost thawing through increasing temperature levels is also a concern as it is projected to release carbon dioxide currently captured in Arctic soil.[48] Although little is known about how such thawing would affect the environment, some scientists estimate it has the potential of releasing carbon greatly exceeding current anthropogenic emissions.[49]

All of these impacts are not derived from human activity in the Canadian Arctic yet they affect domestic and international politics. Decreasing sea ice coverage opens Arctic waters for shipping at a far greater extent and generates interest in seabed exploitation, thereby affecting international relations in that there has been a re-emergence of "long-standing disputes over territorial and sovereignty claims."[50]

Persistent organic pollutants

Persistent organic pollutants are defined by the United Nations Industrial Development Organization as "chemical substances that persist in the environment, bio-accumulate through the food web, and pose a risk of causing adverse effects to human health and the environment".[51] These substances are of significant environmental concern in that "the migration of persistent organic pollutants (POPs), such as pesticides, from thousands of miles away taints the food chain in the Arctic, on which the Inuit depend for survival, with surprisingly dangerous intergenerational consequences".[52] These chemicals travel by air to the Polar Regions because of a particular chemical property in the major POPs, such as PCBs and dichlorodiphenyl trichloroethane (DDT).[53] Specifically, they have a "propensity to accumulate preferentially in environments at higher latitudes; because they are semi-volatile, they volatilize in warmer climates and travel pole ward in a cycle of volatilization and condensation, sometimes referred to as the 'grasshopper effect' because of the way emissions 'hop' north in a series of steps before they condense systematically in colder regions".[54]

Global warming increases the impact of POPs to the Arctic. Since the Polar region is projected to become "wetter" due to sea ice melting, hydrological processes "will affect the pathways and increase the loading of pollutants" such as POPs and mercury to Arctic aquatic systems.[55] This leads many scientists to believe that changes in food webs and aquatic trophic structures has the potential to further modify the accumulation of bio-magnifying of chemicals.[56]

Meyer and Wania have studied POPS in the Arctic to determine what environmental process has the strongest influence on their accumulation.[57] Their study found that the factors having the most impact on potential accumulation were temperature, sea ice cover, and macro-diffusive atmospheric coefficients (as well as the precipitation rate, but to a lesser extent).[58] They hypothesize that "the global transport and distribution behaviour of many persistent organic chemicals, and in particular their accumulation in polar marine ecosystems, may be significantly impacted by global climate change. There is now also experimental and modelling evidence that climate fluctuations can impact on the atmospheric transport of persistent organic pollutants".[59]

POPs are especially dangerous as they progress up the food chain and bioaccumulate in the fatty tissue of living organisms.[60] This bioaccumulation process is exacerbated due to the long length of Arctic marine food webs, which include high-fat species.[61] As a result, comparatively high levels of hazardous chemicals occur when the concentrations increase through food webs at each trophic level.[62]

Due to the evidence of long-range transport of POPs to regions where they have never been used or produced, and the consequent threats they pose to the global environment, the international community has called for action to reduce and eliminate the release of POPs.[63] The Stockholm Convention on Persistent Organic Pollutants and Convention on Long-range Transboundary Air Pollution focuses on twelve of the most hazardous POPs, dubbed the "Dirty Dozen".[64] These include: Aldrin,

Chlordane, DDT, Dieldrin, Endrin, Heptachlor, Hexachlorobenzene, Mirex, Toxaphene, Polychlorinated Biphenyls, Dioxins, and Furans.[65] The Stockholm Convention received nine additional chemicals to this list in May of 2009, which includes those commonly used as pesticides, fire retardants, and other commercial purposes.[66] Exposure to POPs is dangerous to humans in that they can lead to "serious health effects including certain cancers, birth defects, dysfunctional immune and reproductive systems, greater susceptibility to disease and even diminished intelligence".[67]

Ozone depletion

Ozone depletion is caused primarily by a release of human-made chemicals, such as "chlorofluorocarbons (CFCs), hydrofluorocarbons (HCFCs), and other ozone depleting substances (ODS)".[68] These chemicals are found in widely used products such as "refrigerants, insulating foams, and solvents".[69] When these chemicals reach the Earth's stratosphere, the sun's ultraviolet (UV) radiation causes them to break apart releasing chlorine atoms that react with ozone and start to create chemical cycles that deplete the ozone layer.[70] The chlorine atoms are extremely powerful in that one atom can "break apart more than 100,000 ozone molecules".[71] Even more serious are some of the other ozone damaging chemicals, such as methylbromide (employed as a pesticide); halons (used for fire extinguishers); and methyl bromide (used as a solvent in industrial processes).[72] For example, when methylbromide and halons break apart, "they release bromine atoms, which are 60 times more destructive to ozone molecules than chlorine atoms".[73]

The natural levels of UV rays in the Arctic are low, however, ozone level depletion caused by these chemicals travel to the stratosphere and trigger a chain reaction resulting in little or no ozone in the spring.[74] When levels of ozone are depleted, it permits an increase of UV radiation to reach the surface of the Earth with greater impact in the Polar Regions as they are more directly affected by atmospheric deterioration with the potential to cause "photochemical damage in the DNA of living organisms".[75] Increased UV radiation may also "harm the eyes and skin of animals, damage the surfaces of plants, and cause genetic damage".[76] Initial research found that "mosses grew more, but leaf and shoot growth decreased in evergreens".[77] The Montreal Protocol on Substances that Deplete the Ozone Layer, which entered into force on January 1, 1989, has had a profound impact upon the production and release of ozone-depleting chemicals, yet more monitoring is needed to assess long term impacts and ensure more rigorous adherence not only to the original Treaty but also to its subsequent amendments.[78]

Air- and water-borne contaminants

There are a number of air borne contaminants and air pollution issues that impinge upon the fragile Arctic ecosystem. CAFF states that pollution in the Arctic originates both regionally and internationally from the "long-range transport of industrial and agricultural pollutants from temperate and tropical areas".[79] Acidification from precipitation is also a serious problem for the Arctic, especially for freshwater systems.[80] Furthermore, long-range chemical pollutants, such as POPs and some heavy metals that travel through oceanic and atmospheric transport, have been found in whales, seals, polar bears and migratory birds.[81]

The Arctic Monitoring and Assessment Program (AMAP) is one of the Arctic Council's working groups that reports on pollution and related issues including Arctic acidification and Arctic haze. The former is only a major concern in geologically sensitive areas where acid deposition has increased to a point in which it is higher than what the system can inherently neutralize.[82] Sulphur is the most significant source of acidification followed by nitrogen due to anthropogenic sources.[83] Arctic haze is the visible manifestation of air pollution composed primarily of long-range transported contaminants, such as sulphate aerosol particulate organic matter.[84] The chemicals accumulate over the winter months and become noticeable in the spring in large areas of the Arctic resulting in reduced visibility.[85] Although studies have been unable to illustrate important health effects attributed to sulphur emissions from smelters (sulphur's largest source), epidemiological studies have indicated differences in population health in communities with some of the highest acidifying air pollutant levels.[86] These areas, specifically by the Norwegian–Russian border, indicate causes associated with socio-economic factors more than environmental degradation, however, the precise causes have not yet been determined.[87]

The overall trend of air pollution is that sulphate concentrations are decreasing in some subarctic areas, such as Fennoscandia and northwestern Russia, but "levels of nitrate aerosol are increasing during the haze season at Alert (Canada), and possibly also at Barrow (Alaska)".[88] More data needs to be collected to confirm this reported increase and map its sources. International cooperation in the implementation of international agreements is critical if the improvements of acidification in Arctic terrestrial and freshwater ecosystems are to continue as expected until 2020.[89]

The presence of mercury and other heavy metals are naturally of great concern to the Arctic, as even at low doses, serious health risks are caused by methylmercury; and levels of mercury have skyrocketed.[90] Studies have been reporting the presence of mercury and heavy metals since the 1960s, and elevated levels in humans in the 1970s.[91] Despite a decline within North America and Europe in mercury emissions starting in the 1980s, countries such as China have increased their emissions due to rapid industrial growth and the use of coal-fired power plants.[92] Specifically, Asia is causing approximately 50% of mercury emissions globally.[93]

As with air pollution, water pollution arrives to the Arctic through long-range transportation, which then accumulates in the food chain and poses threats to humans and animals.[94]

Changes in biodiversity

Biodiversity refers to the abundance and range of living organisms,[95] including the "number, variety, and variability of living organisms" and how these change from one area to another and over time.[96] Biodiversity is affected by many factors, including temperature, moisture, hydrology, contaminants and climate change that can help determine the growth, reproduction, distribution, and productivity of plants and animals.[97] For example, hydrological changes can influence species due to moisture content affecting metabolic and reproductive processes.[98] Keeping biodiversity in the Arctic vibrant is vital as many species only exist there and have evolved to adapt to the climate and its unique habitats and ecological processes.[99]

CAFF identifies some of the current threats to biodiversity as: climate change, ozone depletion, physical disturbance/habitat fragmentation, chemical disturbance, new species introduction, consumptive use/overexploitation, and other land uses such as tourism.[100]

Fragmentation occurs when portions of a geographic area, such as a natural ecosystem, are subdivided thereby severing the landscape into isolated sections.[101] This can start through the simple division of land by roads but end in devastating changes to native species and an overall change to the landscape, which in turns destroys the area's wilderness heritage.[102] Pressure to build roads and engage in logging and mining activities, as well as constructing petroleum pipelines, are significant causes of fragmentation.[103] Impacts include blocking the movement of smaller animals, exposing larger animals to hunting and poaching, causing sedimentation of rivers due to erosion and leading to increased development, which in turn leads to more fragmentation.

Species introduction is a concern for biodiversity in that one species that is not natural to a habitat can have the ability to take over another and reduce the diversity of the original species.[104] For example, the Nootka lupine introduced in Arctic Iceland grows invasively and has overtaken the native vegetation, which reduced the diversity of plant cover[105].

Lastly, the issue of overexploitation from human over consumption is a threat to biodiversity.[106] Overfishing has caused the decline in some species, which has triggered changes in marine food systems that have caused a crisis in many Arctic areas.[107] The seal starvation is thought to have been associated with the overexploitation of capelin.[108]

Nuclear remnants

Nuclear remnants cannot yet be considered an environmental threat of the past as historic nuclear tests from around the world are still having effects. Radionuclides (atoms with an unstable nucleus often called radioactive isotopes) from atmospheric nuclear weapons testing in the 1940s to early 1960s, as well as from accidents such as Chernobyl in 1986, have infiltrated the Arctic.[109] The radioactive contaminants accumulate in plants, such as lichens, mosses and cushion plants within the tundra and polar desert, that then accumulate easily. Since these plants draw on air as well as precipitation for nutrients, a chain reaction starts as animals consume the plants and are exposed to the contaminants.[110] Caribou and reindeer rely on lichens for their primary winter diet, which leads to these species having "some of the highest concentrations of radionuclides among Arctic animals".[111] Heavy reliance by the Inuit and the Saami, the Indigenous peoples of northern Scandanavia, on caribou and reindeer respectively as a major component of their diet then directly threatens human health.

II. The Impact of Northern Environmental Threats on Arctic Peoples

Not surprisingly, the health and livelihood of Arctic peoples is directly affected by the profound changes occurring. As previously stated:

> "The imbalance among countries, with wealthy nations producing most of the emissions and developing ones bearing much of the burdens, becomes particularly acute for Indigenous communities that are living primarily 'off

the land'. They face the brunt of climate change exposure as well as other forms of environmental degradation."[112]

This global experience has wide-ranging implications in the Canadian portion of the Polar Region. The Inuit have traditionally depended on reasonably predictable patterns in sea ice formation and break up, wind and snowfall, the shift in seasons, and animal migrations for their very survival. The ability to rely on TEK to interpret the landscape and predict the rhythm of the seasons is crucial to endure the winter and to make the most of the abundance of the summer season. Increasingly, the Inuit are facing unpredictable wind patterns, extreme weather events, seasons altered, including longer, hot summers, and warmer, shorter winters. The earlier melting of snow and ice and deeper thaw of the active layer has caused subsidence problems and shorter durations for ice roads essential for safe transport of goods and people among many northern communities.[113] The warming trend implies that recent patterns of ice freezing later and breaking up earlier on rivers and lakes, as well as reductions in ice thickness, will continue, thereby altering the thermal dynamic of waterways and the habitat beneath the ice. Increased river-ice jamming and related flooding creates new hazards to adjacent settlements and the health of the riparian ecosystem, which depends on sediment, nutrients and the spring inundation of water.[114]

Changes to seasonal timing and the flow, including the magnitude of available surface water, will be important to northern communities that rely on this as the main source for drinking water, which is often untreated.[115] Another risk of contamination is from the northward movement of species and their related diseases as well as contamination resulting from sea-level rises, particularly for coastal communities.[116] Hydrological freshwater systems will also be affected by changes due to development and infrastructure due to the "loss of containment security of mine wastes in northern lakes underlain by permafrost; and increased flow and ice hazards to instream drilling platforms and hydro-electric reservoirs".[117] Significant movement in habitat and migration patterns of animals, such as caribou, polar bears and sea mammals, obviously impact heavily on northern Indigenous communities that have relied upon these animals for food and clothing for millennia. The dramatic recent decline in caribou population is especially frightening. As analyzed in a recent research study across the circumpolar region:

"Of 43 major herds that have been monitored during the past decade, at least 34 are declining … Herds for which census data have been gathered have declined an average of 57% from historical population maxima."[118]

The impact in some areas of the Canadian Arctic due to climactic shifts is even starker as "Three successive winters (1995–1997) of unfavourable winter conditions resulted in mass starvation – a 98% decline of Peary caribou on the Western Queen Elizabeth Islands."[119]

All of these events obviously impact negatively upon the cultural identity of the Indigenous peoples of the North in that they have a spiritual connection with the land, animals, and waters around them. Therefore, "when water is unavailable, unsafe or contaminated, it means that these people will suffer spiritual injury and mental health harm as they are unable to protect their environment."[120] As stewards of their traditional territory, they "carry explicit obligations that can be characterized as cultural and religious imperatives to protect their historic and natural space as its physical custodians or stewards to protect spiritual beings."[121] These obligations have been widely understood by the Supreme Court of Canada, the United Nations, the Inter-American Commission on Human Rights, and the International Labour Organization.[122]

These profound developments result in new threats to population health compounded by inadequate access to comprehensive local health care facilities and declining drinking water quality for people whose life expectancy is already ten to fifteen years shorter than that enjoyed by non-Aboriginal Canadians.[123] The implication of further health problems brought on by climate change is extremely serious through more travel related accidents and respiratory stress associated with higher summer temperatures.[124] The increasing presence and persistence of contaminants, such as heavy metals and POPs, that are carcinogenic, may have immunological and neuro-toxic effects, and may influence foetal and infant development concerns all northerners.[125] As Boyd indicates, "Levels of persistent organic pollutants are five to ten times higher in the breast milk of Inuit women than in women from southern Canada."[126]

The influx of unfamiliar birds and insects generates concerns regarding possible spread of disease and allergic reactions. As Thériault notes, the safety and abundance of "country foods" has been damaged by abnormally increased levels of POPs, heavy metals and other toxic substances within the Arctic food chain.[127] Global warming impacts upon food security through "changes in distribution, abundance, behaviour and health of wildlife, changes in hunting patterns and routes, and increasingly unpredictable weather, snow and ice conditions."[128]

Alterations to biodiversity affecting food source quality and availability influences northern residents for social and cultural reasons. The United Nations Millennium Ecosystem Assessment in 2005 stated that biodiversity benefits humans as it contributes to social relations, health, security, resiliency, and freedom of choices and actions.[129] The loss of certain species, therefore, affects human well-being beyond earning a living or obtaining food.

Canada's sovereignty assertions in the Arctic are also under renewed challenge due to the disappearing polar ice cap. This in turn generates internal questions between Canada as a nation and the Indigenous peoples of the north who want to regain a significant level of domestic self-determination over their traditional territory. The Inuit already dominate the Government of Nunavut, have successfully launched the Nunatsiavut Government in northern Labrador and are seeking to establish regional governments within northern Quebec and the Northwest Territories. These governments already are key to land management in the North, however, they seek to acquire a major voice in shaping Canada's pursuit of protecting its territorial boundaries claims and its involvement in the protection – or exploitation – of international areas. The non-voting permanent representative status guaranteed to Indigenous peoples in the Arctic Council has provided one forum in which to articulate their concerns,[130] however, other bilateral and multilateral negotiations are underway among the Arctic nations in which Indigenous participation is absent.

III. The Changing Human Face of the North and Environmental Protection

The political landscape of governance in the North has been dramatically transformed though land claims and reformations of public governments over the last two decades.[131] Despite immense socio-economic, health and environmental problems facing the North, many First Nations and all of the Inuit enjoy "considerable political and institutional autonomy in comparison with their southern counterparts."[132] The creation of the Nunavut Territory was directly inspired by the Nunavut Lands Claims Agreement (NCLA), which was signed on May 25, 1993 by Inuit representatives with the Canadian government, and the Government of the Northwest Territories. Less than one month after the NLCA was signed, the Parliament of Canada passed the *Nunavut Act* and the *Nunavut Land Claims Agreement Act* on June 10, 1993.[133] After six years of work and planning led by the Nunavut Implementation Commission, the *Nunavut Act* finally came into force on April 1, 1999.[134]

The vast majority of land claim and self-government negotiations in northern Canada have been concluded, including all above the Arctic Circle, over the past three decade.[135] The agreements reached, which are constitutionally protected by virtue of s. 35(3) of the *Constitution Act, 1982*, have significantly improved Aboriginal peoples' influence over land, wildlife and other resource decision-making.[136] Since the major environmental threats derive from locations far removed from the Canadian Arctic, the capacity of northern governments to minimize local effects is limited and their ability to regulate the source is non-existent.

IV. The Federal Government's Role

The Government of Canada is slowly coming to grips with environmental degradation and its impact on the North. Although a signatory to the Kyoto Protocol on Greenhouse Gas (GHG) emissions and a leader in the Montreal Protocol on Substances that Deplete the Ozone Layer, the Commissioner of the Environment and Sustainable Development, which is Canada's environment auditor, decried that despite five plans aimed at ameliorating climate change being approved by different federal governments over many years, none had reduced GHG emissions.[137] The current government agreed and released *Turning the Corner*, "a plan aimed at reducing greenhouse gas emissions by 20% from 2006 levels by 2020."[138] The National Round Table on the Environment and the Economy commented that the plan either overestimated the effectiveness of GHG reduction measures or did not properly substantiate how the goals would be achieved.[139] Little has changed federally since then.

Other legislation has been initiated to help improve efforts at avoiding environmental degradation anywhere within federal jurisdiction. Parliament enacted the *Species at Risk Act*[140] (*SARA*) in 2002 to respond to its obligations under the international Convention on Biological Diversity of 1992. *SARA* could aid in protecting several important northern mammals, such as the polar bear and beluga whale, if added to its list of protected species.[141] Canada also protects some vital natural habitat through creating national parks and protected areas such as Migratory Bird Sanctuaries. Furthermore, Canada's *Ocean Act*[142] includes specified Marine Protected Areas while the *Canada Wildlife Act*[143] protects listed National Wildlife Areas and Marine Wildlife Areas.

The *Canadian Environmental Protection Act, 1999*[144] is being invoked to undertake a review of 23,000 chemicals to assess for any human health risks.[145] Newer POPs, like polybrominated diphenyl ethers and perfluorooctanoic acid are also being investigated while Canada became the first country in the world to prioritize Persistent, Bioaccumulative and Toxic compounds.[146]

Arctic-specific environmental legislation includes the *Arctic Waters Pollution Prevention Act*,[147] which prohibits the deposit of any waste directly or indirectly in Arctic waters. More recently, the federal government announced in August 2008 that it is extending the range of the Act's jurisdiction from 100 to 200 nautical miles offshore.[148]

Despite these initiatives and the presence of progressive legislation for some sectors, POPs, GHGs and many other threats are global and no one country can effectively addressing them no matter how much political will, money or human resources it may be prepared to devote to the task. Only strong international co-operative efforts backed up by concrete action and fundamental moves in human behaviour have any chance of overcoming these environmental predations.

Conclusion

International environmental and human rights laws, which are increasingly giving respect to indigenous rights, appear at present to be humanity's only firm hope to bring about the fundamental transformations required to avoid the growing likelihood of catastrophe. While new technologies may yet be found to reverse global warming, restore the ozone layer, eliminate POPs and other airborne contaminants, cleanse nuclear remnants and prevent further pollution of our oceans, no one can rely upon such discoveries being found.

On the other hand, our current international law instruments are clearly inadequate to meet such a massive task as they have been largely designed to protect individuals from state abuse and prevent local sources of contamination. Further, there simply are no effective enforcement mechanisms for this order of magnitude. In addition to efforts by civil society to bring sufficient pressure to bear on states and large multinational corporations to induce a radical realignment of international law and corporate social responsibility, international law is one method to produce at least a modicum of action in prodding states for more effective environmental protection. Indigenous rights, where recognized, can also be a vehicle for transformative change. The United Nations Permanent Forum on Indigenous Issues has contributed in this regard by making climate change the focal point of its 2008 session.

Indigenous peoples are on the frontline in facing the tragic consequences of environmental degradation around the world. The Inuit in the circumpolar region live the changes that have been wrought in recent years on a daily basis. They are seizing the limited opportunities available to them, however, the global nature of so many of these threats transcend their capacity to impose the measures necessary to prevent disaster. It is up to the rest of us to induce the action needed to avoid catastrophe. The timeline to do so in the Arctic, as for many small island states around the world, is almost gone.

Notes

[1] David W. Schindler and John P. Smol, "Cumulative Effects of Climate Warming and Other Human Activities on Freshwaters of Arctic and Subarctic North America" (2006) 35 AMBIO 160 at 161.

[2] Conservation of Arctic Flora and Fauna, "Arctic Flora and Fauna: Status and Conservation" (2004), online: Arctic Portal http://arcticportal.org/arctic-council/working-groups/caff-document-library/arctic-flora-and-fauna [hereinafter CAFF]. .

[3] Intergovernmental Planet on Climate Change, "Climate Change 2007: Synthesis Report, (2007), online: Intergovernmental Panel on Climate Change www.ipcc.ch/publications_and_data/publications_ipcc_fourth_assessment_report_synthesis_report.htm 26 at 30 [hereinafter IPCC].

[4] Ibid.

[5] Ibid.

[6] Francois Cote and Tim Williams, "The Arctic: Environmental Issues" (2008) PRB 08-04E Parliamentary Information and Resource Service, Library of Parliament, online: <www.parl.gc.ca/information/library/PRBpubs/prb0804-e.pdf> at 1.

[7] IPCC, note 3, supra, at 39.

[8] CAFF, note 2, supra, at 95

[9] "Climate Change and Water: IPCC Technical Paper VI" (2008), online: Intergovernmental Panel on Climate Change <http://www.ipcc.ch/pdf/technical-papers/climate-change-water-en.pdf> at 106.

[10] Ibid.

[11] Ibid.

[12] Ibid at 107.

[13] Ibid.

[14] Ibid.

[15] Ibid.

[16] Ibid.

[17] Ibid.

[18] Ibid at 108. See, also, Terry V. Callaghan, L.O. Björn, F.S. Chapin III, Y. Chernov, T.R. Christensen, B. Huntley, R. Ims, M. Johansson, D.J. Riedlinger, S. Jonasson, N. Matveyeva, W. Oechel, N. Panikov and G. Shaver, 2005: "Arctic Tundra and Polar Desert Ecosystems" in C. Symon, L. Arris and B. Heal, eds., *Arctic Climate Impact Assessment (ACIA): Scientific Report*, (Cambridge: Cambridge University Press, 2005) at 243-352.

[19] Ibid.

[20] Ibid.
[21] Ibid.
[22] Ibid.
[23] CAFF, note 2, supra, at 86.
[24] Ibid.
[25] Ibid at 95.
[26] Ibid at 86.
[27] Sheila Watt-Cloutier, "Testimony before the Inter-American Human Rights Commission, Washington, DC" (2007a),online: The Center for International Environmental Law <www.ciel.org/Publications/IACHR_WC_Mar07.pdf>.
[28] IPCC, note 3. supra at 30.
[29] Bradford W. Morse, Samantha Reynolds and Alison Ronson, "A Canadian View on the Northern Dimension" (2008) at 9; translated from Englishas "Anchi gli inuit hanno una voce" with Samantha Reynolds and Alison Ronson in (2008) *Limes, Rivista Italiana di Geopolitica* 63–72 (also published in its original form in the special issue, *The Polar Game*, as "The Voice of the Inuit" in the sister online journal, *Heartland* at pp. 51–59 www.heartland.it/_lib/_docs/2008_02_the_popolar_game.pdf).
[30] CAFF, note 2, supra, at 95.
[31] CAFF, note 2, supra, at 86.
[32] Arctic Climate Impact Assessment (ACIA), *Impacts of a warming Arctic: Arctic Climate Impact Assessment* (New York: Cambridge University Press, 2004) at 26.
[33] Morse, et al, note 29, supra, at 6 relying upon IPCC, note 4, supra.
[34] Garrett W. Brass Ed., Arctic Ocean Climate Change, (Arlington, VA.: US Arctic Research Commission, Special Publication No. 02-1, 2002) at 10.
[35] L. Illeris and S. Jonasson, "Soil and plant CO_2 emission in response to variations in soil moisture and temperature and to amendment with nitrogen, phosphorus, and carbon in northern Scandinavia" (1999) 31 *Arctic, Antarctic and Alpine Research* at 264–271.
[36] Ibid.
[37] J.P. Schimel, K. Kielland and F.S. Chapin III, "*Nutrient availability and uptake by tundra plants*" edited by Reynolds and Tenhunen (Berlin: Springer Verlag, 1996).
[38] Natural Resources Canada, "Taking the chill off: Climate change in the Yukon and Northwest Territories" (2006) www.adaptation.nrcan.gc.ca/posters/wa/index_e.php.
[39] ACIA, note 32, supra, at 71.
[40] Ibid at 70.
[41] Francois Cote and Tim Williams, "The Arctic: Environmental Issues" (2008) PRB 08-04E Parliamentary Information and Resource Service, Library of Parliament, online: < http://www.parl.gc.ca/information/library/PRBpubs/prb0804-e.pdf> at 2.
[42] Ibid.
[43] Ibid.
[44] Ibid.

[45] Ibid.
[46] Ibid.
[47] Ibid..
[48] Ibid.
[49] Ibid. r
[50] Morse, et al, note 29 at 6.
[51] "What are Persistent Organic Pollutants (POPs)?" online: United Nations Industrial Development Organization http://www.unido.org/index.php?id=5167.
[52]D. Boyd, "Unnatural Law: Rethinking Canadian Environmental Law and Policy" (2003) at 67.
[53] Torsten Meyer and Frank Wania, "What environmental fate processes have the strongest influence on a completely persistent organic chemical's accumulation in the Arctic?" (2007) 41 Atmospheric Environment 2757 at 2762.
[54] Ibid at 2763.
[55] IPCC, note 9, supra, at 108.
[56] Ibid.
[57] Meyer and Wania, note 53, supra, at 2758.
[58] Ibid at 2765.
[59] Ibid at 2766.
[60] Ibid.
[61] Ibid.
[62] Ibid.
[63] "Persistent Organic Pollutants", online: United Nations Environment Program <www.chem.unep.ch/pops/>.
[64] "What are Persistent Organic Pollutants (POPs)?" online: United Nations Industrial Development Organization http://www.unido.org/index.php?id=5167
[65]"The Stockholm Convention" online: United Nations Industrial Development Organization <www.unido.org/index.php?id=5279>.
[66] "Governments unite to step-up reduction on global DDT reliance and add nine new chemicals under international treaty" (9 May 2009), online: Stockholm Convention on Persistent Organic Pollutants <http://chm.pops.int/Convention/Media/Pressreleases/COP4Geneva9May2009/tabid/542/language/en-US/Default.aspx>. The additions include:
Alpha hexachlorocyclohexane to Annex A; Beta hexachlorocyclohexane to Annex A; Hexabromodiphenyl ether and heptabromodiphenyl ether to Annex A;
Tetrabromodiphenyl ether and pentabromodiphenyl ether to Annex A;
Chlordecone to Annex A; Hexabromobiphenyl to Annex A; Lindane to Annex A;
Pentachlorobenzene to Annnex A and C; Perfluorooctane sulfonic acid, its salts and perfluorooctane sulfonyl fluoride to Annex B.
[67] "About the Convention" online: The Stockholm Convention on Persistent Organic Pollutants http://chm.pops.int/Convention/tabid/54/language/en-US/Default.aspx.
[68] "Brief Questions and Answers on Ozone Depletion" (2008) online: United States Environmental Protection Agency http://www.epa.gov/ozone/science/q_a.html .
[69] Ibid.

[70] Ibid.
[71] Ibid.
[72] Ibid.
[73] Ibid.
[74] CAFF, note 2, supra, at 95.
[75] Ibid., at 86.
[76] Ibid., at 95.
[77] Ibid.
[78] United Nations Environment Program Ozone Secretariat (2006) online: <http://ozone.unep.org>.
[79] CAFF, note 2, supra, at 95.
[80] Ibid.
[81] Ibid.
[82] Arctic Monitoring and Assessment Programme, "Arctic Pollution 2006: Acidification and Arctic Haze", (2006) at v.
[83] Ibid.
[84] Ibid.
[85] Ibid.
[86] Ibid.
[87] Ibid.
[88] Ibid.
[89] Ibid
[90] Henrik Selin and Noelle Eckley Selin, "Indigenous Peoples in International Environmental Cooperation: Arctic Management of Hazardous Substances" (2008) 17 R.E.C.I.E.L. 72 at 74.
[91] Ibid.
[92] Ibid.
[93] Ibid.
[94] Ibid.
[95] Cote and Williams, note 41, supra.
[96] Greenfacts, "Scientific Facts on Ecosystem Change", (2009), online: <http://www.greenfacts.org/en/ecosystems/>.
[97] IPCC, note 3, supra, at 104.
[98] Ibid.
[99] Ibid.
[100] CAFF note 2, supra, at 86.
[101] Ibid., at 98.
[102] Ibid.
[103] Ibid.
[104] Ibid. at 86.
[105] Ibid.
[106] Ibid, at 253-254.
[107] Ibid, at 86.
[108] Ibid.
[109] Ibid at 152.

[110] Ibid.
[111] Ibid.
[112] Bradford Morse, "Indigenous Rights as a Mechanism to Promote Environmental Sustainability," in Laura Westra, Klaus Bosselmann and Richard Westra, eds., *Reconciling Human Existence with Ecological Integrity: Science, Ethics, Economics and Law*, (London: Earthscan, 2008) at 159.
[113] Natural resources Canada, note 38, supra.
[114] IPCC, note 9, supra at 108.
[115] Ibid.
[116] Ibid.
[117] Ibid. at 108-9.
[118] Liv S, Vors and Mark S. Boyce, "Global declines of caribou and reindeer," (2009) 15 *Global Change Biology* 2626 at 2626.
[119] Ibid., at 2630.
[120] Bradford Morse and Samantha Reynolds, " Indigenous Peoples and Water Rights: Does the United Nations' Adoption of the *Declaration on the Rights of Indigenous Peoples* Help?" (unpublished, paper presented at Global Ecological Integrity Group Conference in Berlin, Germany (July 2008), copyright protected) at 2.
[121] Morse, note 113, at 160.
[122] Ibid., at 160-163.
[123] Statistics Canada, "Aboriginal Peoples in Canada in 2006: Inuit, Métis and First Nations, Census 2006." at 19, www12.statcan.ca/english/census06/analysis/Aboriginal/pdf/97-558-XIE2006001.pdf.
[124] Christopher Furgal and Jacinthe Seguin, "Climate Change, Health and Vulnerability in Canadian Northern Aboriginal Communities" (2006) 114 *Environmental Health Perspectives* No. 12, at 1966.
[125] Arctic Monitoring and Assessment Program, "Fact Sheet: Effects of contaminants on Arctic humans and biota" Arctic Monitoring and Assessment Program (AMAP, 2002), www.amap.no..
[126] David Boyd, *Unnatural Law: Rethinking Canadian Environmental Law and Policy* (Vancouver: UBC Press, 2003) at 67.
[127] Sophie Thériault, 'Northern Frontier, Northern Homeland": Inuit People's Food Security in the Age of Climate Change and Arctic Melting' (2009) 15 *Southwestern Journal of International Law* 223 at 224.
[128] Ibid.
[129] Cote and Williams, note 6, supra at 2.
[130] Arctic States participating as members in the Arctic Council are Canada, Denmark/Greenland/Faroe Island, Finland, Iceland, Norway, Sweden, The Russian Federation, and the United States of America. The Inuit Circumpolar Council is a permanent non-voting representative. See Arctic Council, "Member States", http://Arctic-council.org/section/member_states.
[131] Tonina Simeone, "The Arctic: Northern Aboriginal Peoples" (2008) Library of Parliament: Parliamentary Information and Research Service online:

<www.parl.gc.ca/information/library/PRBpubs/prb0810-e.pdf> at 1.
[132] Ibid.
[133] *Nunavut Land Claims Agreement Act*. S.C.,1993, c. 29.
[134] André Légaré "Canada's Experiment with Aboriginal Self-Determination in Nunavut: From Vision to Illusion" (2008) 15 *International Journal on Minority and Group Rights* 335 at 347.
[135] Bradford Morse, "Indigenous Peoples of Canada and their Efforts to Achieve True Reparations" in Federico Lenzerini. ed., *Reparations for Indigenous Peoples: International and Comparative Perspectives*, (Oxford University Press, 2008) at 271
[136] Ibid.
[137] 2006 Report of the Commissioner of the Environment and Sustainable Development www.oag-bvg.gc.ca/internet/English/osh_20061005_e_23460.html.
[138] Government of Canada, "Turning The Corner: Regulatory Framework for Industrial Greenhouse Gas Emissions" online: Environment Canada <www.ec.gc.ca/doc/virage-corner/2008-03/pdf/COM-541_Framework.pdf>.
[139] Cote and Williams, note 6, supra, at 5.
[140] *Species at Risk Act*, S.C. 2002, c. 29.
[141] Cote and Williams, note 6, supra, at 5.
[142] *Ocean Act* S.C. 1996, c. 31.
[143] *Canada Wildlife Act* R.S.C. 1985, c. W-9.
[144] *Canadian Environmental Protection Act, 1999*, S.C. 1999, c. 33.
[145] Cote and Williams, note 6, supra, at 5.
[146] Rhittu Chatterjee, "Hunting for persistent chemicals that might pollute the Arctic" (2008) 42 Environmental Science and Technology 5034 at 5034.
[147] *Arctic Waters Pollution Prevention Act* R.S.C. 1985, c. A-12.
[148] Cote and Williams, note 6, supra, at 5.

CHAPTER EIGHT

POWER AND THE RECOGNITION OF INDIGENOUS LAND RIGHTS

MICHAEL COYLE

Introduction

One of the direct, and intended, consequences of international colonialism was the dispossession of indigenous peoples around the world from the vast majority of their traditional lands. Whether effected through military force, political negotiation (treaty-making) or through the legal and economic rights granted to settlers and settler corporations, the results across the world have been strikingly similar: disconnecting indigenous peoples from their traditional relationships with the land and limiting their access to the resources on which their economies were once based. In the past 50 years, however, a new international consciousness has emerged concerning the plight of indigenous peoples. In many cases countries have come to grips with the realization that the methods used to dispossess indigenous peoples did not comply with domestic legal norms, much less broader notions of equality among peoples and respect for cultural difference. Among the first nations to develop formal processes aimed at offering redress for takings of indigenous lands were the United States, which established an Indian claims commission in 1946, Canada, which created an Office of Native Claims in 1974 and New Zealand, which passed legislation in 1975 to review alleged state breaches of the Treaty of Waitangi.

A small number of states are currently implementing formal processes to address indigenous land rights claims.[1] In assessing the effectiveness of these processes in reversing the most damaging results of colonialism, one might look to a variety of criteria. The remedies available to indigenous peoples under each process, the efficiency of the process in resolving disputes, the extent to which concepts of fairness play a role in determining outcomes, the degree to which each process departs from colonially imposed norms, and the amount of resources actually transferred back to indigenous peoples under the process – all of these are important considerations that might be used to assess the quality of justice delivered by domestic indigenous rights processes. Existing state-sponsored land rights processes might also be measured against the consensus ideals reflected by the 2007 United Nations Declaration on the Rights of Indigenous Peoples, a non-binding document endorsed by all but three state members of the United Nations.[2] Article 27 of the Declaration provides as follows:

> "States shall establish and implement, in conjunction with indigenous peoples concerned, a *fair, independent, impartial, open and transparent* process, *giving due recognition to indigenous peoples' laws, traditions, customs and land tenure systems*, to recognize and adjudicate the rights of indigenous peoples pertaining to their lands, territories and resources which they have traditionally owned or otherwise occupied or used…"
> (United Nations Declaration on the Rights of Indigenous Peoples, 2007) [emphasis added]

Beyond fairness, independence and inclusion of indigenous values, the Declaration also stipulates that domestic land rights processes should encourage the return of land and resources to indigenous peoples. Thus, Article 28(2) provides:

> "Unless otherwise freely agreed upon by the peoples concerned, compensation shall take the form of lands, territories and resources equal in quality, size and legal status or of monetary compensation or other appropriate redress."

One overarching determinant of the capacity of any domestic process to achieve the ideals set out in above is its responsiveness to pre-existing power imbalances between the state and indigenous groups. It was power imbalance, after all, that permitted states to dispossess and marginalize indigenous peoples in the first place. Only through significant adjustments to those power relations will indigenous groups be able to disturb property interests that have been allocated under state laws. Likewise, the recognition of indigenous values and legal traditions as norms that should have significant sway in the resolution of land disputes will necessarily involve limiting the dominance of established modes of legal reasoning. And for states to establish independent processes for resolving indigenous rights disputes (disputes in which the state is a defendant) requires those same states to relinquish a large measure of control over the outcome of those disputes. Put simply, states must be willing to re-examine, in the context of indigenous rights disputes, the manner in which they have historically exercised power.

In this chapter I propose to examine the capacity of contemporary indigenous land rights processes to counteract the influence of external power relations between the parties on the outcomes produced by those processes. "Power", of course, is a protean concept that can be analyzed through a variety of lenses. Power relations are ensconced in established institutions, in the legal frameworks that now operate in formerly colonial regimes, in the financial and technical resources available to disputants, and even in the language and dialogical norms that prevail in a given society. Further, as thinkers like Foucault have pointed out, power is relational: that is, the effective power relationship between any two parties is determined by the particular context in which they find themselves interacting as well as their perceptions of that background. This analysis will focus on two particular contexts in which indigenous land right disputes are currently being addressed: the formal claims resolution processes established in Canada and Aotearoa/New Zealand. Both countries' land rights processes are well developed, having been established in the 1970s. Both processes, as will be seen, have produced a significant number of outcomes that have resulted in the transfer of hundreds of millions of dollars' worth of land and monies to indigenous groups. Further, the challenges that must be addressed by their processes are roughly similar to those present in most countries seeking to address the injustices suffered by minority indigenous populations. Because both countries rely primarily on direct negotiations with the state to determine the outcome of indigenous rights disputes, this analysis will draw on the insights of negotiation theory in examining the sources and predicted effects of power relations between the parties on the outcomes of indigenous claims.

All of Gaul was once divided into three parts, so too this chapter. First, it will briefly describe the historical and legal context of indigenous treaty claims in Canada and New Zealand. Second, it will examine the determinants and effects of power in negotiations to resolve treaty disputes between the state and indigenous groups. Finally, it will review the treaty processes and legal regimes currently in place in both countries to assess their robustness in counteracting the influence of power relations on outcomes.

The Nature of Treaty Disputes in New Zealand and Canada: Historical and Legal Context

Any assessment of the fairness and effectiveness of a process designed to resolve rights disputes must begin with a consideration of the nature and context of the disputes in question. In both Canada and New Zealand, as with most countries addressing issues arising from their colonial past, disputes over indigenous land rights generally reflect grievances that have been pursued and remained unresolved for a long time. In this regard there are striking parallels between the experiences of state–indigenous relations in Canada and New Zealand. Both countries were originally settled on the basis of political agreements between the British Crown and indigenous peoples. In what is now Canada the treaty-making process began early in the 18th century and continues to this day in areas not covered by historical treaties. In New Zealand the relationship between the settler state and the *tangata whenua* was established by one treaty, the Treaty of Waitangi, a document whose terms were agreed in 1840. In both countries, the Crown deliberately chose treaty-making as its preferred method of dealing with indigenous peoples' rights. On both continents treaty-making allowed the British to procure peaceful relations with indigenous peoples and international credibility for Crown claims of sovereignty over the "new" lands. For indigenous peoples, an important *quid pro quo* was formal recognition by the British Crown that further occupation of indigenous lands would occur only with the consent of the indigenous peoples involved.[3] In both New Zealand and Canada, the Crown negotiated treaties with great solemnity and with assurances that the Crown's treaty promises were intended to endure for generations ("as long as the sun goes round and the water flows" was the formula used in several 19th century treaty negotiations in Canada).[4]

Both states, however, very soon limited the ability of indigenous peoples to enforce the Crown's treaty obligations. Only thirty seven years after the signing of the Treaty of Waitangi, the New Zealand Supreme Court issued a ruling against the Maori that would limit their ability to enforce the Crown's treaty promises for more than a century. In its decision in *Wi Parata*[5], the court dismissed a Maori claim to enforce the terms of a land transfer that had been made on condition that the transferred lands were to be used to set up a school for Maori children. Further, the court concluded that insofar as the Treaty of Waitangi purported to provide for the assumption of British sovereignty in New Zealand it was a mere "nullity." After all, the court reasoned, the Maori were "barbarians", incapable of negotiating such matters of state with the

Crown. In Canada, Privy Council rulings in two cases decided in the 1880s[6] had a similar effect in severely limiting the ability of First Nations to force the Crown to honour its previous treaty commitments. These rulings focused on the legal result of Treaty 3, by which the Anishinaabe signatories appeared to relinquish their property interest in some 55,000 square miles of land, now largely in north-western Ontario. The Crown's reciprocal engagement – to establish Anishinaabe reserves within that same territory – was held to be unenforceable against the province of Ontario (which eventually took 40 years to confirm the reserves that the treaty required). In the trial judgment of the first of these cases, Chancellor Boyd used language strikingly similar to that used by the New Zealand court eight years earlier in *Wi Parata*. For Boyd, the Anishinaabe were "the rude red-men of the North-West", and, for the most part, "a more than usually degraded Indian type."[7]

In both Canada and New Zealand until well into the 20th century, the treaty promises that had been made to indigenous peoples remained generally unenforceable in the courts and unprotected by any legislation that would implement the Crown's obligations. Indigenous peoples' access to their traditional land and resources were left vulnerable in the face of ever increasing waves of immigration and government decisions to allocate resource rights that paid little or no heed to the special situation of indigenous peoples. Further, in both countries legislation was passed that expressly limited indigenous groups' access to the courts for the advancement of land claims.[8] The effects of this shunting away of indigenous grievances can be seen today. In New Zealand more than 1400 claims have now been filed that allege breaches by the Crown of the Treaty of Waitangi, while in Canada some 1300 claims have been submitted against the federal government alleging that its past actions resulted in illegal interferences with First Nations' lands or monies.[9]

The two countries also followed similar trajectories in recognizing in the 1970s that the Crown's dealings with indigenous lands and resources should be subject to scrutiny as involving *legal* rights of indigenous peoples. The Canadian Supreme Court's decision in *Calder*[10] saw six members of that court acknowledge that indigenous peoples' historical occupation of their traditional lands could give rise to a justiciable claim for aboriginal title. Shortly thereafter, the Canadian federal government established a formal claims policy to negotiate aboriginal title and treaty claims, as well as other claims that the Crown had failed to fulfill its legal obligations to particular First Nations. In 1982, amendments to Canada's Constitution included a formal recognition of "existing aboriginal and treaty rights."[11] For its part, in 1975 New Zealand enacted the *Treaty of*

Waitangi Act 1975[12], an Act that (according to its subtitle) was to provide for the observance of the principles of the Treaty of Waitangi. The Act provided for the creation of the "Waitangi Tribunal", empowered to investigate alleged breaches of the Treaty, to hold hearings, to review the alleged breaches, and to make recommendations as to how particular breaches should be remedied by the Crown.

The result is that both Canada and New Zealand today have formal processes designed to investigate and resolve indigenous land and treaty claims. New Zealand's is essentially a two-track model, whereby Maori claims are generally subject to a hearing overseen by an independent statutory tribunal, followed (in the case of almost all claims relating to land and resources) by direct negotiations between the Crown and the *iwi* making the claim. As for the assessment of Maori claims, the *Treaty of Waitangi Act* stipulates that the norms to be used are the "principles of the Treaty", to be derived by reference to both the English and Maori versions of the Treaty. Canada has also recently enacted legislation authorizing the establishment of a special tribunal to hear and adjudicate First Nation land claims.[13] The Canadian tribunal has yet to be appointed and, as we shall see, will be significantly limited in its jurisdiction. At present, direct negotiation between the Crown and First Nation claimants is the only process available outside the courts for resolving treaty and land claims in Canada. Treaty claims are negotiated under the federal Specific Claims Policy, which provides that Canadian domestic legal principles are to govern both the assessment of each claim as well as the principles of compensation to be applied.[14]

Both countries, in short, have opted, in the formal processes they have established to address indigenous claims, to combine the application of legal norms to the dispute with direct dialogue between the Crown and indigenous groups. In the end, however, negotiation remains the primary method of resolving Maori treaty claims, while under Canada's formal land claim processes direct negotiation is the sole option currently available to First Nations. To assess the appropriateness of relying on negotiations to determine the outcomes of indigenous land rights disputes and, in particular, the extent to which power might be expected to influence their outcomes, it will be helpful to turn to negotiation theory.

Negotiation and Power Theory: The Risks for Indigenous Claimants

Given the nature and context of indigenous land and resource grievances, negotiation between the parties appears to be a particularly valuable tool for resolving those claims. This is so particularly in the case of treaty claims. Dialogue, after all, was the approach originally taken by the parties to regulate their respective rights. Resorting to the simple application of one party's legal norms to resolve such disputes seems inappropriate, particularly given that colonialism's imposition of state norms played such a key role in marginalizing indigenous grievances. Relying on dialogue, on the other hand, offers hope that the parties will be able to find enduring solutions to their disputes based on respect for the viewpoints of both sides. Further, the complexities involved in refashioning the parties' relationship in modern terms, and in crafting outcomes that assure indigenous land rights while recognizing the accrued rights of the public and third parties, call for a flexibility that negotiation can offer to a much greater extent than judicial or administrative orders. That same complexity makes interest-based negotiation potentially more efficient than adversarial proceedings in producing creative settlements that meet the overarching goals of both parties. Finally, depending on the ground rules of the negotiation, dialogue offers an opportunity for the state to demonstrate its respect for indigenous values and its willingness to create a healthier partnership for the future.

If negotiation seems to be a promising option for resolving indigenous land claims there are nonetheless obvious risks in relying upon negotiation alone. Historically, treaties were negotiated between peoples with very different worldviews and correspondingly different perspectives on the meaning of the negotiated terms and how they should be implemented. Those same differences in perspective can easily lead to disagreements on the same issues today and, ultimately, to negotiation impasses. Arriving at a negotiated settlement requires a willingness on both sides to find reasonable and respectful ways of bridging those differences. In the absence of neutral criteria and impartial mechanisms for resolving disagreements as to the validity of a claim and appropriate redress, direct negotiations may reach no outcome at all, leaving the *status quo* in place. Of particular concern here is the potential role of power imbalances in conditioning each side's assessment of a reasonable negotiation outcome.

In negotiation terms, the "bargaining power" of each party may be defined as its capacity to influence the outcome of the negotiation.[15] While ability to influence outcome may depend on a number of factors, including the abilities and tactics of each negotiator, the most significant concern for negotiating parties focuses on factors *external* to the dynamics at the bargaining table that may limit their ability to achieve an outcome that advances their interests. This concern, that external circumstances such as imbalances in resources may unduly influence negotiation outcomes, has led to considerable debate among legal scholars as to whether unassisted negotiation should ever be relied upon to resolve disputes over legal rights.[16] It seems worthwhile, therefore, to examine the potential effects of power imbalance on negotiations to address indigenous land rights.

Negotiation theory and empirical studies both indicate that the most important determinant of the parties' relative bargaining power is the attractiveness of their respective alternatives to reaching agreement at the table.[17] Simply put, if one party enjoys more attractive options for achieving its goals independently of the bargaining process it will tend to be more demanding when it considers possible options for negotiated agreement. Thus, the negotiation literature suggests, parties with superior negotiation alternatives tend to achieve better negotiation outcomes.[18] What constitutes a desirable alternative for each party will, of course, depend not only upon the external environment but also upon each party's subjective goals and values. Those goals and values (the parties' negotiation preferences) will not only shape their assessment of the alternatives to negotiation; they will also condition the parties' pursuit of particular settlement options that are clearly superior to their alternatives. Thus, while the interaction of the parties' negotiation alternatives will establish the boundaries of an acceptable settlement agreement, the parties' preferences will tend to shape their choice of a particular outcome within that settlement range.

There is evidence that imbalance between the parties' negotiation alternatives will affect the quality of their discussions as well as the outcomes they obtain. Empirical studies suggest that negotiators who believe they enjoy the stronger bargaining position tend to spend less time uncovering and discussing the interests of the other party. Further, particularly in situations where the parties lack a history of trust, negotiators who perceive themselves to be in a weaker bargaining position are tempted to employ adversarial tactics to resist the perceived effects of power. This is the so-called "power paradox": the greater the perceived imbalance in bargaining power, the more intensely a subordinate party will be tempted to adopt face-saving tactics and positional negotiation tactics.[19] The result is that, in situations of perceived power imbalance combined with distrust, both parties are tempted to avoid the types of open communication and collaboration that lead to outcomes that maximize both sides' interests. In the context of indigenous land rights negotiations, then, the presence of power imbalances between the state and an indigenous claimant will create a risk that the quality of their dialogue will not be conducive to creative and collaborative outcomes. Equally importantly, power asymmetries will present a danger that one party will enjoy a disproportionate ability to influence settlement outcomes regardless of the merits of the rights claim. To assess the significance of these risks in the context of indigenous–state negotiations requires a reflection on the political and legal alternatives available to each side in those negotiations.

Consider the legal alternatives available to each party in indigenous land rights negotiations. To simplify our analysis for the moment, let us consider those alternatives outside of any special dispute resolution mechanisms created by each country's current formal land and treaty claims process. For the indigenous claimant one conceivable alternative to negotiating with the Crown is to pursue its rights claim in the courts. In this regard, the common law in New Zealand offers little solace. The treaty rights of the Maori enjoy no special constitutional status and the New Zealand courts have recently confirmed that treaty promises in themselves cannot form the sole basis of a claim against the Crown.[20] Nor, to date, have the New Zealand courts confirmed in any decided case that specific Maori customary or aboriginal rights persist. In all of these regards, the Canadian case law is much more favourable to indigenous groups. Both aboriginal title claims (where no historical treaty has been signed) and treaty right claims now benefit from the recognition of aboriginal and treaty rights in the Canadian constitution. Further, for both types of claims, the Canadian Supreme Court has ruled that the Crown's behaviour must be measured against the "honour of the Crown" and, in appropriate cases, against the exacting standards of a fiduciary.[21] Regarding the implementation of treaties, the Court has ruled that the Crown must avoid any appearance of sharp dealing, resolving doubts as to the meaning of treaty terms in favour of the Indians.[22] Finally, since the Supreme Court decision in *Haida Nation*[23] in 2002, the Crown has been required to consult with aboriginal groups whenever it proposes to undertake an activity that may interfere with their aboriginal or treaty rights.

For a Canadian First Nation pursuing a land claim there are nonetheless significant limitations in the degree of rights protection offered by the courts. The rights guaranteed in s. 35 of the Constitution are not absolute; thus, the courts have ruled that both the Crown and the federal and provincial legislatures may intrude upon indigenous rights provided that they prove that they had a valid objective in mind and that in the overall context of their responsibilities the intrusion can be characterized as consistent with the honour of the Crown. Second, the Crown's obligation to *consult* aboriginal groups requires that the Crown give serious consideration to their rights; it does not require that the Crown ultimately refrain from interfering with the affected treaty or aboriginal rights. Third, in part perhaps because of the difficulties of mounting treaty rights litigation, there is no reported appellate decision in Canada that has granted a civil remedy for the Crown's historic breach of First Nation treaty rights.

At any rate, regardless of the substance of domestic law, where indigenous groups do not have financial resources to match those of the state, a number of factors tend to make pursuing rights claims in the courts a relatively unattractive alternative to negotiation. Pursuing litigation to the appellate level is a time-consuming and costly process. The cost of litigation is more difficult to bear for an indigenous group than for the state, which will have access to greater financial resources and to a full-time staff of litigation lawyers. Further, domestic courts are required to apply the state's legal norms – norms which have developed with little reference to indigenous values or to the legal traditions of an indigenous claimant. Finally, the remedies available in court, at least in the common law world, are much more restrictive than those available through negotiation. Generally restricted to awarding monetary compensation, ordering parties to cease particular actions, and making general declarations of the parties' rights, the courts are not in a position to fashion the sophisticated outcomes that are frequently required to adjust the parties' contemporary rights based on complex historical injustices. They are typically unable, for example, to order the return of lands, to interfere with third party rights, or to design co-management arrangements in relation to public resources. In all these respects, the litigation process tends to be ill suited to effecting significant changes in the *status quo* of rights over land and resources. For all of these reasons, therefore, even *where an indigenous claimant possesses an unrecognized right that might by confirmed by domestic law*, the litigation alternative will tend to be less attractive to the claimant than to the state.

The second legal avenue that an indigenous claimant might pursue is to seek shelter under the protection of domestic legislation that implements indigenous customary or treaty rights. Here the situation is significantly more favourable to the *tangata whenua* of New Zealand than to First Nations in Canada. Over the past 35 years New Zealand has enacted more than 20 statutes that make express reference to the principles of the Treaty of Waitangi.[24] Such legislative provisions, ranging from the regulation of property transfers by state-owned enterprises, to the management of Crown resources to health care, require Maori values or interests to be considered by government decision-makers. While these statutes might be amended at any time, given that Maori rights are not constitutionally protected, there is no doubt that ensuring Maori concerns are taken into account prior to specific government decisions breathes meaningful life into the historic treaty relationship in New Zealand. Apart from the *Treaty of Waitangi Act, 1975*, however, none of these statutes facilitates the independent pursuit of treaty promises by a Maori claimant.

Canada, for its part, has enacted no federal or provincial legislation that purports to give effect to the Crown's historic treaty promises. Nor, to the writer's knowledge, is there any Canadian legislation that requires a government body to take historic treaty promises into account when making decisions that might affect First Nation interests.

We have seen that the legal alternatives to negotiation in Canada and New Zealand provide only limited opportunities for indigenous rights claims to be resolved in accordance with indigenous values. What about political alternatives to negotiation? The relevant issue here is the extent to which an individual indigenous claimant has the capacity to vindicate a particular land rights claim through political means. Where indigenous peoples represent a minority of the population in a democratic state and particularly where a land claim appears to conflict with the interests of the public at large, the relative political influence of a particular indigenous group will be necessarily limited. Where indigenous peoples form a very small minority, as in Canada where they represent 3 per cent of the overall population and an even smaller proportion of the country's elected representatives[25], the capacity of an individual indigenous community to resolve a land rights dispute through political influence alone will be very limited indeed. In New Zealand, where fully 14 per cent of the population is Maori, and where the "mixed-method" proportional electoral system has led to an even high proportion of Maori members in the legislature[26], the opportunity to exercise political influence regarding Maori interests is significantly greater. The relative preponderance of New Zealand statutes that refer to treaty principles is surely a reflection of this fact. However, the national political influence of an individual *iwi* pursuing its land rights will of necessity be much more restricted. In the end, neither country's legislatures have ever intervened to resolve an essentially local indigenous land rights claim.

Analysing the Current Claims Processes from a Power Perspective

We have seen that, in general, for an indigenous land rights claimant in New Zealand or Canada there are relatively few available political and legal alternatives to seeking a negotiated settlement of its claim. And maintaining the *status quo*, by which the state has already allocated property rights, will inevitably be a less tolerable option for the rights claimant than for the state. The question arises: to what extent do the formal claims processes established in New Zealand and Canada alter the existing landscape of alternatives to ensure that power imbalance does not influence negotiation outcomes. Those claims processes have, in fact, resulted in a significant number of settlements, involving the transfer of a sizeable amount of compensation, land and resource rights to First Nation and Maori claimants.[27] To assess the risk that many of these outcomes may reflect the limited options available to indigenous claimants requires a consideration of the extent to which the national claims processes provide mechanisms to counteract the expected effects of power imbalance.

As noted earlier, Canada's national policies for addressing aboriginal title and treaty claims rely solely on direct negotiation between aboriginal groups and the Crown. Neither of these policies currently provides any process alternative should the parties fail to reach a negotiated agreement, nor any independent mechanism for resolving disputes between the parties as to the validity or value of an indigenous claim.[28] Both policies were developed unilaterally by the federal government; both permit the federal government to determine unilaterally whether a claim is valid; and neither policy requires the government to stay at the negotiation table when disputes arise between the parties. While Canada's Specific Claims Policy calls for legal principles to be used in the negotiation of a claim, the absence of mechanisms for resolving impasses and ensuring timely outcomes led in 2007 to a Commission of Inquiry criticizing the timeliness and neutrality of the federal claims process, and to the federal government itself describing the process as "not sustainable."[29] First Nations' frustration with Canada's land claim processes, reflected in a number of recent high-profile public confrontations, is arguably a direct result of the underlying and uncorrected power asymmetry between the parties.

New Zealand, by contrast, has created in the Waitangi Tribunal a robust alternative to direct negotiation of treaty claims with the Crown. With powers invested by a statutory mandate, the Tribunal is authorized to investigate historic treaty claims as well as claims that current or proposed Crown activities would violate the principles of the Treaty. Although the Tribunal's power is generally limited to making recommendations (its limited jurisdiction to make binding rulings has been exercised on only one occasion) its recommendations and general conclusions about the nature of the treaty relationship have had considerable influence over individual claims negotiations as well as over the development of general government policies.[30] The Treaty of Waitangi Act's requirement that at least one Maori member sit on each hearing panel, combined with the fact that for most of its history the Tribunal has been led by a Maori chairperson, also give it considerable credibility as an impartial body. On the other hand, in the end, once the Tribunal has made its recommendations on a claim, the claimant must seek its redress through unassisted negotiation with the Crown. The negotiation process is not ordinarily monitored by the Tribunal and it has few resources to hear complaints of unfairness in the Crown's negotiation tactics. Further, as we have seen, in general the Maori claimant will at this stage have no opportunity to resolve negotiation disputes in the courts. Finally, unlike the Canadian government, the New Zealand government does not purport to consider legal norms in negotiating the value of a claim settlement; in the absence of such norms, the government has made clear that each settlement should roughly conform to an overall "budgetary envelope" pre-determined by the state.[31]

Turning to the aspects of the current claims processes that might ensure that indigenous values are taken into account in a land or treaty claim, New Zealand has taken significant steps in this direction unmatched by any similar initiative in Canada. In addition to the Tribunal's bicultural membership requirements, appointments to the Tribunal must take into account each candidate's knowledge and experience of the matters likely to come before it. The Act requires Tribunal members, in ascertaining the meaning of the Treaty, to consider both the Maori and English texts of the Treaty. The Act expressly authorizes the Tribunal to develop hearing proceedings that take into account the customs of each Maori claimant. Witnesses can give evidence in the Maori language and the Tribunal's procedures call for hearings to be held whenever possible on the Maori claimant's territory. As for New Zealand's Crown negotiation policy, it expressly requires that government negotiators consider Maori values in determining appropriate settlements, and requires its negotiators to consider including a special component in each settlement to provide "cultural redress" for the Crown's actions.[32] All of these measures reflect a conscious effort by New Zealand to minimize the risk that power imbalance could lead to the eclipsing of indigenous values in the claims settlement process. By comparison, Canada's Specific Claims Policy makes no reference to the cultures of the parties and establishes no negotiation ground rules that would take into account First Nation values. Indeed, the Policy expressly provides that compensation for takings of land must not include any amount based on "special value to owner" (i.e. reflecting the indigenous group's particular attachment to its land and the environment).[33] Nor will the newly enacted *Specific Claims Tribunal Act* improve matters in this regard. The Act makes no provision for bicultural appointments of Tribunal members, nor does it require that members have any expertise in indigenous rights issues. The criteria to be applied by the Tribunal follow closely the rules of the current Specific Claims Policy and the Act expressly provides that compensation for a First Nation claim shall not include any non-pecuniary loss, "including loss of a cultural or spiritual nature."[34] Finally, the Act permits, but does not require, that the Tribunal take into account "cultural diversity" in developing its procedures.

Conclusion

In any country where indigenous peoples are a minority there is a significant risk that power imbalances between the state and an indigenous group will affect the outcomes of land rights negotiations. Where negotiation plays a significant role in the state's claims resolution framework the adequacy of that framework will depend largely upon the viability of legal and political alternatives to the negotiation process. If the existing alternatives to negotiation create a significant risk of power imbalances favouring the state, the design of a fair and timely indigenous claims framework requires a deliberate effort on the part of the state to ensure that the process includes neutral dispute resolution mechanisms and that the process itself leave space for indigenous values. We have seen that the land rights processes currently available in New Zealand and Canada do not completely address the concern that state power will unduly influence the outcomes of indigenous land rights disputes. On both counts, however, New Zealand's treaty claims processes appear to provide a much more substantial counterweight to power than the equivalent processes in Canada.

Notes

[1] These include Australia, Brazil, Canada, the Philippines and New Zealand. The courts in certain other countries, such as Argentina, Belize, Malaysia, India and Japan, have recently recognized that indigenous collective rights are justiciable domestically, while the Inter-American Court of Human Rights has ordered that mechanisms be created to recognize indigenous land rights in Nicaragua and Paraguay. It may be hoped that such countries will also establish formal indigenous land rights processes.

[2] The countries that have not endorsed the Declaration are Canada, United States and New Zealand.

[3] In Canada this was effected through the *Royal Proclamation of 1763* and the Treaty of Niagara in 1764. In New Zealand the assurance was provided in Article 2 of the Treaty of Waitangi.

[4] For examples of government negotiators in Canada assuring First Nations that the treaties would bind them forever, see A. Morris, *The Treaties of Canada With The Indians of Manitoba and The North-West Territories Including The Negotiations on Which They Were Based, and Other Information Relating Thereto* (Toronto: Prospero Books, 2000) at 51, 94, 202 and 275.
[5] *Wi Parata v. Bishop of Wellington* (1877), 3 NZLR 72 (S.C.).
[6] *St. Catherine's Milling And Lumber Company v. The Queen* (1888), 14 A.C. 46 (P.C.), *Ontario Mining Company, Limited and Attorney-General for Canada v. Seybold et al. and Attorney-General for Ontario*, [1903] A.C. 73 (P.C.).
[7] *St. Catherine's Milling*, *supra* note 7, at paras 24 and 64.
[8] The *Indian Act*, R.S.C. 1927, c. 98, s. 141 made it a summary conviction offence for a lawyer to take money from an Indian band for the advancement of any claim without the consent of the federal government. In New Zealand legislation specifically enacted to prevent the advancement of Maori claims included the *The Land Titles Protection Act 1902 and the Maori Land Claims Adjustment and Laws Amendment Act 1904:* see D. Williams, "*Wi Parata* is dead, Long Live *Wi Parata*", in A. Erueti and C. Charters, eds. *Maori Property Rights and the Foreshore and Seabed: The Last Frontier.* (Wellington: Victoria University Press, 2007) at 36–40.
[9]As of June 2008, the Waitangi Tribunal had received 1448 registered claims: F. Cooke, Assistant Registrar, Waitangi Tribunal (Personal communication, June 22, 2008). The Canadian figure comes from Indian and Northern Affairs Canada, "Specific Claims – A Statistical Snapshot" online: www.ainc-inac.gc.ca/al/ldc/spc/pubs/rpts/sss-eng.asp.
[10] *Calder v. A.G.B.C.*, [1973] S.C.R. 313.
[11] s. 35(1) of *The Constitution Act, 1982, Schedule B of the Canada Act 1982 (UK)*, s. 35(1).
[12] *Treaty of Waitangi Act, 1975* No 114, Public Act, Schedule 1.
[13] *Specific Claims Tribunal Act*, S.C. 2008, c. 22, s. 6 ["SCTA"].
[14] Canada began a policy of negotiating indigenous claims in 1974. For the current policies in relation to treaty claims and aboriginal title claims, see DIAND, *Outstanding Business: A Native Claims Policy – Specific Claims* (Ottawa: Minister of Supply and Services, 1982)["*Outstanding Business*"] and DIAND, *In All Fairness: A Native Claims Policy* (Ottawa: Minister of Supply and Services, 1981.
[15] See R. S. Adler and E. M. Silverstein, "When David Meets Goliath: Dealing with Power Differentials in Negotiations" (2000) 5 Harv. Negotiation L. Rev. 1, R. Fisher, "Negotiating Power: Getting and Using Influence" (1983) 27 The American Behavioral Scientist 149 and R. Korobkin, "Inertia and Preference in Contract Negotiation: The Psychological Power of Default Rules and Form Terms" 51 Vand. L. Rev. 1583.
[16] See, e.g., O. Fiss, "Against Settlement" (1984) 93 Yale L.J. 1073; R. Delgado et al., "Fairness and Formality: Minimizing the Risk of Prejudice in Alternative Dispute Resolution" (1985) Wis. L. Rev. 1359 and T. Grillo, "The Mediation Alternative: Process Dangers for Women" (1991) 100 Yale LJ. 1545.

[17] For a general discussion of the role of alternatives in negotiation, see D.A. Lax and J. K. Sebenius, "The Power of Alternatives of the Limits of Negotiation" (1985) 1:2 Negotiation Journal 163.
[18] R. J. Wolfe and K.L. McGinn, "Perceived Relative Power and Its Influence on Negotiations" (2005) 14 Group Decision and Negotiation 3 at 6., citing among others, a study by Mannix and Neale 1993; see also P.H. Kim and A.R. Fragale, "Choosing the Path to Bargaining Power: An Empirical Comparison of BATNAs and Contributions in Negotiation" (2005) 90:2 Journal of Applied Psychology 373 at 374 and Lax and Sebenius, *supra* note 18 at 171-72.
[19] E. Lawler, "Power Processes in Bargaining" (1992) 33 Sociological Q. 17, P. H. Kim, R. L. Pinkley and A. R. Fragale, "Power Dynamics in Negotiation" (2005) 30 Academy of Management Review 799 at 813-814, J. Z. Rubin and B. R. Brown, *The Social Psychology of Bargaining and Negotiation* (New York: Academic Press, 1975) at 213-233. Compare J.-S. Rioux, "Crisis negotiation outcomes among equals and unequals, democracies and non-democracies" (1999) 36 International Politics 25.
[20] See *New Zealand Maori Council v. Attorney-General*, [1987] 1 NZLR 641 at 655-656. A 1941 Privy Council decision ruled that the Treaty of Waitangi has the status of an international treaty, but the domestic effect of this is that Maori treaty rights are enforceable only if confirmed by New Zealand legislation: *See Hoani Te Heuheu Tukino v. Aotea District Maori Land Board,* [1941] NZLR 590 (P.C.).
[21] *Haida Nation v. British Columbia (Minister of Forests)*, 2004 S.C.C. 73 at para. 19 [*Haida Nation*]. *R. v. Sparrow*, [1990] 1 S.C.R. 1075.
[22] See, e.g., *Haida Nation, ibid*. at para. 19, *R. v. Sioui*, [1990] 1 S.C.R. 1025.
[23] *Haida Nation, supra* note 22, at para 35.
[24] Te Puni Kokiri, *He Tirohanga Ō Kawa Ki Te Tiriti O Waitangi* (2001) Appendix 1 at 111. Available online:
http://www.tpk.govt.nz/en/in-print/our-publications/publications.
[25] In Canada, aboriginal members represent 1.65 per cent of seats in the House of Commons. See
www2.parl.gc.ca/Parlinfo/Compilations/Parliament/Aboriginal.aspx?Menu=HOC-Bio&Role=MP&Current=False&NativeOrigin.
[26] See www.elections.org.nz/democracy/electorates/ethnicity-mps.html.
[27] Since 1973 there have been more than 300 settlements under Canada's Specific Claims Policy, with a cumulative value of more than Cdn$2 billion: Canada, Senate Standing Committee on Aboriginal Peoples, Proceedings (13 July 2006), online: Senate Standing Committee on Aboriginal Peoples
<http://www.parl.gc.ca/39/1/parlbus/commbus/senate/com-e/abor-e/03ev-e.htm?Language=E&Parl=39&Ses=1&comm_id=1 >. In New Zealand the Office of Treaty Settlements has reached 25 final agreements with Maori claimants, worth close to NZ$750 million: Office of Treaty Settlements, *Four Monthly Report: March-June 2009,* online: www.ots.govt.nz.
[28] In 2008 Canada enacted the SCTA, *supra*, note 14. The Act authorizes a tribunal (yet to be appointed) to arbitrate disputes over the validity and compensation owing on certain Specific Claims. The tribunal will be unable to hear claims for a

value greater than 10 million dollars or claims seeking the return of land or resources. For a description of the mandate of this tribunal and the limits of its jurisdiction, see M. Coyle, "Addressing Aboriginal Land Rights in Ontario: An Analysis of Past Policies and Options for the Future – Part II" (2006) 31 Queen's Law Journal 796.

[29] The Hon. S. B. Linden, *Report of the Ipperwash Inquiry*, vol. 2 (Toronto: Queen's Printer, 2007) at 77-104. See Indian and Northern Affairs Canada, "Backgrounder: Canadian Centre for the Independent Resolution of First Nations Specific Claims"; online: www.ainc-inac.gc.ca/ai/mr/nr/j-a2004/02462abk-eng.asp

[30] C. J. Iorns Magallanes, "*Reparations for Maori Grievances in Aotearoa New Zealand*" in F. Lenzerini, ed., *Reparations for Indigenous Peoples: International and Comparative Perspectives* (Oxford: Oxford University Press, 2008) at 536-542.

[31] In 1994 the New Zealand government announced that it proposed to settle all treaty claims in accordance within an overall budget of NZ$1 billion. The Crown appears to have been guided by this budget in subsequent claim settlements: see M. Palmer, *The Treaty of Waitangi in New Zealand's Law and Constitution* (Wellington: Victoria University Press, 2008) at 196-7.

[32] See Office of Treaty Settlements, *The Negotiation Process*, online: <http://www.nz01.2day.terabyte.co.nz/ots/DocumentLibrary/RedBookPart2.pdf> at 42. The relevant provisions of the Treaty of Waitangi Act are ss. 4(2A), 5(6)(c), 5(9) and 6(2).

[33] DIAND, *Outstanding Business*, *supra* note 15.

[34] SCTA, *supra* note 14, s. 20(1).

CHAPTER NINE

HAUDENOSAUNEE GREAT LAW OF PEACE: A MODEL FOR GLOBAL ENVIRONMENTAL GOVERNANCE?

JACK P. MANNO

This chapter is a speculative thought experiment: what might global environmental governance look like if it took as its model the founding constitution, known as the Great Law of Peace, of one of the oldest continuing democracies in world, the Confederacy of the Six Nations of the Haudenosaunee (Iroquois)? This is not meant to be an introduction to Haudenosaunee history nor contemporary Haudenosaunee democracy. I am not an expert in either. I am not a citizen of the confederacy and I do not have Native heritage that I know of. I have, however, done considerable research and writing on environmental governance and decision-making, especially in the Great Lakes basin of North America and have written on the role of non-governmental organizations in international environmental politics. I have also served as organizer and Chair of a number of special environmental advisory groups to the Council of Chiefs of the Onondaga Nation. I have learned much through reading (Parker 1968, Wallace 1994 [1946], Schaaf and Swamp 2004, Grinde and Johansen 199[illegible], Jemison and Schein 2000, Barreiro 1988, Annunziata 1995) and even more from listening to my Onondaga friends and neighbors, Chief Irving Powless Jr., Chief and Faithkeeper Oren Lyons, Chief Jake Edwards, Clan Mother Audrey Shenandoah, Jeanne Shenandoah, Freida Jacques, Ada Jacques, Faithkeeper Wendy Gonyea, the late *Tadodaho* Leon Shenandoah and the current *Tadodaho*[1] Sid Hill. I am convinced that while our environmental efforts have most often focused on healing the earth, it's not the earth that needs healing but our relationship to it.[2] The best way to make certain that this healing happens and is sustainable is to incorporate into our governing institutions practices that restore relationships of respect, gratitude and reciprocal care between humans, our nations and the rest of the natural world.

The Great Law of Peace as a Model

The need for global environmental governance to steer humanity's collective responses to the gathering planetary crises has brought forth a number of proposals for reforming existing international institutions and/or creating new ones. Accomplishing the goals of these proposals seems unlikely yet inevitable. Unlikely that enough of humanity will come to identify as global citizens responsible for the stewardship of a shared and fragile biosphere and inevitable that the task will be ours. As a result there is a flurry of speculation, including in this book, about what environmental governance ought to look like; what authorities it should have, how it can be made democratically accountable and, most importantly, how it can effectively reform the ecologically destructive political economy that presently predominates on the planet and transform it into one based on a relationship of reciprocity and mutual benefit between humans and the earth, between Nature's productive capacity and humans' dependence on it. If our stay on earth has reached the point at which we are required to live and earn our livelihoods in ways that no longer undermine the ecological foundation on which all prosperity ultimately depends, then our task is to organize and govern ourselves to make this possible. It is most likely that global environmental governance will emerge incrementally, built on reforms to existing multilateral environmental and economic institutions that have imbedded in them neo-liberal cultural assumptions and practices, part of the hegemony of western culture that may well be part of the legacy that has wrought the environmental calamity we face. In this chapter I propose that another model of governance be considered, the Great Law of Peace (*Gayanashagowa)*, the Constitution of the Haudenosaunee (Iroquois) confederacy.

As Denevan (1992), Cronon (1983). Mann (2005), Sale (1990) and the oral history of Indigenous Peoples have revealed, the Americas prior to European colonization was not the "pristine" sparsely populated wilderness of American romantic myth but rather a settled landscape based on highly developed agriculture and resource management systems, with elaborate trading systems encompassing thousands of miles. After first encountering the lands and peoples of the Haudenosaunee, early explorers, fur traders and missionaries frequently reported fertile and productive lands, abundant food supply, remarkable health of the people and seemingly strong and stable social conditions. What is it about the Haudenosaunee confederacy that brought it through periods of internal strife, has helped it survive the genocide of Indigenous peoples unleashed on the Americas, and supports the continuity of its basic structure and purpose into the 21st century? This is not to suggest that survival of traditional Haudenosaunee governance has been easy or complete. The Six Nations (Tuscarora was added in the 18th century) presently govern only remnants and, in the case of Cayuga, none of their vast original territory. Their populations are greatly reduced yet increasing. And in many communities a variety of governing bodies operate outside the context of the current Grand Council of the Confederacy. Yet the Confederacy continues to deliberate at its capital at the Longhouse in Onondaga making it one of the longest continually functioning democratic institutions in the world today. It may be that those who hope to construct the architecture for global institutions to deal collectively with climate disruption and the many other environmental challenges would be served well by considering the model created a thousand or more years ago by the Peacemaker and leaders of the Haudenosaunee confederacy.

There have been a number of case studies of local co-management systems in which resource management is shared between indigenous peoples drawing on traditional environmental knowledge (TEK) and practices, and resource management agencies whose decisions and practices are based on western scientific environmental knowledge (SEK). The evidence in these cases generally suggests that the integration of TEK governance principles into environmental decision-making has reduced conflict and improved decision-making (Berkes at al. 2003; Chapin et al 2009). The Great Law of Peace is one of the most widely known and documented examples of traditional governance principles. It is the intention of this chapter to consider how the Great Law of Peace could serve as a model for environmental governance on a global scale.

The Great Law of Peace is the constitution of the Haudenosaunee ("People of the Longhouse"), more commonly known as the Iroquois Confederacy, one of several confederations of Indigenous Nations that existed in North America at the beginning of European conquest, many of which continue as distinct and sovereign Indigenous Nations today. Johansen and Grinde (1991, 21) report that Indian confederacies were widespread and had "evolved to co-ordinate governance across geographic distances that seemed huge to European eyes at the time, and to permit maximum freedom to nations within the confederations, and individuals within nations." While the Great Law was a comprehensive international agreement involving all aspects of Haudenosaunee collective life, for our purposes most importantly it specifically included and reinforced an ethic of responsible resource management, a perspective of respect and gratitude toward the natural world, a requirement to consider the impacts of decisions on future generations (those "whose faces are coming from beneath the earth") and clear assignment of stewardship responsibilities. Presently throughout the Lake Ontario Plain, western Lake Erie, the St. Lawrence and Niagara Rivers and the Finger Lakes Watersheds of northern, central and western New York State in the United States and southern Ontario, Canada; the six Haudenosaunee Nations continue to be represented in the Grand Council of Chiefs[3] at its capital, the Longhouse of the Onondaga Nation in Central New York.

The relevance and influence of the Great Law of Peace is not ancient history. Contemporary Haudenosaunee leaders have been instrumental since the 1970s in publicly warning of the coming environmental calamities. Haudenosaunee intellectuals and spiritual spokespeople carried the message from the Great Law in a collection of essays titled Basic Call to Consciousness (Akwesasne Notes 1977). They helped organize a highly visible presence at the United Nations Conference on Environment and Development, the Earth Summit, in 1992 and they were influential in the decades-long effort that led in 2007 to United Nations General Assembly adopting the Declaration on the Rights of Indigenous Peoples that includes the recognition that, "respect for indigenous knowledge, cultures and traditional practices contributes to sustainable and equitable development and proper management of the environment." According to the authors of "Basic Call",

> "When the United Nations of the world was proposed toward the end of World War II, researchers were dispatched to find models in history for such an organization. For all practical purposes, the only model they found concerned the Constitution of the Five Nations, whose author had envisioned exactly that."

According to the Haudenosaunee telling of how the Confederacy began, the five independent nations, the Mohawk, Oneida, Onondaga, Cayuga and Seneca, whose people had long been in brutal conflict with each other, were convinced by a Huron prophet known as *Deganawidah* "The Peacemaker" to bury their weapons beneath the tree of peace on the shore of Onondaga Lake. The roots of that tree spread out in the four directions and any nation or people in the world would be welcome to follow those roots back to their source and, if they were willing to live by the Great Law of Peace, they would find shelter there. In the contemporary search for a model for how for international environmental cooperation we might consider the path to where those roots lead. What can we learn from the Great Law of Peace to guide global environmental agreements and institutions?

That the contemporary efforts to agree the structure and functions of environmental governance institutions should be influenced by the Great Law is not surprising. There is a long history of western political thought being inspired by Haudenosaunee philosophy, experience and practice. Schaaf and Swamp (2004) report that, "Just before the founding fathers were about to make their declaration of independence, a delegation of the Haudenosaunee came before Congress." In 1992, The United States Congress passed a joint (House and Senate) resolution concluding that, "the Confederation of the thirteen colonies into one republic was explicitly modeled upon the Iroquois confederacy as were many of the democratic principles which were incorporated into the Constitution itself." What seems to have impressed liberal philosophers of the 18th century was how the Haudenosaunee confederacy was able to balance individual liberty, national sovereignty and international cooperation. If, indeed, the US constitution was in part modeled after the Great Law of Peace[4], they left out key aspects of it especially the political importance of the clan system and the powerful role of women.

At the end of the first decade of the 21st century the need for improved institutional mechanisms for effective cooperation has become increasingly urgent (Walker et al 2009). Ron Engel, a theologian and philosopher, a drafter of the Earth Charter and one of the editors and authors of the IUCN volume, *Governance for Sustainability* (Bosselmann et al 2009, p. 47), argue that global institutions must rest on foundational political agreements with shared yet differentiated responsibilities, as seems to have been the case with the Great Law of Peace. According to the authors of *Governance for Sustainability*:

> "a sustainable society demands a people with a strong covenantal identity. Sustainability requires what a covenantal commitment alone can offer: sacrifices for the common good, including the survival and well-being of all members, self-limitations on individual and collective behavior, and responsibilities that reach across generations."

Bosselmann et al emphasize the importance of democratic relations between the state and its people, the people with each other and, a commitment to equality before the law, articulated often in covenantal constitutions. Without such commitment to the community as a whole, including as Aldo Leopold famously added, the community of life, democratic societies can have the tendency to degenerate into selfishness, greed, inequality and wasteful excess, the kind of environmentally destructive behavior that led to high profile and urgent international efforts to invent and adopt more sustainable forms of development, beginning with the Bruntland Commission (1987), the World Conference on Environmental and Development (1992), the UN Commission on Sustainable Development, and more recently the UN initiated Millennium Development Goals and the Millennium Ecosystem Assessments. The clear conclusion of this movement toward Sustainable Development is that success depends on incorporating environmental responsibility and accountability into all government decisions. This was also the original intent of US and other nation's original environmental laws that often included requirements that governments carry out "environmental impact assessments" of all major decisions. Instead, environmental assessments came to be expected only for development activities with clear potential impacts on local environmental amenities. This led, in effect, to ignoring the environmental consequences of wasteful, inefficient and profligate patterns of economic development that wrought environmental and climate havoc through the accumulated stresses and unintended consequences now broadly acknowledged. Despite the difficulties in agreeing to substantial changes in our models of governance, the situation now calls for a significant overhauling, one that expects and requires a new emphasis on personal and collective responsibility for care of the environment.

The Haudenosaunee Confederacy was founded during another time of crises. According to oral historians, the confederacy was founded to solve the problem of blood feuds, where families felt compelled to avenge the death or capture of a relative. The Condolence Ceremony whose ritual of acknowledgement of loss and the need for healing the mind are prescribed in the Great Law. To curtail warfare among the nations the Peacemaker sought and secured from each Nation an agreement to treat the natural world in a manner that required cooperation and conservation.

Many historians and commentators of contemporary Indigenous governance systems have remarked that the power of governance rests on the internalization of social norms involving collective and participatory decision processes, demands on personal duties and the high regard placed on the powers of persuasion, a skill of Native people that numerous settler chroniclers remarked (Grinde and Johannssen 1991). Taiaiake Alfred (1999), a Mohawk scholar of Indigenous government, notes that:

> "The Native concept of governance is based on .....the 'primacy of conscience.' There is no central or coercive authority, and decision-making is collective. Leaders rely on their persuasive abilities to achieve a consensus that respects the autonomy of individuals, each of whom is free

> to dissent from and remain unaffected by the collective decision. The clan or family is the basic unit of social organization, and larger forms of organization, from tribe through nation to confederacy, are all predicated on the political autonomy and economic independence of clan units through family-based control of land and resources."

The clan-based system works to internalize a sense of family between members of the same clan even in separate nations, a self-identification even with people with which one is not related by shared ancestors, and whose relationship is through a shared identity with a part of the animal kingdom, to whom one is also related in a web of reciprocal respect.

Fundamental Principles

Some of the Haudenosaunee fundamental principles are of the kind that Ron Engel calls covenantal, involving basic agreement of the core responsibilities of a human being. Many of these have their origins in the Great Law of Peace, others like the Thanksgiving Address come from the original instructions of the Haudenosaunee Creator. They include[5]:

- *Righteousness.* A form of consciousness also known as the "Good Mind," an alignment of one's mind with natural law and the welfare of one's clan and nation.
- *Diversity within unity.* The point of unity is to create the conditions within which individual and national uniqueness can flourish.
- *Freedom.* Freedom from want, from coercion and violence, freedom of speech and freedom of religion.
- *Consensus.* Extensive public debate. The expectation and often requirement that consensus be achieved.
- *Checks and Balances.* Checks on power through balances between branches of government, (representatives of the younger brothers [Oneida and Cayuga] and older brothers [Mohawk and Seneca]), and the "executive," the Onondaga, keepers of the central fire of the confederacy. An especially important matter that threatens the confederacy must be submitted back to the people. People can propose their own laws even when the leaders fail to do so by a form of popular referendum. Balance between genders (women select the chiefs [*Honoye*] to represent the clans and have the authority to remove errant leaders, a women's council as a form of judiciary where people can appeal for justice.

Balance between war chiefs and peace chiefs. And a complex process of decision-making designed for thorough vetting of decisions and the achievement of consensus. There is also a clear delineation of powers between the confederacy, its member nations and the clan system. The Grand Council of the Confederacy does not interfere with the internal affairs of any nation and likewise Nation councils have no say in matters that are traditionally the concern of the Clan.

- *Fundamental human rights.* Natural rights of all people, direct public participation: freedom of speech, freedom of religion and the rights of women.
- *Gratitude.* The requirement that all meetings and important events begin with a ritual oration known as the "Thanksgiving Address" or sometimes called "Greetings to the Natural World" and "The Words That Come Before All Else" where all the elements of nature , including and beginning with the people that have gathered, are given thanks and respect for continuing to perform their duties (or "original instructions") that have allowed the world to go on as a place where life, and human life in particular, can flourish. Part of the "original instructions" with which humans were endowed by the Creator was the mental capacity, known as intelligence, to notice, think about, recognize and appreciate ones own dependence on all other non-human persons carrying out their duties. It is one of human beings main duties to give thanks and respect.
- *Management of the Commons.* The settled areas, known as the "clearings" were under the responsibility and management the nation of which the settlement was a part. The other areas, the "woods" were understood to be open for access and use by all Haudenosaunee people but the Nation within which the woods existed had the responsibility to manage and protect these Commons.
- *Open borders.* People have full rights in any of the Nations, except the right to hold high office and that too can be granted to an individual through the designated process of raising leaders.
- *Families and Clans as fundamental social structure.* Clans exist in each nation and form political groups that have little to do with where people live. All authority springs from the clans. Issues are discussed within the clans and presented to a clan council made up of women. The clan council determines who is the clan leader (Clanmother). The process of selecting *Hoyane* begins very early

and is part of a young boy's upbringing of being observed, especially for how he treats women and girls.

Lessons for Global Governance

As scholars, practitioners and negotiators explore plans for new institutions or reforms to existing ones, it could be helpful to consider the lessons one might draw from the model of the Great Law of Peace. The following I consider some of the most important:

- *Gratitude, duties and healing.* Because most of the important declarations and manifestos of "western" political history have been composed in response to a long history of harsh treatment and oppression it is understandable that they emphasize primarily individual liberty and human rights. Yet central to solutions to the environmental crises of our time would be a broad internalization and spread of an understanding of reciprocal responsibility and care toward the natural world, an awareness that it is each person's duty to nurture and be stewards of the ecosystems and natural resources on which human well-being ultimately depends. Instilling within the institutions of global governance a ritual expression of gratitude as part of all the "conferences of the parties" and meetings of decision-making bodies would help make gratitude, responsibility and duty a part of everyone's awareness. A covenant based on an expression of the reciprocal ecological duties required of all human beings/citizens should be a foundational part of any set of agreements creating the new institutions of global environmental governance. Such a covenant would place appropriate emphasis on the relationship between human beings and the rest of nature and foster an understanding that it is not just the land, air and waters that need healing but our relationships to them and to each other.
- *The central role of female authority.* The balance of power designed by the Peacemaker includes the balance between women and men. Their roles are distinct and complementary. While menhave the role of Chief, women traditionally are the property owners.. A person inherits her or his Clan identity from the mother, and property traditionally passed down through her family. A man was traditionally expected to join the longhouse of his wife. It is believed that because women watch over the raising of the children they are also watching over the future leaders.

Women bring forth life and with that comes a fuller awareness of the future generations the Great Law requires be considered in all decisions. For those of us who have survived the harsh upbringing typical in "western" "modern" cultures and their oppressive gender stereotypes, any talk of women's power being expressed in particularly gendered ways is difficult to stomach. It is certainly plausible that what the Haudenosaunee consider women's authority can be embedded in our global institutions and be yielded by both women and men. However it is done women's "standpoint" needs to be formally recognized and represented in the new global environmental institutions.

- *Commons and ecosystem services.* The Great Law of Peace distinguishes between who has responsibility for the land and resources in "clearings," or the sellteld areas, and who has responsibility in the "woods." Proper behavior when in each area is clearly spelled out. It is essential that the new institutions recognize and support local collective stewardship and in particular recognize traditional and native peoples with special relationships that may or may not include legal titles..
- *Sovereignty in an interdependent world.* The Great Law makes clear distinctions about which decisions are the prerogative of the Clans and families, which belonged to the Nation and which the Confederacy. Outside of a fairly narrow but critically important range of decisions that affected all the Nations collective wellbeing, the confederacy has little or no authority over the individual nations. The new global institutions under discussion will need to narrowly carve out its mandate, setting legally enforceable environmental duties while creating diverse opportunities for international cooperation in environmental restoration and resource management.
- *Clan identification.* Perhaps the most important of the social institutions for stability and sense of identification with non-human nature is the clan system. Through the Peacemaker's incorporation of the Clan system in the Great Law, international relations were made literally family relations. This may be the most difficult aspect of the model to incorporate into our new governance institutions. It is conceivable that we could design institutions that would link certain groups to specified environmental responsibilities, develop rewards or rights attached to membership in these groups, make membership inheritable, and develop team or clan-based training programs to train

members in the skills to carry out duties associated with membership. Conceivably this could be similar to various scouting programs for children and professional affiliations for adults but these new clan-like identities would be lifelong and associated with specific fields of knowledge and ecologically based responsibilities such as river-keepers, advocates for particular species, soil builders, foresters, tree doctors, etc. While certain groups could be bioregionally specific, there could be enough geographic overlap to create social relationships between members in different countries, facilitated through clan-based social networks operating online and in person for specific environmental projects. A portion of a person's political representation would function through these groups. Each could have representatives in the new institutions of global environmental governance.

Declaration on the Rights of Indigenous Peoples

However the new institutions are designed I hope they will include a commitment to making all their actions consistent with the United Nations Declaration on the Rights of Indigenous Peoples. While this chapter has focused on the Great Law of Peace, there are many indigenous governance practices from which the international community can learn. Simply by deciding to recognize and honor the rights of indigenous peoples in all its deliberations, the new environmental institutions will open up the possibility of communication and collaboration with Indigenous nations based in mutual respect, equity and empowerment.

References

Annunziata, J (ed) 1995. *Haudenosaunee Environmental Restoration: An Indigenous Strategy for Human Sustainability.* Cambridge: Indigenous Development International. Akwesasne Notes 1991 (1978). *A Basic Call to Consciousness: The Haudenosaunee Address to the Western World.* Sumnertown, TN: Book Publishing Co.

Barreiro, J. 1988. *Indian Roots of American Democracy.* Special Constitution Editions of the Northeast Indian Quarterly Vol. IV, # 4 (Winter 1987) and Volume V, No. 1 (Spring 1988).

Berkes, F., Colding, J., Folke, C. 2003. Introduction. In: Berkes F, Colding J, Folke C (eds). *Navigating Social-Ecological Systems.* Cambridge: Cambridge Univ. Press. pp. 1–30.

Bosselmann, K., Engel, R. and Taylor, P., 2009 *Governance for Sustainability – Issues, Challenges and Successes*, IUCN – The World Conservation Union, Environmental Law and Policy Series Vol.70, Bonn/Germany

Chapin, F. S., Folke. C, and G. P. Kofinas. 2009. A Framework for Understanding Change. In: Chapin, F.S., Kofinas, G. P., and C. Folke (eds), *Principles of Ecosystem Stewardship: Resilience-Based Natural Resource Management in a Changing World.* New York: Springer. Pp. 3–28.

Cronon, W. 1983. *Changes in the Land: Indians, Colonists, and the ecology of New England.* New York: Hill and Wang.

Denevan, W. D. 1992 "The Pristine Myth: The Landscape of the Americas in 1492," *Annals of the Association of American Geographers*, 82:3, pp. 369–385.

Elazar, D. 1996. *Covenant and Constitutionalism,* New Brunswick, NJ: Transaction Publishers.

Grinde, D. and Johansen, B. (1991) *Exemplar of Liberty: Native America and the Evolution of Democracy.* Los Angeles: American Indian Studies Center 1991

Jemison, G.P. and Schein, A. M. (eds) (2000) *Treaty of Canadaigua 1794: 200 Years of Treaty Relations between the Iroquois Confederacy and the United States.* Santa Fe: Clear Light.

Mann, C, 2005 *1491: New Revelations of the Americas before Columbus.* New York: Knopf.

Parker, A. C. (1968) *The Constitution of the Five Nations* In: *Parker on the Iroquois*. Syracuse: Syracuse University Press.

Sale, K. 1990. *The conquest of Paradise: Christopher Columbus and the Columbian legacy.* New York: Knopf.

Schaaf, G. and Swamp, J. (2004) *The US Constitution and the Great Law of Peace*. Santa Fe, NM: CAC Press.

Wallace, P. 1994 (1946) *The White Roots of Peace: Iroquois Book of Life*. Santa Fe, NM: Clear Light Publishers

Walker, B. et al. 2009. "Looming Global-Scale Failures and Missing Institutions," *Science,* Vol. 325, 11 September 2009, pp.1345–1346.

Notes

[1] The *Tadodaho* carries responsibilities from the time of the founding of the Confederacy as spiritual leader and facilitator of meetings of the Grand Council of Chiefs of the Haudenosaunee.

[2] I credit the writer and cultural botanist, my friend and colleague, Robin Kimmerer (*Anishinabe*), the director of the Center for Native Peoples and the Environment at SUNY College of Environmental Science and Forestry with this formulation.

[3] "Chief" is an English translation of the Onondaga word, *Honoye* (*Rotiane,* in Mohawk and similar words in the other Haudenosuanee languages) meaning "one with a good mind."

[4] While, there remains considerable controversy over the degree to which key US constitutional framers consciously modeled their document on the Great Law of Peace, it is widely accepted that Benjamin Franklin, Thomas Jefferson and others had extensive contacts with Iroquois representatives and knew of their governance practices. It is likely that their knowledge of Native governance affected their thinking about political systems along with their own ideological background and the influence of enlightenment thought.

[5] This summary is drawn largely from Paul Wallace's "White Roots of Peace" (1994), Geoffrey Shaaf and Jake Swamps, "The US Constitution and the Great Law of Peace (2004) and many conversations and interactions with Haudenosaunee elders, faithkeepers, chiefs and Clan Mothers.

CHAPTER TEN

INDIGENOUS LAND RIGHTS IN MALAYSIA

RAMY BULAN

Introduction

Malaysia has a large indigenous population consisting of Malays, Aboriginal people (Orang Asli) and the natives of Sabah and Sarawak. Although the government has a strong indigenous base, with the political power in the hands of the Malays, the outworking of its obligation to the different indigenous groups, governed by different land laws, presents a particular challenge for the government.

The existence of a government policy of positive discrimination in favor of indigenous peoples, has not insulated certain parts of indigenous communities from blatant inequalities, partly as a result of laws inherited from the British colonial administration. While protection measures have been put in place for Malay land rights other sections of the indigenous population, particularly the Orang Asli remain exposed to certain vulnerability, exploitation and domination. This paper discusses the issue of indigenous identity in Malaysia and its relevance to land rights with a particular emphasis on the situation on peninsula Malaysia. It compares the relevant statutes that provide for the creation and extinguishment of aboriginal land rights and Malay reservation lands, highlighting some of the disparities in the legal provisions. It then looks at landmark judicial decisions that have pushed the frontiers of aboriginal land rights at common law.

Malaysia: The Demography of the Indigenous Population and the Application of Personal Laws

The Federation of Malaysia is made up of thirteen states, eleven states of which are in the Malay Peninsula, and two states, Sabah and Sarawak are part of the island of Borneo.[1] The ninth schedule of the *Federal*

Constitution, divides the executive and legislative powers between the federal and state governments. The states' exclusive powers of any significance are Islamic law, Malay and native customs, forestry, land, including land tenure, Malay reservations and native reservations, cadastral land survey, agriculture, local government and the constitution and procedure of native courts. Sabah and Sarawak are especially concerned with the preservation of native law and customs, which relate to safeguards of the special interests of their indigenous peoples.[2]

Malaysia's multiethnic, multicultural, multi-religious population of nearly 24 million[3] comprise an indigenous population of about 65 per cent of which Malays form the majority. The Chinese who comprise 26 per cent and Indians at 7.7 per cent, are the immigrant population. In Peninsular Malaysia the indigenous Orang Asli (literally, "original peoples") or Aboriginal people form about 0.5 per cent of the population, numbering a close to 150,000 individuals.[4] In Sarawak, indigenous groups comprise about 69.1 percent and in Sabah, the overall indigenous population is 60.7 percent.[5]

Befitting the multicultural ethnicity of the nation, the Malaysian legal system is a plural legal system with an integration of the common law, *syariah* law and customary law tradition. The indigenous groups in Sabah and Sarawak, officially called "natives" traditionally occupy the interior areas of the state where their communities continue to be governed by customary laws, which apply as "personal laws."[6] Customary laws, the aboriginal and native customary laws and to some extent, Malay customary laws have continued to be of significance in matters of family laws and customary land rights.[7]

Customary land tenures are recognized and encapsulated in different legislations. In Peninsula Malaysia, section 4(2) of the National Land Code provides for customary land tenure, along with the National Land Code (Penang and Malacca Titles) Act 1963 and the Customary Tenure Enactment (FMS). The Aboriginal Peoples Act 1954 ("APA") provides for the creation of Aboriginal reserves, and designates other areas as aboriginal areas, and aboriginal inhabited areas. In Sabah, the Land Ordinance (Cap 69) ("SLO") makes provision for native customary tenure under the Land Ordinance whilst in Sarawak, the Land Code 1958 (Cap 81) ("SLC") makes provision for native customary lands. Recent decisions of the highest court in Malaysia have now firmly declared that aboriginal or customary title and native customary rights to land exist based on native law and customs. These rights are recognized by common law and co-exist with statutory rights. They are to be viewed as complementary rights under the Federal Constitution ("Constitution"). The entitlement to these rights are determined by proof of indigenous identity. It is thus instructive to know the definition given to each indigenous group.

Definition of "Indigenous" Peoples in Malaysia: "Malay", "Aborigine" and "Native"

Indigenous peoples in Malaysia fall under a general class called "bumiputera" which means "sons of the soil", and includes Malays, Natives and Aborigines or Orang Asli, meaning "original people". They are defined for this purpose by the Federal Constitution ("Constitution") as well as regulation by state laws. Bumiputera is a "political" term that is used a basis of preferential treatment and affirmative action based on indigeniety. A legally verified indigenous status is essential as it affects rights entitlement including property interests.

a. Malay

Article 160 (2) of the FC defines a Malay as one who professes the religion of Islam, habitually speaks the Malay language and conforms to Malay customs; and was born or domiciled in the Federation or Singapore before *Merdeka*[8] or born of parents one of whom was born in the Federation or in Singapore or is the issue of such person.[9]Practically, for the purpose of the Constitution, a person may not strictly be of Malay ethnic origin or of a Malayan race. A person of Indian, Persian or Turkish origin might be a Malay if he or she professes the Muslim religion, habitually speaks Malay, and conforms to Malay customs. Conversely, a Malay by birth, who does not profess the Muslim religion does not satisfy the condition in Article 160(2).[10]

Malay identity is crucial for entitlement to special Malay privileges and the right to Malay Reservation Lands ("MRL") or customary lands. Characteristically, MRL are registered titles, and regulated by various state enactments. Every state may have its own definition in the Malay Reservation Enactment of the particular state.[11] For former Federated Malay States (Negeri Sembilan, Pahang, Perak and Selangor),[12] a Malay must belong to "any Malayan race" and "speaks Malay or any Malayan language."[13]The Kedah[14] and Perlis[15] Malay Reservation Enactments include Arabs as Malays, as long as they speak Malay and are Muslims. In both Kedah, Perlis and Kelantan, if there is any doubt as to whether a person is a Malay (or Siamese),[16] the question would be referred to the relevant State Ruler in Council whose decision on the matter would be final, and "shall not be questioned or revised by any court".[17]

b. Aborigines or Orang Asli

Article 160 states that an "aborigine" means "an aborigine of the Malay Peninsula".[18] They are also known as Orang Asli ("original peoples"), a collective term used for eighteen sub-ethnic groups who each have their own language and culture and social organization. The Orang Asli are the descendents of the early inhabitants of the Malaya Peninsula before the establishment of the Malay kingdoms, and include the Temuan, Jakun, Semai, Semelai, Cewong and the M' Betsi.[19]

Section 3 (2) of the Aboriginal Peoples Act 1954 "APA" then defines an Orang Asli as a person whose parents are both aborigine or one parent, male or female, is or was, a member of an aboriginal ethnic group. He or she can be a person of another race who is an adopted child. The criteria is, that he or she speaks an aboriginal language and habitually follows an aboriginal way of life and aboriginal customs and beliefs and includes a descendent of that person. In the case of *Sagong Tasi and Ors v Kerajaan Negeri Selangor (Sagong 1)*[20] the indigenous status of plaintiffs was challenged. They proved that they were from the Temuan community, lived in a organized society and follow an aboriginal way of life. They had their own system of adjudicating disputes, maintain their own language, and culture relating to marriages, inheritance, burial practices, religion and practised their own system of land tenure, and fulfilled thereby, the requirements of s 3 APA.

c. Natives

"Native" is defined in the Constitution, Article 161A, as well as the Interpretation Ordinances of each state of Sabah and Sarawak. In Sarawak, the main criteria for native identity is a person listed in clause 7 or a child born of native parents "exclusively" from the races specified under Article 161A clause (7).[21] Under state law however, it is possible for a person who was not born a native to apply to the Native Courts to be "deemed" a native and be subject to its personal law under s 20 of the Native Courts Ordinance 1992.[22] It is possible under this section for a person who is an issue of a mixed marriage, to claim native status.[23]This gives rise to a contradiction between the Constitution and the state law. Should the state law be read to conform to the FC as is normal, in cases of conflict? Since native status falls under the jurisdiction of the Native Courts, might it not be argued that this is a matter for the state to legislate upon? At best, it becomes a matter of concurrent jurisdiction. Given that this matter that could affect the fundamental rights of a person, it should not be interpreted in a pedantic manner. Rather, an approach that leans towards a liberal interpretation should be given due weight.[24]

In relation to Sabah a person which is a citizen, is the child or grand child of a person of a race indigenous to Sabah, and was born either in Sabah or to a father domiciled in Sabah at the time of the birth. Native groups in Sabah are not enumerated in the Constitution although there are at least 38 known major groups.

It must be noted that the definition of "natives" in Sabah and Sarawak include Malays. Be that as it may, a person who is not a Muslim cannot be declared to have become identified with or be subject to the personal law of a native Islamic community. In the Federal Court case of *Datuk Syed bin Kecik v Government of Malaysia and Anor*[25] a Malay from Peninsular Malaysia, was declared a native of Sabah when Suffian LP held that he was rightly a member of, resident in and connected with the state of Sabah and "belongs" to Sabah. His Lordship declared, "Who could belong to Sabah more than a native of Sabah?" This case illustrates the intricacies of proof of identity or membership of an indigenous group in Malaysia.

One the most compelling reasons for proving indigenous identity is the entitlement to rights in land reserved for indigenous peoples, whether as Malays, Orang Asli or Natives. The basis for recognition lies in the idea of indigeneity, pre-existence and first possession. Of the three groups, the rights of Malays to Malay Reservation Land ("MRL") appears to have the strongest safeguards and is, in fact one of the most entrenched rights in the Constitution. This paper will only look at Malay land rights under MRL and juxtapose it Orang Asli rights in aboriginal reserves.[26]

Malay Reservation Land

Lands reserved for Malays are provided for under the Federal Constitution as well as state legislation. Article 89 (6) of the Federal Constitution defines Malay Reservation as "land reserved for alienation to Malays or to the natives of the state in which it lies". A Malay includes any person who under the law of the state in which he is resident, is treated as a Malay for the purpose of the reservation of land. While article 89 lays the constitutional foundation of the reservation policy, article 90 stipulates special provisions relating to customary lands in Negeri Sembilan and Malacca and Malay holdings in Trengganu. These constitutional provisions have entrenched the rights that have already been provided for in several state enactments. Historically, one uniform law was enacted for Selangor, Negeri Sembilan, Perak and Pahang, known as the Federated Malay States Malay Reservation Enactment 1933. There are separate state enactments in each of the former Unfederated Malay States namely, Johor Malay reservation Enactment 1936, Kelantan Malay Reservation Enactment 1930, Malay reservation Enactment of Kedah 1931, Malay Reservation Enactment of Trengganu 1941, and the Perlis Malay Reservation Enactment 1353. There is no equivalent law for the Federal Territory nor is there any for Penang and Malacca.

The ostensible reason for the Malay Reservation policy was the protection and preservation of Malay land rights and to assist Malays economically, socially and politically,[27] based on the assumption that historically they were disadvantaged.[28] The After the signing of the Pangkor Treaty 1874, the British were involved in administration of Malay states through the Residential system, where a British Resident would advise the Malay states on all matters except matters relating to Islam and Malay customs. The British economic policy resulted in massive flow of immigrants from China and India, and the liberal alienation of land caused land to become a valuable and marketable commodity as the demand for land for commercial cultivation, mining and settlement increased. Thus as early as 1874, the Perak Resident J WW Birch voiced the need to safeguard the welfare of the Malay race against being "bled" by immigrants who had begun to acquire Malay land.[29]To prevent them from becoming vagrants[30] and "dispossessed wanderers" in their own country and to protect them lest they alienate and "divest themselves of interests of ancient standing, … depriving themselves and their children of their future livelihood," restrictions or *syarat nyata* (special conditions) were put on their lands. In order to keep the value of lands from being commercially attractive to non-Malays, no rubber trees were to be planted on Malay lands.[31] It was felt that "the end justified the means".[32] They were prevented from mortgaging these lands lest the lands will be taken over in default of payment by cash rich Chinese or Indian traders and others.

These policies mirrored the British perception of themselves as guardians of the Malays, in much the same way that native peoples elsewhere were viewed.[33] It might have been truly altruistic, but today, that British policy is questioned as being a marginalization of the Malays and is perceived as one that kept them from participating in the economic benefits that had flooded into Malaya at the time. Some writers see the MRL policy more as a safeguard for British political interests,[34]and the non-conversion from padi to rubber lands as a design to keep the Malays as an agricultural or a peasant class. Arguably, the policy has helped to preserve lands in a considerable area of lands in the hands of the Malay community.

a. Entrenching Malay rights to land through creation of Malay reservation

Although land falls under the legislative jurisdiction of the states, the importance given to MRL is underscored by the fact that the federal government is empowered to monitor the implementation of the Malay Reservation Enactments. The constitutional provisions entrenched the rights of Malays to their lands making that protection a matter of national policy. This enables the Federal government to intervene in a state matter, and gives it the power to advise and to formulate land policies affecting the states through the National Land Council.[35]

Prior to merdeka,[36] except for the Federal Territory, Sabah, Sarawak, Penang and Malacca, there were already MRL in existence in the peninsula. What Article 89 (1) of the Constitution does is to ensure that the lands under MRL will not be excised, revoked, acquired and the size not be altered unless it goes through the prescribed procedure. All MRL remains so unless otherwise provided by an enactment which is passed by a majority of the total number of members of the state Legislative Assembly and by the votes of not less than two-thirds of the members present and voting. Any laws that relate to the change of status of Malay Reservation Lands must also have the consent of the Ruler in Council. For added protection such an enactment would also need to be approved by resolution of each House of Parliament, namely the House of Representative and the Senate, and passed by a majority of the total number of members of that House and by the votes of not less than two thirds of the members present and voting. The stringent procedure is to safeguard against hasty and arbitrary amendments on a matter that is the subject of public trust.[37]

The provisions for MRL is by far one of the most entrenched rights in the Federal Constitution. In comparison, amendments to the constitution itself under Article 159, cannot be made in either houses of parliament unless it has been supported on second and third reading by vote of not less than two thirds of the total number of members of the house.

After Merdeka, the government in any state may acquire lands for the purpose of declaring it as Malay reservation. A proprietor may also apply for his land to be declared as a reservation. However, where any land ceases to be a Malay reservation land, another land of similar area and character shall be declared as Malay reservation in its place. Nevertheless, no such declaration may be made in respect of any land which at the time of the declaration is owned or occupied by a person who is not a Malay.[38] However, where a new reservation is declared, Article 89 (2) provides that an equal area of undeveloped and uncultivated land in that state shall be made available for general alienation; and a total area of land in a state for the time being declared as a Malay reservation shall not at any time exceed the total area of land made available for general alienation.

b. Revocation of MRL

MRL may lose its character only by way of an express and not implied revocation, and such revocation has to be gazetted to be effective. As such an issue of a mining lease over a piece of land does not take the land out of the Malay reservation.[39] As Raja Azlan Shah J (as he then was) said,“ To hold that an agricultural land in Malay reservation when converted into mining land loses its character as Malay reservation land would I think, be contrary to the intention of the Enactment which aims at preserving and protecting Malay interests.”[40]

Despite the “entrenched” status, MRL have not been spared from “disappearance” as a result of being acquired, or degazetted. Although Article 89(3) of the Constitution specifies that any MRL that has been revoked must immediately be replaced, large areas of land had been acquired under the Land Acquisition Act 1960 (“LAA ”) have not been replaced by the state governments.[41] Article 89 (3) provides that the replacement land has to be similar in character, and an area not exceeding the area revoked, and replaced immediately.[42] It has been suggested that the phrase “land of similar character” ought to be of the same economic value, and that means, land planted with coconut cannot be replaced by padi land. The location and the soil of the land and the potential development of the land should be the same.[43] This mandatory requirement puts the obligation on the state to get replacement for the land prior to approving the revocation of the MRL. Be that as it may, it often happens that land is often acquired for public purposes under the LAA 1960, but are not officially degazetted. As such, they are not replaced. For that reason it has been suggested that the LAA must not be used to acquire MRL to prevent further losses of Malay lands.

Clearly, MRL is well entrenched in the Constitution. The actual implementation of law and policy on the ground may be less than perfect, but it is still much better compared to the land rights of Orang Asli which falls at the opposite end of the spectrum.

Orang Asli Land Rights

The welfare of the Orang Asli comes under the Federal List, and they are a federal concern. Article 8 expressly authorizes the federal government, to undertake affirmative action in favour of Orang Asli, to make laws for the "protection, well being or advancement of aboriginal peoples of the Malay Peninsula (including the reservation of land) or the reservation to aborigines of a reasonable proportion of suitable positions in the public service". Such positive discrimination in favour of Orang Asli will not be invalidated or prohibited by the provision guaranteeing equality before the law to all persons. There are a number of legislation that affect the status of Orang Asli but a legislation that concerns Orang Asli specifically is the Aboriginal Peoples' Act 1954 (APA).

a. The Aboriginal Peoples Act 1954 (Revised 1974)

Under the APA all matters pertaining to land, including the gazetting and degazetting of aboriginal reserves come under the purview of the state, who may by notification in the gazette, declare any area exclusively inhabited by aborigines to be declared an aboriginal reserve[44] or an aboriginal area, [45]or aboriginal inhabited area.[46] Within an aboriginal reserve or aboriginal area, no land shall be declared as MLR, as a sanctuary for protection of animals or birds, or as forest reserve, or be alienated, granted or leased except to aborigines normally resident in the area. Furthermore no licence for collection of forest produce or for any commercial undertaking, is to be issued in an aboriginal area without consulting the Commissioner for Aboriginal affairs.[47]

The State authority may grant rights of occupancy, free of rent or subject to conditions in a grant within any aboriginal area or reserve to Orang Asli individually or communally but such rights would not be "any better than that of a tenant at will".[48] In each case, the state authority may revoke wholly or in part or vary any declaration of an aboriginal area or aboriginal reserve. There appears to be no provision or obligation imposed on the state authority for replacement of any land taken or degazetted as such. This is the only plausible explanation for the decline in acreage of gazetted Orang Asli reserves between 1990–1994 from 20,666,96 ha to 17,963,00 ha. Besides, approval for gazetting has been withdrawn from 7,443.8 hectares of 36,076 hectares originally approved before 1990.[49]

Where an aboriginal community establishes a claim to fruit trees or rubber trees on land which is alienated, granted or leased for any purpose, the state authority *may* grant compensation to the persons entitled in his opinion, or may if he thinks fit, pay the funds to the Commissioner to be held on common fund for that person or community in a manner prescribed by the Minister. This broad discretionary power gives the Minister concerned, or his representative, the Director General of the Department of Orang Asli Affairs (JHEOA), the final say in matters concerning the administration of Orang Asli.[50]

In matters concerning land, the state authority has the final say. This gives rise to problems because, administratively, the Department of Orang Asli Affairs can only request the State Executive Authority, to exercise its statutory powers to declare Orang Asli lands as reserves. Delays in processing and approval of such requests is a common complaint and to some extent, has provided a convenient excuse for disowning responsibility for the land woes of Orang Asli. Thus by placing Orang Asli under federal authority and having their lands rights subject to state authority, the constitution has created a conundrum in respect of division of powers, as to how aboriginal rights are to be safeguarded.[51]

This problem should not be as intractable as it appears to be because, there are adequate constitutional provisions that the federal government could utilize. As Rachagan[52] suggests, that Article 83 read with Article 160 (2) and Article 91, allows the federal government through the National Land Council to secure compliance of APA by the state governments with the APA.[53] There is also nothing to preclude the state governments from alienating, granting or leasing state land to Orang Asli but this is simply not done.[54] It is in the interests of the states to maintain control over their forest lands and that the creation of Orang Asli reserves in the valuable land banks would deny the states of a revenue-generating facility as that reserve would now come under the Federal government control.[55]Be that as it may, the Federal government's have a policy statement on dealing with Orang Asli which should be a good guideline.

b. Statement of Policy Regarding the Administration of the Orang Asli of Peninsula Malaysia of 1961 ("Policy Statement")[56]

The Policy Statement is an enlightened one. It reads:

> "The special position of the Orang Asli in respect of land usage and land rights shall be recognised. Thus, every effort will be made to encourage the more developed groups to adopt a settled life and this to bring them economically into line with other communities in the country. Aborigines are not to be moved from their traditional areas without their full consent."

Recognizing the nomadic lifestyle of some of the groups the Policy Statement relaxes forest policies in favour of Orang Asli. Nonetheless the ultimate aim of the policy is settled agriculture and that entailed providing "the necessary land … where they are willing to settle." If the policy was taken as a guiding principle, it is ironical that Orang Asli communities are excluded from the schemes such as the Federal Land Development Authority (FELDA) schemes[57] for the development of oil palm under the Land Group Settlement Areas Act 1960.[58]

On the contrary, notwithstanding the declared policy of recognizing Orang Asli usage of land, Orang Asli today find themselves living as squatters on State or alienated land. The statutory mechanism for gazettement of lands for their benefit has not been effective in protecting their rights to land.[59] In recent years, they have brought these grievances to the courts and the Malaysian courts have recognized their rights in consonant with the growing jurisprudence on indigenous title internationally.

c. Case law on Orang Asli rights

One of the earliest cases on Orang Asli rights was in *Koperasi Kijang Mas and 3 Ors v Kerajaan Negeri Perak*. The High Court[60] held that the state government of Perak had breached the APA when it gave a concession to Syarikat Samudera Budi Sdn Bhd (the company) to log certain areas which included lands approved by the State government as Aboriginal Reserves, for the regroupment schemes of RPS Sungei Banun and RPS Pos Legap. The company had no rights to carry on logging activities in those reserves. The fact that the lands were yet to be gazetted did not preclude the Orang Asli from exclusive rights to forest products in those areas. Justice Wira opined that gazettement was not a mandatory requirement. Rather, the approval by the state government was the crucial requirement. The decision in *Koperasi Kijang* augurs well for aboriginal rights and hopefully offered some measure of protection against the encroachment and displacement of Orang Asli communities.

Another milestone was created in the case of *Adong bin Kuwau v State of Johore (Adong 1)*[61] where the High court awarded compensation to 52 Jakun families for the loss of 53,273 acres of ancestral lands which the Johor state government had taken for the construction of a dam to supply water to Singapore and Johor. The court held that the Orang Asli had customary rights over their traditional and ancestral lands which they had depended upon to forage for their livelihood. The court applied indigenous title cases from the United States, Canada, Australia, New Zealand and other commonwealth jurisdictions and held that in relation to common law the rights of the aboriginal people

> "[I]nclude the rights to move freely about their land without and disturbance and also to live form the produce of the land itself but not to the land in the modern sense…since they have been in continuous and unbroken occupation and / or enjoyment of the rights from time immemorial.'[62]

Until the decision in *Adong 1*, it was presumed that the only rights that the Orang Asli had were the limited rights provided for by the APA. Mokhtar Sidin JCA (as he then was) held that the APA does not limit Orang Asli common law rights. In order to determine the extent of Orang Asli rights under the law, their rights under common law, and statute have to be looked at conjunctively, for both these rights are complementary.[63] Furthermore, the court declared that the Orang Asli right is protected under the Constitution for both the common law as well as the statutory rights are proprietary rights protected under Article 13 of the Malaysian Federal Constitution.[64] Thus the Orang Asli must be compensated for the deprivation of their heritage lands, freedom of habitation or movement, produce of the forest, future living for the plaintiff and his family and descendants.[65]The state's appeal against the decision was dismissed by the Court of Appeal which fully endorsed the view of the Mohtar Sidin JCA at the High Court. A further appeal to the Federal Court was again dismissed by the Federal Court.[66]

Another case arose in *Sagong Tasi v Kerajaan Negeri Selangor (Sagong 1)*[67] when a Temuan settlement in Selangor was given notice by the State Authority to vacate their land to make way for the construction of a highway to the Kuala Lumpur International Airport. They were given 14 days notice to move, and when they failed to do so, they were forcibly evicted from the land. Their fruit trees, crops and buildings were destroyed. They accepted under protest, compensation which was calculated under s 11 of the APA and filed a suit contending that they were customary owners of the land based on common law, statutory law and the Federal Constitution.

The High Court followed *Adong (1)* but went further to hold that the Orang Asli had an interest in and to the land. Having regard to Article 13, the court held that the Orang Asli right is a proprietary right protected under the constitution but "that this conclusion is limited to the are that forms their settlement but not to the jungles at large where they used to roam and forage for their livelihood in accordance with their tradition." This pronouncement is *obiter*, given the fact that the case was brought only for compensation for settled areas and there was no claim for foraging lands. The court held that the common law rights have not been extinguished by the APA and further declared that the compensation payable under APA based only on "things on the land" was not adequate, within the meaning of Article 13 of the Constitution. The payable compensation should be based on the LAA 1960, and based on market value. Since the plaintiffs have established possession, the second and third defendants, the contractors and the highway authority respectively, were ordered to pay damages for trespass.[68]

On appeal, the Court of Appeal in *Kerajaan Negeri Selangor* v *Sagong Tasi (Sagong Tasi II)* [69]upheld the proprietary rights and the communal nature of the plaintiffs interests. The Court of Appeal went further to emphasize the fiduciary nature of the state's relationship to the Orang Asli. That breach of fiduciary duty arises by reason of Article 8 (5) (c)[70] and the Ninth Schedule (item 16) of the Constitution and their failure to positively discriminate in favour of Orang Asli. Apart from those provisions, the court, held that the 1962 Policy Statement put a duty on the government to protect the Orang Asli, by reserving land for their benefit, protecting them against unscrupulous exploitation and to safeguard their tribal organization and way of life.[71] The Federal and the State governments have also breached their fiduciary duty by failing to gazette the Orang Asli settlement, through the deprivation of Temuan property rights without adequate compensation under sections 11 and 12 APA, and for the unlawful eviction by giving unreasonable and insufficient notice to vacate the land. As Gopal Sri Ram JCA (as he then was) said "They cannot take advantage of their on their own wrong." These principles have now established Orang Asli customary title and proprietary rights to land.[72]

The Federal Court has subsequently reaffirmed those principles in *Superintendent of Lands and Surveys, Miri Division and anor v Madeli bin Salleh (suing as administrator of the Estate of the deceased (Madeli 3)*.[73] Although that case concerned native customary land rights, the principles applied equally to aboriginal customary title. The full bench held that common law respects existing property rights of inhabitants and their rights to compensation for loss of those rights.

Madeli bin Salleh (Madeli 3)

Madeli bin Salleh a Malay, sought declaration that he was a licencee of a piece of approx. 6 acre land situated in Miri, Sarawak upon which he and his father before him had acquired native customary rights ("NCR") to land. Madeli alleged that the act of the Superintendent of Lands, without prior extinguishment of his NCR and without payment of compensation was wrong in law and constituted a wrongful interference with his enjoyment of his NCR.

The court affirmed that the plaintiff had created and acquired NCR in the disputed land and these rights have survived and had not been extinguished by reservation of land for concession (in this case (Shell Concession) or for a public purpose. There cannot be extinguishment of any rights without a clear intention to the effect.[74]On the facts Madeli had proven occupation of the land. At the Court of Appeal it was held that occupation need not necessarily be equated with actual occupation. There must be something done on the land, not necessarily on the whole, and there was evidence that the plaintiff had done that.[75]

The state had contended that the courts below should not have relied on *Adong, Sagong Tasi* and *Nor Nyawai*[76] as those cases had relied on the Australian case of *Mabo (No.2)*[77] and the Canadian case of *Calder v AG of British Columbia*.[78] It was contended that the Civil Law Act s 3 only provided for the application of English common law. The Federal Court explicitly rejected that contention, ruling that those cases "reflected the common law position with regard to native title throughout the Commonwealth." And as to the crown's radical title, applying *Mabo* (No.2) and the Privy Council case of *Amodu Tijani v Secretary v Southern Nigeria*,[79] the Federal Court held that the Crown may acquire a radical or ultimate title to the land but did not thereby acquire absolute beneficial ownership of the land that displaced the presumptive of the natives.

The case of *Madeli bin Salleh* is significant for common law native title in Malaysia.[80] It affirmed the earlier decisions in *Adong, Sagong Tasi* and *Nor Nyawai* and has affirmed principles that are drawn from other common law jurisdictions, that have now become part of the Malaysian jurisprudence which applies to the aboriginal as well as native land rights. The salient features are summarized below:[81]

a. Native customary rights were not created by statute and do not owe their existence to any statute or executive or judicial declaration.[82] They are pre-existing rights.[83]
b. The common law recognizes and protects the pre-existing rights under native laws and customs,[84] and the Courts will assume that the Crown intends that the rights of property of the inhabitants are to be fully respected.
c. Common law has the same force and effect as written law.[85]
d. By common law the Crown may acquire a radical title to the land. The Crown's right or interest is subject to any pre-existing native

rights over the land. The doctrine of acquired rights – change of sovereignty or rule does not affect acquired property rights of the inhabitants.[86]

e. Native title has its source in prior occupation – ownership of land within a territory in exclusive occupation of a people is vested in that people[87]
f. Proof of native title is by way of continuous occupation[88] and oral histories of the plaintiffs relating to their customs and traditions and on their relationship with their lands may be adduced as evidence within the confines of the Evidence Act 1950 Act 50 (Malaysia). [89] Occupation of land does not require physical presence, but there must be evidence of continued exercise of control over land.[90]
g. Native title has its source in traditional laws and customs.[91] The particular nature or rights associated with native title is a question of fact to be determined by the customs, practices and usages of each individual community.[92] It is pertinent to note that customs and usages come within the definition of "law" under Article 160 (2) of the Federal Constitution. As such customs and usages enjoy the status of being "law".
h. Native title under common law is inalienable.[93]
i. Native title may be held communally or individually.[94]
j. The rights may be taken away or extinguished by way of clear and unambiguous words of legislation[95] or an executive act authorized by such legislation.[96] A reservation of land for a public purpose may not necessarily extinguish native title unless it is inconsistent with the continued enjoyment of native title.[97]
k. Native title is a proprietary right protected by the Constitution. If native title is extinguished, adequate compensation must be paid in accordance with article 13 of the Constitution.[98]

While indigenous rights are contained in legislation, they co-exist with common law rights. The enforcement of those rights necessarily requires taking into account the statutory rights, the common law, native law and customs (and Orang Asli customary laws) and the provisions of the Constitution.

A Comparative View of Rights under Malay Reservation and Aboriginal Reserves

It is clear from the above discussion that there is a lot of difference between the laws that apply to the various groups of indigenous people in Malaysia. Much of the differences and disparity is a result of the historical and political background that they have been subject to. On one end of the pendulum, the rights of the Malays under the Constitution and the various Malay Reservation Enactments is assured and is very well entrenched.[99] As discussed earlier, there cannot be any change in the status of MRL unless it goes through a stringent legislative process. Under the state enactments, Malay reservation lands are given land grants, and thus indefeasibility of title to the owners. Despite the argument that the motivation of the colonial administrators was self-serving and perhaps not so altruistic, there appears to have been more care by the colonial government to create reservations to protect Malay lands through proper legislation.[100] To a large extent they have managed to preserve much of the lands in Malay hands today.

On the other end of the pendulum, is the provision for Orang Asli land rights. The APA, provides for creation and declaration of any area exclusively inhabited by aborigines to be an aboriginal reserve by notification in the gazette. The state authority may in like manner revoke wholly or in part vary any declaration of an aboriginal reserve. There does not seem to be any proper guidelines or other protection against such revocation. Orang Asli may be given occupancy rights that are "deemed not to confer on any person any better title than that of a tenant at will."[101] The effect is that they may stay on the land at the pleasure of the authorities and in the event of a relocation and displacement, there is strictly no obligation under the statute to pay compensation.

Stemming from the fact that Orang Asli are seen as tenants at will, the compensation for any taking of land provided under the APA ss 11 and 12 are only for "things on the land" and the payment of compensation is discretionary. As the case of *Sagong Tasi* has shown, this is inadequate compensation and breaches article 13. Gopal Sri Ram JCA (as he then was) declared that s 12 must be brought into conformity with the Constitution, making it yield to article 13, by interpreting the word "may" for shall and by introducing "adequate" before compensation. By giving a liberal interpretation he was merely giving effect to Article 8(5) of the Constitution which sanctions positive discrimination in favour of Orang Asli. Despite the positive outcomes and the declarations of aboriginal rights of the Orang Asli through the courts, there is yet to be any change or amendments made to the APA.

Concluding Remarks: Towards a More Just Land Policy

The positive discrimination measures and the protection of the special rights of the Malays had their roots in that period when their customs, religions, traditional leadership were recognized. Their personal laws and customary lands were specially granted protection to prevent unscrupulous exploitation of the Malays in their own lands. These protections were then entrenched in the Constitution. And now, as the courts have firmly recognized aboriginal rights, the states should move in tandem by creating, affirmative action policies for all indigenous groups along the lines of the laws introduced for the protection of Malay rights. Article 8(5) must be read as not just in terms of a power and discretion but a duty on the federal government to protect aboriginal rights.

Indeed the Policy statement is a clear evidence of the assumption of the duty on the government to protect their land rights. There is a fiduciary duty on both the Federal and the state governments to gazette lands for the benefit of Orang Asli. That duty must require the government to find a long term solution to their land woes. There must be a commitment on the part of the federal agency entrusted with the protection, wellbeing, and advancement of Orang Asli in creatively applying the law and the Constitution as a bulwark against their territorial dispossession.

Perhaps the National Land Council can play a greater role in securing adherence by states to the Federal government's policy regarding Orang Asli land rights. As earlier suggested, under article 83, state governments may be asked to grant land for federal purposes. Since the welfare of the Orang Asli is a matter within Federal List and is a federal matter, a requisition by federal government for state lands which will be gazetted for aboriginal reserves and aboriginal areas is perfectly in order.

In recent months the State of Selangor has announced the formation of a Task Force comprising Orang Asli, to propose an Orang Asli land policy. The details of the proposal for a land policy is yet unclear but it is hoped that adequate provisions would be made not only to confer land titles to them, and also to provide land areas where Orang Asli might continue to practice their culture and traditions as well as to sustain their livelihood. Since Malaysia is a signatory to UNDRIPs, those standards might be used in "the spirit of partnership and mutual respect" to come up with equitable policies for the benefit of its most vulnerable citizens.

Notes

[1] Sarawak is the largest of the thirteen states that make up the Federation of Malaysia. It is an area that is almost equal to the area of the peninsula. The eleven states in the Malay Peninsula, namely Johor, Kedah, Kelantan, Melaka, Negeri Sembilan, Pahang, Perak, Perlis, Selangor, and Terengganu. Two federal territories, namely Kuala Lumpur and Putrajaya are in the peninsula while the third, Labuan is in north-west Borneo.
[2] Report of the Inter-Governmental Committee on the participation of Sabah and Sarawak in Malaysia 1962.
[3] Population Distribution and Basic Demographic Characteristics Report: Population and Housing Census 2000.
htpp://www.statistics.gov.my/English/census/pressdemo.htm
[4] Colin Nicholas puts the figure at 147,412 individuals in 2003. See Colin Nicholas. Centre For Orang Asli Concerns, www.coac.org.my. Cited 2 November 2009.
[5] Malaysia, Department of Statistics, State/District Data Bank 2005.
[6] Principles of English common law and equity applied in so far as the religions, manners and customs of the local inhabitants permitted, to prevent the common law from operating unjustly and oppressively. See the Application of Laws Ordinance 1928 and 1949, Sabah Application of Laws Ordinance 1938, and the Civil law Act 1956. See also the case of *Khoo Hooi Leong v Khoo Chong Yeok* [1930] AC 346, 355.
[7] Much of the Malay customary laws relating to family law has been merged into Syariah laws and are administered by the Syariah courts. Malay customary laws apply mainly in Negeri Sembilan with regard to election of traditional rulers.
[8] Or Independence day. The Federation of Malaya became independent from the British on 31 August 1957 and in 1963, formed Malaysia, together with Sarawak, Sabah and Singapore. Singapore seceded in 1965 to form the Republic of Singapore.
[9] The reference to Singapore was introduced in 1963 when Singapore joined Malaysia but has not been removed even after Singapore's secession from Malaysia in 1965. Prior to independence in 1963 the only legal definition of a Malay appeared in para (b) of sub-clause (3) of clause 124 of the Constitution of the Federation of Malaya 1948 which defined a Malay simply as a person who habitually speaks Malay, and professes the Muslim religion, and conforms to Malay customs.
[10] The requirement of connection to the Federation and Singapore means that an Indonesian who satisfies the first limb would not satisfy the test unless he or she was born in the Federation or Singapore before independence or was domiciled there on Independence Day. See Tun Mohamed Suffian bin Hashim, *An Introduction to the Constitution of Malaysia*, Government Printers, 1976, 283–288.
[11] There are a number of state legislation in respect of Malay reserved lands in Peninsula Malaysia. Malay Reservation Enactments include the FMS Malay Reservation Enactments (Cap 142) which is one uniform law for the Federated

Malay states of Selangor, Negeri Sembilan, Perak and Pahang, and a separate enactment for each of the former Unfederated Malay states, the Kelantan Malay Reservations Enactment (No.18 of 1930), The Kedah Malay Reservations Enactment (No 63), The Perlis Malay Reservations Enactment (No 7 of 1353), the Johor Malay Reservations Enactment (No 1 of 1936), the Trengganu Malay Reservations Enactment (no 17 of 1360)

[12] These states formed the Federated Malay States and were considered British Protectorates.

[13] FMS Cap 142. See also Kelantan Malaya Reservation Enactment (No.18 of 1930), Trengganu Malay Reservations Enactment (No 17 of 1360.

[14] Kedah Malay Reservation Enactment (No .1 of 136).

[15] Perlis Enactment (No.7 of 1353).

[16] In both states (Kedah and Perlis), a Siamese agriculturalist who is a permanent resident, is given the right to own Malay reserved land.

[17] See *Hanisah v Tuan Mat* [1970] 1 MLJ 213 where the Federal Court would therefore not make a ruling on the Malay identity.

[18] Aboriginal is a common usage in many parts of the world, to refer to 'first' nations or peoples, for example, in the settler states of the USA, Canada, Australia.

[19] The eighteen sub-ethnic groups generally classified for official purposes under Semai, Negrito, Senoi and Aboriginal-Malays. Each group has its own language and culture.

[20] [2002] 2 MLJ 591.

[21] The races to be treated for the purposes of the definition of 'natives' in clause (6) as indigenous to Sarawak are Bukitan, Bisaya, Dusun, Land Dayak, Sea Dayak, Kadayan, Kelabit, Kayan, Lugat, Kenyah (including Sabups and Sipeng), Kajang (including Sekapans, Kejamans, Lahanans, Punans, Tanjongs and Kanowits), Lisum, Malay, Melanau, Murut, Penan, Sian, Tagal, Tabun, Ukit.

[22] The states of Sabah and Sarawak have established Native Court systems with specific jurisdictions to try breaches of native law and customs. The respective courts also have jurisdiction to decide whether s person is a native or is deemed to be a native.

[23] In *Law Tanggie v Untung ak Gantang* [1993] 2 MLJ 537, the plaintiff, was born of a Chinese father and native Iban mother was able to claim native status. His statutory declaration and the decision of the Native Court recognizing his status was held as conclusive evidence. Law bought a piece of native customary land. It was registered in the name of his uncle, the defendant (who was an Iban) until he attained native status. The plaintiffs father had assimilated into the community, had taken on a native name, was accepted as a member of the Iban community, lived and was buried an Iban, and his children had carried on as Iban.

[24]At the time of writing, there is conflict brewing in the state over this very issue where few students who excelled in their studies have not been considered for scholarships because of mixed parentage and thus did not fall clearly in the category of "native".

[25] [1979] 2 MLJ 101. He proved that he was ordinarily resident in Sabah, lived as a member of the native community for a continuous period of five years immediately

preceding his claim, was of good character and was not limited by the Immigration Act.

[26] Native customary land rights in Sabah and Sarawak deserve a separate study. For a critique of the Sarawak Land Code, see Ramy Bulan, "Statutory Recognition of Native customary Rights under The Sarawak Land code 1958: Starting at the Right Place", (2007) 34 *Journal of Malaysian and Comparative Law* 21.

[27] Bashiran Begum Mobarak Ali, "The Federal Constitution – A Shield for the Protection of Malay Reservation Policy, (2008) *The Law Revi*ew 53.

[28] Ibid at 53

[29] Copy of a Minute by the British Resident, Perak Dated the 7 Sptember 1910". SS BA Office 1910, File No.3170/10, cited by in Bashiran Begum Mobarak Ali, "The Red-Ink Grant: Tracing Legitimacy in History", (2007) 34 *Journal of Malaysian and Comparative Law* 159.

[30] "Minute by Attorney General JW Bonser, 28 April 1891 on Acting British Resident Selangor to Colonial Secretary, " SS BA Office (1875–1955), File No 294/1891, in Lim Teck Ghee, *Peasants and Their agriculture Economy in Colonial Malaya 1874–1897* (Kuala Lumpur: OUP, 1977) at 18.

[31] In a letter to the Resident on 17 April 1912, RJB Clayton, the District Officer Ulu Langat recommended that "no rubber" should be endorsed on the lands and the number of fruit trees planted on the lands should also be limited. This would have a direct effect on the value of the land and it would be less attractive to non-Malays. SS BA Office 1910, File No. 3170/10.

[32] RG Watson, Acting Chief Secretary of the Federated Malay States, in debating the Malay reservation Enactment, Federated Malay States, *Annual Report* (1913) at 22.

[33] For instance in *Cherokee Nations v The State of Georgia* 30 U S 1, Justice Marshall spoke of the relationship between the Indians and the state as one of "pupilage. Their relation to the US resembles that of a ward and a guardian". He said "They look to our government for protection; rely upon its kindness and its power ; appeal to it for relief to their wants".In *Worcestor v Georgia, 31 U S 515 (1832* he said policies were made for the Indian "comfort and benefit" or 'for the prevention of injuries and oppression".

[34] On arguments that the British did not want the Chinese to acquire so much land that they become a threat to British political and economic predominance. It is also argued that whatever lands were given as MRL were rural agricultural lands which were occupied by Malays. See Bashiran Begum Mobarak Ali n 36 at 92. See also Mohamed Azmi Abdul Hamid, "Malay Reserve Land and Malay Land – the fate of the Malays in Future Development," in Tanah Air Ku- Land Issues in Malaysia, Consumer Association of Penang.

[35] The National Land Council consists of a Chairman and a representative from every state. Article 91 provides:

> "It shall be the duty of the National land Council to formulate from time to time in consultation with the Federal Government, the State Government's and the National Finance Council, a national policy for the promotion and control of the utilization of land throughout the federation for mining,

agriculture, forestry or any other purpose, and for the administration of any of any laws relating thereto; and the Federal and State Governments shall follow the policy so formulated."

The state governments may also consult the National Land Council on any matter and it is the duty of the Council to advise the state. The states of Sabah and Sarawak are invited to sit in the National Council but its decision does not bind the two states whose land laws are different. There is no provision for Malay Reservation lands in those two states. Since Malays are considered as 'natives' in those two states, the laws relating to native customary rights apply to them.

[36] Merdeka refers to the independence of the Federation of Malaya from the British. This is marked on 31 August. This is now distinguished from Malaysia day which is on 16 September, the day Sarawak and Sabah joined Malaya (and Singapore) to form Malaysia.

[37] Nik Abdul Rashid, *Land Law and Land Administration*, Kuala Lumpur: University of Malaya Press (1971) at 143.

[38] Article 89 clause 4.

[39] *Mohamed Isa and Ors v Abdul Karim and Ors* [1970] 2 MLJ 165.

[40] The Perak Malay Reservation Enactment s 4(i) (b) states: "The Menteri Besar (Chief Minister) may at any time, with the approval of the Ruler of the State in Council by declaration published in the Gazette revoke any declaration whereby any land has been declared to be Malay Reservation, either as to the whole or any part of the area therein referred to."

[41] Mohd Azmi Abdul Hamid, "Malay Reserve Land and Malay Land – the Fate of the Malays in Future Development", in *Tanah Air Ku, Land Issues in Malaysia, Consumer Association of Penang,* at 147.

[42] In *VSP Suppiah Chettiar v KS Navaradnam* [1972] 2 MLJ 60, immediately means "allowing a reasonable time to do it."

[43] Muhammad Said Bin Abdul Kadir al-Haj, *Enakmen-Enakmen Rizab Melayu – Satu Kajian Perbandingan*, Pejabat Daerah dan Tanah Sabak Bernam, Selangor Darul Ehsan (1987) at 8.

[44] Section 7 (1) APA. Any area exclusively inhabited by aborigines and where the aborigines are likely to remain permanently may be declared and aboriginal reserve. The reserve is to be gazette.

[45] Section 6, APA. Any area s predominantly or exclusively inhabited by aborigines and where it appears unlikely that the aborigines will remain permanently in that place may be declared an aboriginal area.

[46] Any place inhabited by an aboriginal community.

[47] Now the Director General of the Department of Orang Asli Affairs (JHEOA). The Department of Orang Asli Affairs is a Federal department set up to manage all the affairs of Orang Asli. Previously it was placed under the Ministry of Home Affairs, but is now under the Ministry of Rural Development. The Orang Asli are the only people in Malaysia solely managed by a government department.

[48] Section 8, APA.

[49] Colin Nicholas, "The law on natural resource management as it affects the Orang Asli" paper presented at the UNDP–RIPP/PACOS workshop on Indigenous

peoples and Natural Resource Management Laws, 10 November 2005, Donggongon, Sabah. www.coac.org.my cited on 12 November 2009.

[50] For an in depth discussion on Orang Asli and resource rights, see Colin Nicholas, *The Orang Asli and the Contest for Resources, Indigenous Politics, Development and Identity in Peninsular Malaysia*, IWGA Document No. 95, Copenhagen, 2000.

[51] Lim Heng Seng (above) at 173.

[52] Sothi Rachagan, "Constitutional and Statutory Provisions Governing the Orang Asli" in Lim Teck Ghee and AG Gomes (eds), *Tribal Peoples and Development in Southeast Asia*, (1990) at 101–111.

[53] Above n 35.

[54] See s 8 APA.

[55] Colin Nicholas, above n 49.

[56] Jabatan Hal Ehwal Orang Asli (1961).

[57] This is a resettlement scheme which received funding from the World Bank and other institutions, where participants were given ten acres of land with housing and all the amenities, to encourage landless peasants to settle down. The government prepared the settlement sites, ploughed the lands, constructed houses, planted the crops which were almost ready by the time the settlers arrived. The settlers would pay back the costs of development as their revenue grew.

[58]The Land (Group Settlement Areas) Act 1960 enables land agencies such as the Federal Land Development Authority (FELDA), the Federal Land Consolidation and Rehabilitation Authority (FELCRA) and other agencies such as the Pahang Tenggara Development Authority (DARA) to take over State land and to develop them for the purpose of land settlement which culminates in the issue of titles to the settlers. Frequently Orang Asli traditional areas have been converted to such land schemes with them not enjoying both the fruits of the programme nor being entitled to such titles for the land. See Colin Nicholas, www.coac.org.my cited on 12 November 2009. For a comprehensive report on the impact of development on Orang Asli and the many ways in which they are marginalized, see Dentan, R K, K Endicott, A G Gomes and MB Hooker, *Malaysia and the Original Peoples, A Case Study of the Impact of Development on Indigenous Peoples*, Allyn and Bacon, 1997.

[59] Mohd Tap wrote about the regroupment schemes for Orang Asli thus: "Even the choice of land to be "declared" Orang Asli reserves or areas is, first and foremost dependent on the land not being considered "commercially important" by the authorities. Land given to out to Orang Asli is usually considerd "second class land". That is land with poor soil quality…unattractive to other communities. …the land needs to be 'well rehabilitated" before it would be used for agriculture" Mohd Tap, " An examination of Development Planning among the Rural Orang Asli of West Malaysia. Unpublished diss. University of Bath, quoted in Dentan et al, (1997) n 60 at 25.

[60] The High Court is the lowest of the superior courts. Appeals will go to the Court of Appeal and further appeals to the highest court the Federal Court.

[61] [1997] 1 MLJ 4_8.

[62] *Adong 1* [1997] 1 MLJ 418 at p 430.
[63] *Adong 1* [1997] 1 MLJ at 431.
[64] Article 13 states: (1) No person shall be deprived of property save in accordance with law. (2) No law shall provide for the compulsory acquisition or use of property without adequate compensation.
[65] *Adong 1* [1997] 1 MLJ 418 at 436.
[66] No written judgment was handed down.
[67] [2002] 2 MLJ 591.
[68] *Sagong Tasi I* [2002] MLJ 59 at pp 620–1.
[69] *Sagong Tasi II* [2005] 6 MLJ 289..
[70] Article 8 (5) (c) states that legislation for the "protection, well being or advancement of the aboriginal peoples of the Malay Peninsula (including reservation of land) do not offend the provision on equality before the law.
[71] Ministry of the Interior, Federation of Malaya (1961). Principle (d) states that Orang Asli are not to be removed from their lands without their full consent. There was also documentary evidence to indicate that the Department of Orang Asli Affairs had taken responsibility for the Orang Asli in the area.
[72] In 2006 the Federal Court granted leave to the defendants to appeal to the Federal Court on points of law but the appeal has since been postponed pending negotiations for settlement between the parties. In the meantime, the state government had also withdrawn its appeal
[73] [2008] 2 MLJ 677.
[74] Quoting *Colonial Sugar Refining Co v Melbourne Harbour Trust Commissioners* [1972] AC 343, Skinner JCA at the Court of Appeal said that it is a well accepted rule of statutory interpretation that a statute should not be held to take away private rights of property without compensation unless the intention to do so is expressed in clear and unambiguous terms.
[75] Reference was made to Lord Denning in the Privy Council which was followed in *Hamid bin Matussin v Superintendent of Lands and Surveys and Anor* [1198] MD 2.
[76] *Nor anak Nyawai v Borneo Pulp Plantations and Ors* [2001] 6 MLJ 241.
[77] The landmark Australian High Court decision of *Mabo and Anor v The State of Queensland (Mabo No 2)* (1992) 175 CLR 1.
[78] (1973) 34 DLR (3d) 145. *Calder* was cited in *Madeli* for the proposition that common law recognizes 'native rights over their lands'. *Madeli 2* [2008] 2 MlJ 677 at 691–2.
[79] [1921] 2 AC 399.
[80] Although the term NCR is generally used in Sabah and Sarawak, the term common law native title is used for ease of reference. Mokhtar Sidin JCA s put it quite clearly in in *Adong 1* when he said *:* "The customary rights of the aborigines in Peninsula Malaysia is a common law right which the natives have and which the Canadian and Australian Courts have described as native titles".[80] Another judge, Abdul Aziz bin Abdul Rahman underlined the connection between the customary title in the Peninsula and that of Sarawak saying that the aboriginal rights are "similar to the native customary rights over communal reserves in Sarawak". See

Amit bin Salleh and Ors v The Superintendent of Lands and Surveys Department, Bintulu and Ors [2005] 7 MLJ 10.

[81] This is not exhaustive but it gives a good summary of the present law.

[82] *Wik Peoples V Queensland* (1996–97) CLR 1.

[83] *Nor Nyawai 2* [2006] 1 MLJ 256

[84] *Nor Nyawai 1* [2001] 6 MLJ 241 , Chin J quoting Brennan J in *Mabo No 2* (1992) 175 CLR 1; and The Wik Peoples v Qurrnsland [1996–1997] CLR 1, 84

[85] *Madeli 3* [2008] 2 MLJ 677

[86] *Madeli 3* [2008] 2 MLJ 677

[87] *Amodu Tijani*[1921] 2 AC 399, Sagong Tasi *(2)* [2005] 6 MLJ 289

[88] See *Nor Nyawai* 2 [2006] 1 MLJ 256 at 269

[89] See *Sagong 1* [2002] 2 MLJ 591 at 610,.621–4

[90] *Madeli 3* (Federal Court) [2008] 2 MLJ 677 at 694–5

[91] See *Nor Nyawai* 1 [2001] 6 MlJ 241 at 268; *Madeli 3* [2008] 2 MLJ 677 at 694

[92] *Sagong 2* [2005] 6 MLJ 289 at 301–2

[93] *Adong 1* [1997] 1 MLJ 418, at430; *Adong 2* [1998] 2 MLJ 158 at 162.

[94] *Sagong 1* [2002] 2 MLJ 591 at 613–4; Madeli 3 [2008] 2 MLJ 677 at 692–3.

[95] *Sagong 1* [2002] 2 MLJ 2 MLJ 591 at 612; Nor Nyawai 2 [2006] 1 MLJ 256 at 270; Madeli 3 [2008] 2 MLJ 677 at 690, 696–7.

[96] *Sagong 1* [2002] 2 MLJ 591 at 612.

[97] *Madeli 3* [2008] 2 MLJ 677 at 697–8 (FC).

[98] *Adong 1* [1997] 1 MLJ 418 at 436. affirmed on appeal to the Court of Appeal and the Federal Court. *Adong 2* [1998] 2 MLJ 158 at 163–4; *Sagong Tasi* 1 [2002] 2 MLJ 591 at 617; *Sagong 2* [2005] 6 MLJ 289 at 309–10; *Madeli 3* [2008] 2 677 at 691–2. Art 13 states : (1) No person shall be deprived of property save in accordance with law. (2) No law shall provide for compulsory acquisition or use of property without adequate compensation. Foraging lands and settlement have been treated differently for the purpose of assessment of "adequate" compensation. Cf *Adong 1* [1997] 1 MLJ 418 and *Sagong Tasi* 2 [2005] . In *Sagong*, the market value was used to assess compensation based on the Land Acquisition Act.

[99] This is not to say that there are no problems with regard to Malay Reservation lands. As the nation move into modernization, there is fear of the "disappearing " Malay Reservation Lands, but not in the ambit of this paper.

[100] Bashiran begum Mobarak Ali, above n 27.

[101] Section 8, APA.

PART III –

ENVIRONMENTAL HARMS, PROPERTY RIGHTS AND RESPONSIBILITIES

Part III Introduction

Laura Westra

After considering democracy in Part I, and other alternative forms of Indigenous governance in Part II, Part III turns to civil society, the possible conflicts between the human rights of individuals and legal entities, and the responsibility engendered by the novel and increasingly destructive powers of humankind. The question of responsibility is the main focus of Prue Taylor's analysis, based on the work of Hans Jonas, but attempting a reconciliation between that work and the existing legal context. Yet Taylor acknowledges that, despite the wealth of "alternative ethics" proposed by many environmental scholars, the legal framework lags far behind.

The disconnect between morality and law is only one of the problems that surface as Taylor considers the interface between environmental responsibility and entrenched property rights. The protection of the "collective good" might work best in states with a favorable written constitution that might encourage the internalizing of environmental protection. Yet it takes a unique soft law instrument like the Earth Charter to incorporate and clarify the "substantive concept of responsibility" in international law.

In contrast, Valentina Vadi considers existing investors' rights as protected by investment treaties, most of which ignore or treat as secondary the responsibility to maintain environmental health. States have an obligation to respect people's health, hence to protect citizens from "health infringements" and provide them with a safe environment. Although the case law does not trace this obligation "to fulfil environmental health" explicitly, the rights to life and to family life, often cited in these cases, connect with environmental health, at least implicitly.

Nevertheless many trade treaties, including NAFTA and other arbitral tribunals, tend to act to "balance" the public interest and corporate rights in favour of the latter, so that the interests of investors, rather than environmental health concerns, reflect the final disposition of most cases.

The next chapter, by Colin Soskolne, reflects on the pressing need for attention to be channeled to eco-epidemiology as the discipline that considers the dynamics between ecosystem change and human health.

Given the hazardous circumstances that beset us today, more attention is needed in support of preventive recommendations based on informed science. An international ecohealth forum held in December 2008 in Merida, Mexico, resulted in a draft call for action. It sets out, among other things, the need for action to achieve sustainability through respect for the principles reflected in the Earth Charter, including ecological integrity and social justice.

The final chapter in Part III takes on one of the gravest problems regarding environmental responsibility: the limits of present democratic governance. Democracy is normally designed to be the form of government appropriate to a limited, specific nation. But the problems we are facing today have been initiated and caused globally, rather than locally; hence a new form of democracy needs to be devised and adopted, such as allows us to approach environmental issues collectively, through a global democracy.

CHAPTER ELEVEN

"THE IMPERATIVE OF RESPONSIBILITY" IN A LEGAL CONTEXT: RECONCILING RESPONSIBILITIES AND RIGHTS

PRUE TAYLOR

Introduction

> "The danger derives from the excessive dimensions of the scientific–technological–industrial civilization. What we could call the Baconian program – namely, to aim knowledge at power over nature, and to utilize power over nature for the improvement of the human lot …" (Jonas 1984: 140)

In the last forty to fifty years, a large body of literature has emerged which investigates the moral and ethical foundations of the ecological crisis that currently confronts human society. Much of this literature, which is both scholarly and popular, identifies a human centred ethic ("anthropocentric") as a motivating cause for behaviour that plunders the Earth's resources to meet the short-term interests of humanity. This behaviour largely ignores the present interests of other life, together with the longer term interests of future generations of humans and non-humans. For centuries it has been possible to conduct human activity with little regard for these other interests. Earth's abundance has been sufficient to accommodate such behaviour,[1] and the potential needs and interests of future generations were both too remote and without voice. But now, as the limits of Earth's biological capacity are over-reached,[2] as present human populations become impoverished by ecological degradation, and species extinction rates threaten the resilience and function of life-supporting systems, fatal flaws are emerging to challenge the "logic" even the rationality, of human behaviour.[3] Humanity is undermining the conditions of its own present prosperity and existence, together with that of other living beings. The long term prospects for the prosperity of humanity and the continued existence of millions of species look grim.[4] This situation has resulted in many calls for a new or alternative "ethic" or morality to guide human activity and to bring humanity into "right" relationship with nature.[5] In response to such calls, scholarly literature is replete with examples of alternative ethics that derive from a rich array of sources.[6]

The debate, within what might broadly be termed the field of environmental ethics, has been of considerable interest to legal theorists for the last four decades. It is now almost trite to say that legal systems recognise and protect the moral values of society together with accepted spiritual, religious, economic and political philosophies. With this in mind, legal theorists have investigated the ethical foundations of current legal systems and found them wanting. They have also suggested alternative ethical paradigms and begun the task of implementing them into law.[7]

As mentioned above, there are a rich variety of sources from which we can draw to develop alternative ethical norms to guide the human–nature relationship from one of utilitarian exploitation, to one of care and respect. Most of the world's great cultures, spiritual and religious traditions, reflect an ancient wisdom that the land and the people are one. Furthermore, this interdependence requires humanity to care for and respect nature, beyond the fulfilment of human interests and needs. However, from the perspective of emerging environmental law theory, the work of two twentieth century scholars has been particularly influential. They are German philosopher Hans Jonas (1903–1993), and American ecologist and forester, Aldo Leopold (1887–1948). Their work, together with Garrett Hardin's famous analysis of our tendency to exploit common resources,[8] reminds us of the ongoing task of environmental law: to understand our place in nature *and* to reconcile individual (legal) rights with responsibilities toward the collective. In this chapter, the "collective" is taken to extend beyond present generations of humanity, to include present and future generations of *all life* on Earth.

In a recent book on the philosophical foundations of environmental law, Coyle and Morrow (2004) argue that environmental law is much more than merely a collection of public laws that incrementally encroach upon the exercise of private legal rights, for the purpose of attaining agreed social outcomes (e.g. pollution control and conservation, biodiversity protection). They contend that from its inception, environmental law has always been (and continues to be) an attempt to address the relationship between humanity and the world, and the limits of our entitlement to the use of its resources (Coyle and Morrow 2004: 1–7).

According to this analysis, environmental law is also an effort to shape the content and substance of the rights of private individuals so as to require the fulfilment of responsibilities *that go far beyond the reciprocal legal duties owed to other individuals.* The continuing rapid decline of Earth's ecological systems, despite some forty years of advanced environmental policy and law, is testimony to the ongoing failure of environmental law to achieve this fundamental objective. Indeed, Joseph Sax recently made this point in an article critiquing the last four decades of environmental law. He noted that our plethora of environmental laws and regulations:

> "still float at the surface of our domestic legal systems … and thus, at the profoundest level, the legal system is handicapped in doing the job it needs to do." (Sax 2008: 9)

This is because our legal systems fail to address the values that create the incentive systems for environmental harm. Our legal systems do not yet "reconstitute themselves to respond to the new values and understandings" that comprise our appreciation of environmental degradation (Sax 2008: 9–10). Sax concludes by noting that it is only when a *duty to protect*, in addition to [property] rights to use, is expanded into a general principle of our domestic law, governing all resources; that we will "be able to say we have truly implanted environmental jurisprudence into our legal system"(Sax 2008: 25).[9]

The task that confronts environmental law is becoming clearer, as our focus on the ethical dimensions of the human–nature relationship deepens. However, implementing human responsibilities for nature into law will not be an easy task. As will be explained below, this is largely because of the traditional "rights" focus of the law. This is a feature of many Anglo-American domestic legal systems that place primacy upon individual freedoms, articulated as legal rights. Overcoming the individual rights focus of the law will not be resolved by simply writing "responsibilities" into the law. The task is more complex than that. The main argument of this chapter is that in looking for ways to deal with this challenge, Hans Jonas's theory on human responsibility for all life, is very helpful. As will be seen, it is a highly developed theory that encourages (and requires) us to step beyond the limitations of traditional interpersonal ethics. Accordingly, this chapter begins by summarizing key aspects of Jonas's theory of moral responsibility. It then discusses why, in general terms, the implementation of moral responsibility poses difficulties for the law. The novel contribution of Hans Jonas to the reconciliation of moral responsibilities and legal rights is then elaborated upon. Finally, it is suggested that despite the difficulties of achieving this reconciliation, there are signs, from the evolving body of environmental law, that this is already beginning to occur.

Hans Jonas and The Imperative of Responsibility

Hans Jonas is considered one of the most widely read philosophers of the twentieth century, due to the popularity and topicality of *Das Prinzip Verantwortung* (1979).[10] The 1984 translation (*The Imperative of Responsibility: In Search of an Ethics for the Technological Age*) made his work available to an English readership. His primary objective was to design a new theory of *moral responsibility*, applicable to both the public and the private sphere. He argued that new technologies have given humanity the power to cause enormous harm to other humans, to future generations and to the balance between humans and nature. Given these changed and novel circumstances, a new theory of responsibility is required. This theory should be commensurate with the powers humanity now exercises and it should be capable of extension into the (unpredictable) future.

Stated in these terms, his objective does not seem so difficult to achieve. However, this conclusion is too simplistic. His task must be appreciated against the background of dominant philosophical concepts. He is considered by many as having provided one of the best formulations of the challenge ecological degradation poses for traditional ethics. His work made many aware that traditional (Kantian) ethics was only designed for interpersonal (human) relationships "within limited horizons of time and space", and was thus unable to help guide current human interactions *with nature*. This is not to say that traditional ethical systems were wrong, rather they were simply "not designed to cope with the current and prospective scope of human agency", in relation to other humans (present and future) and nature (Levy 2002: 84).

Jonas was also alert to the fact that humanities changed circumstances required not only a reassessment of its power/freedom, but also a reassessment of human nature (i.e., a new understanding *of* humans; Westra 1994: 6–7). This deeper ontological or metaphysical understanding of human beings asks: *who* are we, and *where* are we? In addressing these questions, it is possible to develop a metaphysical basis for applied environmental ethics. As will be seen below, Jonas characterised his own efforts to answer these bigger metaphysical questions as a "speculative attempt" (Jonas 1984: x). Furthermore, it is not entirely clear whether his attempts actually create a wholly non-anthropocentric value theory.

Closely related to the theory of moral responsibility is what he called the "heuristics of fear". This heuristics of fear recognises that our technological prowess is now such that we can no longer foresee the causal and temporal reach of our actions and this raises the real possibility of putting human and non-human survival at serious risk (Levy 2002: 84–85). Thus: "in our pursuit of technological innovation we should educate ourselves to imagine always what maybe the worst consequences of what we do" (Levy 2002: 84–85).

The Theory of Moral Responsibility and Environmental Law

As noted above, Hans Jonas began by arguing that the magnitude and novelty (in temporal and spatial scales) of human activity in relation to the environment, and the actual and potential impacts on the conditions of life (for individual human beings, future generations of human beings and nature itself), raises important moral issues. He considered the potential to cause environmental harm, through the application of new technologies (the apocalypse of the "too much"), to be worse than the risk from the atom bomb. This is because averting ecological disaster requires "revocation of the whole lifestyle, even of the very principle of the advanced industrial societies, and will hurt an endless number of interests (the habit interests of all)" (Jonas 1984: 202). In response to the novel dangers faced, *new ethical principles are required.* He argued that past ethics (direct dealings between humans) has left us unprepared: "[t]he lengthened reach of our deeds moves *responsibility ... into the centre of the ethical stage*" (Jonas 1984: x).

With this overview in mind, the purpose of this section is consider (in general terms) why the law finds the task of reconciling legal rights with responsibilities toward the collective for the protection of nature, so difficult. It following section will focus, in more detail, on the content of the theory of responsibility and apply aspects of it to the development of environmental law.

The first problem is the traditional view of many Anglo-American legal systems that morality is separate from law. That it is "in some sense prior to and independent of social practices in general and legal practices in particular" (Cane 2002: 2). While this view has been dominant for some time, it has now become more widely accepted that morality does inform the content of appropriate law (Cane 2002). Of course, democratic law making processes are critical to ensuring that moral values become part of the law, with the necessary consent of society. Despite the growing acceptance of the interrelationship between morality and law, there remains a great reticence to bring them too closely together.

The second problem is the traditional rights focus of the law. This is a feature of many Anglo-American domestic legal systems that place primacy upon *individual freedoms*, articulated as legal rights. The exercise of these legal rights are only limited by reciprocal legal duties, not to interfere with the legal rights of others. According to this very narrow legal equation, an individual is free to do what she/he likes, subject only to a reciprocal duty not to harm the legal rights of others. The predominance of this rights approach, is a reflection the liberal ideology that has dominated many political and legal systems for the last 200 years. In these circumstances, the system of justice has become preoccupied with framing issues in terms of rights and then attempting to balance, or reconcile, conflicting claims to rights. It does this primarily via a system of reciprocal duties.

The legal rules that surround property (use and enjoyment of land and natural resources) are predominantly expressed as "rights". As these property rights have been developed to protect human "freedom" (from interference by the state or other private individuals), and facilitate economic growth via exploitation, they concern only human needs and interests (e.g., natural resources as economic factors, privacy and possession) to the extent that a legally recognised interest exists. Equally, the counterbalancing duties that attach to these rights, are only those required to protect the rights of others. As a result, property rights regimes offer a flawed approach to achieving environmental protection. The whole range of interests that the collective might have, in the maintenance of a healthy and resilient environment, are not legally recognised and protected. There is no legal standing or capacity to seek protection of these collective interests and no liability for harm to the environment itself.[11] Accordingly, ecological systems themselves and the interests of the collective, are mere bystanders in this legal process. They may receive indirect protection at best. An example serves to illustrate. A landowner threatens to destroy a species habitat on his/her land, which will cause a serious decline in specie numbers and threaten their continued survival. Who can take legal action to prevent this occurring? A neighbouring land owner may if he/she can demonstrate that destruction of the habitat will interfere with their property rights to use and enjoy their own land. An example would be if vegetation destruction were to cause flooding and erosion of the neighbours land, impacting on their use of that land for crops etc. But what about the collective interests?; the interests of the public generally to ensure the survival of the species, the interests of future generations of humans to ensure survival of the species, the interests of the species itself, in its own continued survival and the interests of all other members of the web of life, in the continued integrity of the ecological system of which that species is a member? At present, property rights regimes to not directly protect any of these legitimate and important interests. If they do receive any protection, then it is contingent upon, and an indirect result of, the legal action of the neighbouring land owner.

A further flaw in the rights approach is that it is based on a concept of reciprocity. The legal "duties" that attach to the rights only extend as far as non-interference with the rights of others. If broader counterbalancing obligations do exist (via, for example, constitutional provisions), they are limited to protection of the interests of human society (e.g. "Property imposes duties. Its use should also serve the *public weal* [good]."; Article 14(2) *German Grundgesetz*). Accordingly, the German Federal Administrative Court noted: "the law cannot provide for the health of ecosystems per se, but only in so far as required to protect the *rights of affected people*" (Bosselmann 2008: 94–97). Thus far the concept of rights and reciprocity has provided a barrier to the idea of nature (or non-human life) and future generations being the legitimate holders of rights and duties. It could therefore be said that concern for the state of nature, and human–nature relationships, have been revoked by the social contract theory of the founding fathers of modern law. The law has been left blind to concern for the health of ecological systems.

What solutions can be pursued? One approach is to essentially avoid these problems by continuing to expand the body of environmental regulation, that is, by increasing the subject matter of environmental law in an attempt to mitigate the ecological blindness of legal rights. This has largely been the approach of both international and domestic legal systems, for the last forty years. A number of important problems are associated with this approach. It is piecemeal, incremental, and slow. The law is reduced to play endless game of catch-up. More importantly, as Joseph Sax pointed out (Sax 2008), this approach does not go to the roots of the problem because it does not challenge the anthropocentric values inherent in property rights, those core pillars of the legal system that create incentive for ecological harm. Aldo Leopold made the same point in *A Sand County Almanac* (Leopold 1968: 214), when he noted that a system of environmental protection, that does not change the focus away from commodification of land for economic human self-interest, is "hopelessly lopsided". It relegates to government functions that *become too large, too complex and too dispersed to be performed.* Accordingly, a sense of environmental responsibility (or ethical obligation) on the part of a private owner is the only remedy (Leopold 1968: 214). This approach also fails because it does not challenge the traditional negative liberal paradigm of freedoms (articulated as rights) being protections from interference by the state and others. This thinking makes rights extremely resistant to the imposition of any counter-balancing obligations, on the basis that they undermine fundamental individual freedoms – a barrier which a state should not cross. However, if the exercise of these freedoms and rights begin to erode the very ecological (and social systems) upon which they ultimately depend, then they must change.[12] It is not semantics to say that environmental responsibilities would ultimately protect freedoms rather than threaten them.

What other solutions emerge? Another approach, which has been pursued in the last twenty-five years has been to expand the rights approach. This has seen the creation of a whole new catalogue for rights including procedural rights of participation, environmental human rights, rights of future generations, animal rights and nature's rights. It is beyond the scope of this chapter to traverse the literature that promote and critiques this expanded rights approach. However, an example from the critique of nature's rights will serve to illustrate. Here debate has centred on whether the creation of nature's rights distracts us from the more important task of learning to value nature *for itself*. In responding to Christopher Stone's seminal article: *Should Trees Have Standing?*(Stone 1972: 450), which advocates granting rights for nature, two commentators argued:

> "[r]ights [for nature] will not liberate the environment or force us to value it more – what they do is 'allow for' the environment to engage in flawed legal institutions, the arena in which other rights-holders must battle. This is not true valuing." (Taylor: 377)

Baird Callicott argues that nature's rights creates a form of moral pluralism which leads to ridiculous results as we pick and choose between competing moral norms (Taylor: 377). Despite such concerns, the State of Ecuador recently amended its constitution to recognise the rights of nature.[13]

A third approach is to make a conscious choice to extend moral consideration to nature. That is, to locate moral value in the needs and interests of human beings, *and* also in the needs and interests of nature. If we make this conscious choice to extend the scope of our moral concern, then we are acknowledging that nature is worthy of our respect, and *responsibility for nature* will stem from this. It is here that Jonas's theory of moral responsibility is of particular assistance. As shall been seen,

Jonas provides us with a justification (if we must argue in this way) to better co-ordinate the exercise of individual freedoms, with the necessity to protect nature. More specifically, what we find in the *Imperative of Responsibility* is a clear and convincing articulation of: (i) the *source* of a moral obligation to protect nature; (ii) the *scope* (or object) of responsibility; and (iii) a definition of responsibility. Each of these will now be considered in turn.

What is the source (or root) of responsibility?

This question can be rephrased as: what is the source of the "ought" or "obligation to do"? Alternatively, this could be expressed as: what is the source that commands or forbids certain uses of nature? This is, of course, the fulcrum point for Jonas's moral theory. His answer was humanities power or the causal force that is unleashed through free will and knowledge. The "ought" arises as the self-control of consciously exercised power. "In sum: that which binds (free) will and obligation together in the first place, power, is precisely that which today moves responsibility in [to] the centre of morality" (Jonas 1984). As noted above (Levy 2002), this reassessment of human power and freedom must not be separated from Hans Jonas's "heuristics of fear" and "new image" of man (ourselves as among Nature's children). We must also see it in light of our present ecological circumstances including global climate change, biodiversity loss, desertification and massive harm to human health and prosperity. These threats were only distant prospects in the 1970s, now they are our reality.

How can we apply this point to the development of environmental law? Here we find that Hans Jonas has provided us with a valid source of moral obligation towards present and future generations of all life, which can provide a solid foundation for environmental law. His analysis enables us to avoid the difficulties of basing the law solely upon so-called "natural law" theories.[14] Furthermore, his analysis incorporates a reality that earlier moral philosophers failed to see. Put simply, the combination of our power and the scope of our freedom has put us in a novel situation of such peril that there is now a rational basis for humanity to agree to expanded restrictions upon the exercise of our freedoms. These restrictions include, but go beyond, traditional restrictions to protect interpersonal relations and the good order and stability of human society.

From a legal point of view, this analysis has the potential to fundamentally change the way we view freedoms protected as legal rights. While these rights would still be limited by the rights of other human beings, they "ought" also be limited by responsibility to the collective, to ensure that the ecological basis of life is protected and enhanced. With this changed perspective comes the realisation that the exercise of our freedoms need now to be constrained by the prior and pre-emptive interests of the collective. In this manner, we have a situation in which a determination of the collective benefit determines the scope of individual freedoms and not the other way round, as it currently is. Individual freedoms would not be seen in isolation from their ecological context, nor could they be defended at all costs as an ultimate end in themselves. This change could make rights (a social mechanism for establishing moral norms in law) relevant to our current ecological circumstances. At best, legal rights (such as property rights) could then become a positive tool for ecological protection and restoration when *responsible use* is adequately defined. At worst, the legal system is free to move beyond reconciliation of conflicting utilitarian human interests to sanction *irresponsible use*.

Once we can agree that the exercise of individual freedoms need to be qualified by protection of the collective good, as outlined above, then we need to turn to the more difficult tasks of: (i) creating the governance systems to achieve this change, within democratic systems and (ii) creating "complex" decision-making frameworks and defining intended substantive outcomes. The first of these tasks might be easier to achieve in nations that have written constitutions. The second task requires much more attention than it currently receives. However, here it is important to be clear that the focus is on controlling human behaviour (use) and not on attempting to manage nature. Human behaviour can be managed, but nature cannot. Traditional legal rights doctrine has inhibited environmental lawyers from addressing the substance or content of legal rights, to internalise environmental protection.

What is the scope (or object) of moral responsibility?

The concept of responsibility implies that there is an object *of* responsibility. However, traditional moral theory has applied very limited criteria for identifying worthy "objects". Jonas makes it clear that other human beings are the first objects of responsibility. In doing so he states that "existence" (i.e., life) and "reciprocity" are important to his moral theory. This raises a vital question: what about living entities other than human beings – can can they also be objects of responsibility? His answer to this question is not entirely clear. He notes that every living thing is its own end (of intrinsic worth) and can be the object of responsibility. Further, man can have responsibility for guarding the self-purpose of other life. In short, he has not difficulty in acknowledging the intrinsic value of nature. However, he goes on to add a crucial qualification; *to be an "object" in moral terms*, the thing (or "it") must be alive *and itself a potential bearer of responsibility*. This qualifier (*reciprocity* or capacity for moral reasoning) reminds us of how important traditional moral theory was to the work of Hans Jonas. The implication is that the potential to be a bearer of moral responsibility is unique to the human condition or other living beings (if they should prove to have a capacity for moral reasoning; Ott 2008: 47).

To his credit, Jonas recognises the moral dilemma here. Man has great power to harm himself, future generations of humans and the "plenitude of terrestrial life under his dominion" (Jonas 1984: x). He recognises therefore that a philosophy of nature is needed (beyond one of traditional human inter-relations) which can bridge the traditional chasm between an "is" (object) and a morally binding "ought". But how can this chasm be bridged?

It is at this point that his new (metaphysical) "image of man" becomes critical. He acknowledged that a new concept of man must be a secular, cogent and persuasive notion of general being.[15] The construction of such a new concept is no easy task as he admits in these words: "That attempt must brave the veto of reigning analytical theory against all attempts of this kind and indeed cannot hope for more than a tentative result. But dare it we must" (Jonas 1984: x). Later, he refers to his attempt as "speculative" (Jonas 1984: x).

What is his new image of man? It is a human being whose physical survival depends upon the survival of nature (needs) *but also* a human being whose very "humanity" or "human dignity" depends on the survival of nature *beyond his own needs*. He reasons that human beings are so much a part of nature, that nature's protection or destruction beyond human need and interests, will dignify humanity or dehumanise humanity. The latter outcome is described as an atrophy of our essence, even in the lucky case of biological survival. In his words (Jonas 1984: 136):

> ""There is no need … to debate the relative claims of nature and man when it comes to the survival of either, for in this ultimate issue their causes converge from the human angle itself. Since, in fact, the two cannot be separated without making a caricature of the human likeness – since, rather, in the matter of preservation or destruction the interest of man coincides, beyond all material needs, with that of life as his worldly home in the most sublime sense of the word – *we can subsume both duties as one under the heading 'responsibility toward man' without falling into the narrow anthropocentric view. Such narrowness in the name of man, which*

> *is ready to sacrifice the rest of nature to his purported needs, can only result in the dehumanization of man, the atrophy of his essence even in the lucky case of biological survival.*"

And, further, that (Jonas 1984: 137):

> "Ourselves being among (Nature's) children, we owe allegiance to the kindred total of her creations, of which the allegiance to our own existence is only the highest summit. *This summit, rightly understood, comprises the rest under its obligation.*"

In this manner, Jonas has avoided the need to make non-human life the object of moral responsibility. In this manner, he has side-stepped the traditional chasm by constructing a holistic or integrative concept of man which includes all life. Humans can remain the moral "object", but a discharge of moral responsibility toward other human beings (properly understood in terms of both physical needs and metaphysical essence/dignity) *requires care and concern for all life beyond human needs*. In other words, rediscovering our own dignity within nature requires us to care for nature's integrity over and above utilitarian needs. This integrity is not sentimental but includes the laws of life – eat and be eaten. Furthermore, as humans can remain the moral object, the problem of reciprocity is resolved. In taking this approach, he has not strayed too far from Immanuel Kant's argument that destruction of nature and cruelty to animals is morally wrong because it damages the moral character of the person so acting (the so-called *Verrohungs* argument). This has been described as a binding ethical obligation against oneself (Ott 2008: 47). Nor has Jonas created a clearly *non-anthropocentric* metaphysical basis for environmental ethics. In this sense, he remains consistent with Kant's position that "anybody who believes that there are moral obligations with respect to non-humans commits a fallacy …" (Ott 2008: 47). However, it is at this critical point that philosophers such as Laura Westra, and others, part company with Hans Jonas. They cogently argue the need for an *explicitly non-anthropocentric moral theory.* [16]

Of course there have been many efforts to bridge the object ("is") and the "ought", which is viewed as integral to developing an adequate moral theory for the relationship between humanity and nature. Konrad Ott, for example, argued that Kant's apparently inviolable criteria of rationality was not insurmountable, but could be overcome by a transcendental analysis. Ott himself offers a concept of "zoocentrism" (based on prehensions i.e., being alive and able to perceive something) as a basis for recognising the moral value of nature.[17]

Returning to Hans Jonas, what insights does his writing on the "object" of moral responsibility provide for the purposes of legal development? First, he provided us with the basis for creating a "secular, cogent and persuasive image of man", as part of nature. Such efforts will be critical to the fundamental shifts in thinking required to develop moral theory and the law. Second, he did not have a difficulty with valuing nature for its own ends (i.e. recognising the intrinsic value of nature). However, as noted above, he did strike significant difficulties in dealing with the Kantian need for reciprocity, which stood in the way of making nature itself a moral object. He resolved this problem by resorting to a kind of evolved anthropocentrism: discharge of moral responsibility toward other human beings (properly understood in terms of both physical needs and metaphysical essence/dignity) *requires care and concern for all life beyond human needs*. Thus, if we are to fully extend moral consideration to nature itself, we must rely on the body of modern philosophical thinking that goes beyond that of Hans Jonas. Both philosophers and lawyers a like, have argued that human society, and the law, will continually fail to provide adequate responses to the problems faced by human society, if this further step is not taken.[18]

Acting out of care and respect for nature itself is required to motivate the necessary limitations and sacrifices to our human interests. And, as Aldo Leopold reminded us in 1948:

> "Conservation is getting nowhere because it is incompatible with our Abrahamic concept of land. We abuse land because we regard it as a commodity belonging to us. When we see land as a commodity to which we belong, we may begin to use it with love and respect." (Leopold 1968: 214)

Leopold also stated that he failed to see how we could act ethically in relation to land (Nature) if we did not also love and respect it it (Leopold 1968: viii).

What is meant by "responsibility"?

Hans Jonas went to a lot of effort to explain what he meant by "responsibility". Of particular interest was his clear distinction between narrow *legal concepts* of responsibility and a more *substantive concept of responsibility*. Law tends to concern itself with attributing liability or guilt (responsibility) for the consequences of acts or omissions. In this sense, the law is concerned with accountability, punitive and redistributive responses. In contrast, Jonas explained his substantive concept of responsibility as being a *positive duty of power*. This involves a forward determination of what is to be done. It is responsibility *for* the matter that has a claim on our acting. This *substantive concept of responsibility* not only screens present actions but also creates a positive obligation towards actions not yet contemplated. One becomes committed objectively (in the sense of the object that is entrusted) and committed subjectively (in the sense of *feeling* responsible or caring). He also asked; what is acting *ir*responsibly? The answer; disregard for the well-being, interest and fate of those under ones care (Jonas 1984: 140). This concept of positive responsibility invokes an *affirmative call to guard and to protect*. He used the archetype of the parent/child relationship to explore his concept of guardianship (Jonas 1984: ch.5). At this juncture it is important to recall that he sought to develop a moral concept that applied both to the private and the public sphere.

Jonas's substantive concept of responsibility, which he characterised as a positive duty of (human) power, raises some interesting issues and possibilities for the development of environmental law. First, (together with other aspects of his theory) this notion of responsibility creates a *direct* relationship between humanity and nature. Importantly, this direct relationship is based on an understanding of humans as part of nature, but it also acknowledges that humanity has a particular place within nature. As moral agents, we can act to cause great harm *and* we also have the potential to act as guardians or stewards, in respect of our own actions, to protect, respect and not cause harm, to others. In this respect, there is the possibility to base legal rights (such as property rights) on an appropriate ethical foundation (i.e., humans as part of nature and in relationship with nature). The significance of this should not be overlooked. Simply attempting to constrain the exercise of rights by a set of responsibilities towards the collective would be meaningless, if humanity continues to view nature in largely instrumental terms i.e., nature primarily for the purpose of meeting human needs and interests.

Second, Jonas's concept of responsibility invokes a positive or affirmative requirement to guard and protect, that he equates with a form of guardianship. This aspect has parallels with notions of stewardship or trusteeship, all of which are often called into play as a means (moral *and* legal) to implement positive actions of care and respect for "others". This will be explored in more detail below, for now it is important to recognise that guardianship, stewardship, and trusteeship are all important and familiar legal concepts for providing the necessary inter-generational reach of obligations *and* for acting for those that can not act for themselves (e., children, unborn and future generations). Furthermore, acting as a guardian or steward can be exercised either in concert with legal ownership or independently of, legal ownership. In fact, a notion of stewardship in relation to legally owned land and natural resources enables us to ask deeper questions about the meaning of ownership and what can and can not be owned.

Third, a positive substantive concept of responsibility provides a basis for non-legal (i.e. voluntary) measures for the protection and the restoration of degraded ecological systems. As reports such as the UN Millennium Ecosystem Assessment make clear, we can no longer limit our responses to merely ones of "protection" or "mitigation" of further harm. With fifteen out of twenty four ecological systems seriously degraded or being used unsustainably, we need to urgently turn our attention to the active task of restoration.[19] A positive concept of responsibility could create a valuable foundation to support and invigorate voluntary measures to protect and restore ecological systems, beyond self-interest. Indeed, there is a body of thought that suggests that ecological self-responsibility provides a critical component for our overall efforts to respond to ecological degradation (Teubner 1994). The existence of a positive duty of responsibility, beyond that which is directly legally enforceable, may also be important in terms of changing social behaviour and cultural understanding. It could enable communities (through non-legal means) to both sanction irresponsible behaviour and support responsible behaviour.

Fourth, a positive substantive concept of responsibility provides a basis for legal measures for the protection and restoration of ecological systems. Dealing with protection first, as Jonas noted above, acting responsibly requires care, respect and regard for the *well-being, interest and fate* of those under ones care. In doing so, he reminds us that we should "educate ourselves to imagine what may be the worst consequences of what we do, and we must recognise that we can no longer foresee the causal and temporal reach of our actions" (Levy 2002: 84–85). All this argues for a duty to do no harm, implemented via a strong precautionary approach that takes uncertainty seriously. That is, it acknowledges that uncertainty, in an ecological context, is irreducible. We are dealing with enormous complexity, especially across spatial and temporal scales. There simply will never be enough scientific information. A key element of a precautionary approach would be a reversal of the burden of proof. In short, the onus would be on the proponent of activities to prove that they will not cause ecological harm, or that they are consistent with ecological resilience, before undertaking an activity.

Applying this to property rights in relation to land, for example, undertaking new land uses would require first proof that they were either not harmful, or were consistent with maintaining the resilience of ecological systems. This turns current interpretation of property rights to land, on their head. In many legal systems these rights are currently interpreted as including the right to do ecological harm, subject only to the legal rights of others and the regulatory framework for environmental protection. Thus the onus is on others (state or those with legal rights) to prove harm, usually after the fact of its occurrence. The issues that a strong precautionary approach raises stem not so much from the reversal of the burden of proof, but from understanding and defining harm or appropriate notions of ecological resilience or integrity. This is an area where much has been done, but much more is needed. However, it would seem more appropriate to create regimes that focus on the positive rather than the negative. That is, they should attempt to establish what constitutes resilient and healthy ecosystems, and build in margins for uncertainty, with the objective of determining whether activities contribute to the maintenance and enhancement of these ecosystems, rather than focusing on what is or is not harmful. Approached in this way we have the opportunity to ask the right questions; i.e., how can we protect and maintain the health and well-being of ecosystems, while also using them for meeting human needs? How can we do good, rather than how can we merely do less harm? Obviously, this alternative approach will raise extensive issues for the management of science and instruments such as environmental impact assessments.

Turning now to the issue of restoration, when requirements to restore take the form of legal rules, then important issues will arise regarding (for example) financial burden and liability for historic harm. Difficult as these maybe, such issues are not insurmountable and lend themselves to creative use of financial mechanisms such as land taxes.

A fifth, and final point, to be made about the legal implications of a positive concept of responsibility is that it potentially expands the opportunities for legal action and legal standing in cases of *ir*responsible behaviour. If responsibility is given clear legal form, and its exercise is owed to the collective, then there is the possibility of it being enforced by members of the collective, either independently (e.g., present generations of humans) or via legal representation or the agency of the state (nature itself, or future generations of humanity). This would take environmental law enforcement well beyond the narrow focus of harm to human health or the economic interests of identifiable individuals, or harm to the public interest (such as in cases of public nuisance). It also takes the enforcement of environmental law beyond the scope of public officials effort's to ensure compliance with regulatory frameworks. An approach such as this comes closer to the type of fundamental change canvassed by Joseph Sax. He referred to many of the above examples of enforcement as delivering limited results because the environmental remains "at best only a second order legal interest", that is, a bystander in the legal process. He noted (Sax 2008: 22):

> "[T]here is a very significant difference between receiving the benefit of legal support as a matter of contemporary policy or politics, and having a right recognized as one of the *fundamental pillars of the legal system.* Although we may talk about fundamental rights to environmental quality, and even embody those precepts in major declarations or constitutions, we need to understand at the same time *how far we are from embracing any real conception of public rights as legal entitlements* in matters like biodiversity protection."

While the "public rights" to enforce environmental law, that he referred to did not extend to include non-human life, his point remains valid. Fundamental changes to our approach to legal standing will be an important component of reconsidering our general legal principles to reflect new values, "…so that they encourage resource uses consistent with protecting and restoring the natural services of the land and water, rather than giving incentives to disturb and destroy those services" (Sax 2008: 17).

The above paragraphs have made some specific observations, of a legal nature, in relation to Jonas's concept of human responsibility for nature. The objective has been to explore how a moral obligation could be given form in law, particularly, in the context of private law rights relating to land and natural resources. This section concludes by making some points of a more general kind before turning the next task of this chapter. That is, identifying nascent signs of a moral obligation of responsibility, for present and future generations of all life, being given legal form.

The idea of private rights being qualified by positive legal obligations is not a novelty in law. For example, Article 14 (1) and (2) of the *German Grundgesetz* (Constitution) provide that:

(1) Property and the right of inheritance shall be guaranteed. Their content and limits shall be defined by the laws.

(2) Property *entails obligations. Its use shall also serve the public good* [weal].

The problem is, of course, that the broader counterbalancing obligations that exist, are limited to protection of the interests of human society only (Bosselmann 2008: 138–141,195). Nevertheless, a positive obligation to serve the public good introduces a range of obligations that go well beyond the traditional liberal notions of social contract theory. That is, rights are limited only by the reciprocal legal rights of others, and any encroachments by the state are largely treated as a form of expropriation, for which compensation must be paid.

In reconciling rights with responsibilities, it will be necessary to steer a steady course between vague notions of responsibility and specific legal definition of what it means in terms of processes and outcomes. However, as has been demonstrated, traditional legal and moral theory has created significant barriers to this reconciliation. This requires us to argue through a tangle of thought and practice that has developed over the last 200 or so years, demonstrate its limits and argue for an overcoming of these limits. Yet, if we really saw ourselves as a part of nature, and nature as more than the sum of its parts, why would we not acknowledge moral obligations and develop legal frameworks to recognise and implement them?

An Example of Emergent Responsibility in International Environmental Law

This section considers an emergent example of human responsibility for nature, which has been given expression in legal form. This example is not yet hard law, but rather indicates that environmental law is attempting to reshape or create fundamental legal principles to reflect values of care and respect for nature, together with some form of human stewardship. The example used is from the field of international law, whereas the sections above used examples from domestic law (primarily property rights). The parallel rights in international law include the rights attached to territorial sovereignty and permanent sovereignty over natural resources.[20] Furthermore, international law is capable influencing the content of domestic law, as increasing convergence between the two spheres of law indicates.

The example used here is the IUCN Draft International Covenant on the Environment and Development. This is not yet a hard law document. It contains both a restatement of existing law (hard law) and principles and concepts that have not yet been accepted as having binding quality (soft law). It is intended to place the unfulfilled task of creating an internationally binding general framework treaty on the environment, squarely before states, and to have a realistic chance of gaining their acceptance.[21] More specifically, its intention is to state a general international obligation on all states (and individuals) to protect the whole of Earth's environment. The Preamble includes a statement of the basic ethical premises upon which the Covenant's substantive parts are based. It therefore refers to the unity of the biosphere, the dependence of all life on functioning natural systems and to humanity as being a part of nature. It also recognises the need for inter and intra-generational equity. Another specific reference to responsibility is in the invocation of the international principle of "common but differentiated responsibilities". This principle acknowledges the special needs of developing countries, but also that developed countries *share the responsibility* to assist developing countries to progress towards sustainable development. Poverty eradication is one of the core elements for meeting this responsibility.[22]

One of the most significant affirmations in the Preamble provides "[t]he duty of all to respect and care for the environment and promote sustainable development."

The commentary states that this wording is comprehensive as it is intended to address all actors (states, corporations, international organisations and individuals), the object of the duty is the totally of the environment (which includes humans and all life forms), and respect requires not only abstaining from harm but also taking affirmative action to preserve and rehabilitate. Most importantly, this duty is intended to express a legal obligation and not only a moral one.[23]

Moving to the fundamental principles, Article 1 states that the Covenant's objective is to create *"integrated rights and obligations"*. In this sense, it follows the lead of Earth Charter, noted above. The clear intention is to better co-ordinate rights with responsibilities, not by incremental restrictions to rights but by the creation of overriding moral and legal obligations. The key international legal "rights" include territorial sovereignty and permanent sovereignty over natural resources. These international law rights parallel (in many respects) the exercise of property rights in land and natural resources, found in many domestic legal systems. Article 2 restates of the duty of care and respect affirmed in the Preamble, as a legal norm.[24] is headed "

Respect for All Life Forms" and states that:

> "Nature as a whole warrants respect. The integrity of the Earth's ecological systems shall be maintained and restored. Every form of life is unique and is to be safeguarded independent of its value to humanity."

The commentary gives us valuable information about the intention and scope of Article 2. It draws, for example, from existing international conventions that recognise the uniqueness of every form of life and the respect that is warranted regardless of its material worth to man. Some treaties express this as "intrinsic" value and use it as a basis for environmental protection. Further, in Article 2 humanity is treated as a part of Nature, and because *all* parts of the web of life are linked and interdependent, they must *all* be protected and conserved. The phrase "independent of value to humanity" is intended to expressly put to rest the former utilitarian (anthropocentric) approach of the law, which limited economic interests. Finally, as noted above, the commentary to the Preamble specifies that "respect" requires not only abstaining from harm but also taking affirmative action to preserve and rehabilitate.[25]

Article 3 is also highly relevant to implementing moral responsibility into international law:

> "Common Concern of Humanity
>
> The global environment is a common concern of humanity. Accordingly, all its elements and processes are governed by the principles of international law, the dictates of the public conscience and the fundamental values of humanity."

The commentary explains that this Article states the legal basis upon which the international community, at all levels, can and must take joint and separate action to protect the global environment. It is based on the scientific reality that local human activities are causing global degradation, requiring worldwide co-operative responses (e.g., climate change, loss of biodiversity, ozone depletion etc) "This implies both the *right and the duty* of the international community as a whole to have concern for the global environment." [26] The commentary also notes that the "common concern of humanity" (CCH) does not create specific obligations and rules but establishes a general basis to act. In this sense, it can be understood as articulating a *general responsibility* on all to protect and respect the global environment.

The absence of specific rules and obligations to implement a general responsibility inherent in the CCH is a reflection of the current state of legal development. There is still a lack of definition and agreement surrounding this concept, its application and level of acceptance by states. However, the commentary identifies three generally agreed components. All are an indication of the extent to which "responsibility" is intended to fundamentally reshape rights. First, it expresses a shift from classical treaty making notions of reciprocity and sovereign self interests to requiring action in the longer term, and in the interests of all. This is very significant, as constant prioritization and protection of national interests (as an element and assertion of state sovereignty) frustrates the development of international environmental law regimes and drives standards to the lowest common denominator. This reflects the view that the acceptance of legal obligations, beyond the protection of sovereign interests, is purely at the discretion of states. Second, CCH also implies that the state of the global environment, which includes that part of the global environment within national territory of any state, can no longer be considered as solely or exclusively a matter for domestic jurisdiction. Thus, the traditional rights of territorial sovereignty, including to exclusively manage their own environment, now includes a general responsibility to protect and respect that environment. This does not deprive a state of its sovereign rights, including those of permanent sovereignty over natural resources, but it does qualify them significantly and alters their fundamental character.[27] Third, CCH is intended to create a legal right to have and exercise concern for the global environment.[28] The implication is to give members of the international community (including individuals) the legal capacity to take action to protect the environment. This potentially creates and exception to the one of the core articles of the UN Charter; namely non-intervention in matters exclusively within the domestic jurisdiction of states. In short, the right of non-intervention in the affairs of the territorial sovereign is trumped by a general responsibility to respect the global environment.[29]

To conclude, this section has focused on international law because there is growing evidence of a developing responsibilities approach to law, within this field. This may be due, in part, to the lack of consensus concerning the "creation stories" devised to explain the origins of international law (Young 1995: 187). As a result, the strict separation between morality and law (that is said to apply in positivist domestic systems), does not constrain international legal development in the same way. Equally, international law is now more open to influences from other cultures, and the attempt to mitigate the almost exclusive insistence on the superiority of rights can be seen as an attempt to reconcile the differing ideologies of Western and Eastern worldviews. Whatever the reasons and sources of a responsibilities approach, as it strengthens it will certainly begin to influence the content of legal norms in both international and domestic law.[30] Nevertheless, we must be realistic about the pace of any development –if left exclusively to states, it is likely to be very slow and incremental. A recent survey on state practice and international treaties, including those on human rights, war crimes, weapon systems, environment and cultural and natural heritage, identified a new trend in international law; sovereign rights are being injected with duties toward a community beyond national boundaries and to accommodate interests beyond narrow internal interests. However, the same survey gave rise to the observation that (Schrijver 2008: 228):

> "Treaties and general principles of international law often do not receive adequate follow-up in national legislation and in the setting up of a well-equipped implementation and monitoring instrument...Internationally, there is also scarcely any well-organized monitoring of compliance with international (treaty) obligations".

Clearly, there is much hard work to be done. While there may be emerging clarity about the basis for responsibility in law, to respect and care for the environment, how this is to be expressed in specific legal norms and how these are to be monitored and enforced, requires much more attention and effort.

Those closely engaged in the effort, to fundamentally transform law and governance with the objective of protecting and restoring the Earth's ecological integrity, need to be both realistic and hopeful. Realistic in their acknowledgement that traditional liberal rights based concepts still dominate both international and domestic law. But, hopeful in detecting and interpreting nascent shifts such as the growing support for the state as an environmental trustee (Bosselmann 2008: ch.5). Many scholars are also hopeful that modern law will begin to relax its formalism, and open itself up to an "important and fertile source of nourishment" found in the traditional wisdom of humanity (Weeramantry 2004: x). In doing so, modern law has the opportunity to overcome its excessive emphasis on rights by giving much more attention to responsibilities that are inherent to the exercise of rights.

Conclusion

One of the core features of modern environmental law is its almost entire emphasis on rights. These include the rights attached to territorial sovereignty, property, development, natural resources and more recently, to human beings (human rights). This emphasis on rights has been identified as an impediment to the achievement of global sustainable development. When seen through the lens of global ecological degradation, it is apparent that the exercise of legally protected rights to own, manage, and use natural resources, in the absence of the influence of responsibilities, risks applying legal concepts beyond the limits of their own logic. In contrast, many traditional cultures/spiritual beliefs place much more emphasis on responsibilities, reflecting an ancient wisdom that the land and the people are one.

> "If the land withers and dies, so also will the people because the health of a community is dependent on the health of the land. And the health of the land is lost unless you pay due regard and reverence to that land, caring for it as you would any other living thing." (Weeramantry 2004: xi–xii)

How could modern western law incorporate both the positive obligation of care and the restraining influence of responsibility? One way is to open legal systems to the guidance of traditional wisdom. Another route is to open our legal systems to developments in moral philosophy such as that identified in The Imperative of Responsibility. Neither of these options are without difficulty. As has been seen, many modern legal systems have adopted formal rules about the sources and content of new legal concepts and principles, which limit and slow development.

This article has considered the relevance of Hans Jonas's work to the emergence of responsibilities toward nature. Its particular focus has been on how his theory of moral responsibility might inform the concern that the predominately rights based approach of law needs to be reconsidered, within the broader context of human responsibility for all life. The primary benefit rests in the creation of a positive requirement to restore and not to harm, in a manner that goes to the heart of legal principles and brings human affairs and needs into balance with nature. If a legal requirement of responsibility can be created, to implement a moral responsibility, then there is the potential that human activity will fundamentally change so that there is much less call for the law to respond to an endless catalogue of destructive human behaviours and technologies. This is because legal responsibility for ecological systems could create an overarching framework. Legal use (utility) rights would then be interpreted and exercised, subject to fulfilment of responsibilities toward ecological systems. Private land owners, for example, would be transformed from having the right to use the land, subject to the rights of others and an expanding array of environmental regulations, to stewards or guardians of the land with clearly prescribed rights to use but none to consume and to degrade (Freyfogle 2007: 141). To achieve this would be a contribution toward the development of a legal framework that explicitly embeds humans within the larger ecological order and speaks to the overall condition of that larger order (Freyfogle 2006: 236).

We cannot, however, simply wait for governments and the law to make these changes. It is the task of each and every one of us, in all our individual and collective capacities, to define and exercise responsibility for the integrity of the Earth's ecological systems. Only then will our political participation be grounded in an understanding of the tension between individual rights and freedoms, and the experience of caring for the relationships that bind humans and nature as one and create the community of all life.

References

Bosselmann, K., "In Search of Global Law: The Significance of the Earth Charter" (2004) vol 8, issue 1 Worldviews: Environment, Culture, Religion 62. K. Bosselmann and P. Taylor The Significance of the Earth Charter in International Law in P.B. Corcoran (ed) *The Earth Charter in Action* (KIT Publishers, 2005).

—. *Principle of Sustainability: Transforming Law and Governance* (Ashgate, 2008).

Cane, P., *Responsibility in Law and Morality* (Hart Pub, 2002).

Coyle, S. and K. Morrow *The Philosophical Foundations of Environmental Law: Property, Rights and Nature* (Hart Pub, 2004).

Freyfogle, E., *On Private Property: Finding Common Ground on the Ownership of Land (*Beacon Press, 2007).

—. *Why Conservation is Failing and How it Can Regain Ground* (Yale University Press, 2006).

Jonas, H., *The Imperative of Responsibility: In Search of an Ethics for the Technological Age* (University of Chicago Press, 1984).

Kimmerer, R., The Rights of the Land (Orion Magazine, Nov/Dec 2008) http://www.orionmagazine.org/index.php/articles/article/3647/ (last accessed on 4 June 2009.

Leopold, A., *A Sand County Almanac: And Sketches Here and There* (Oxford University Press, 1968).

Levy, David J., *Hans Jonas: The Integrity of Thinking* (University of Missouri Press, 2002).

Mackey, B., "Some Observations on the IUCN, the Earth Charter, and Global Governance" in M. Jeffery (et al, eds) *IUCN Academy of Environmental Law Research Studies: Biodiversity, Conservation, Law and Livelihoods. Bridging the North South Divide* (Cambridge University Press, 2008).

Manno, J., and Chief Irving Powless Jr. (Chawhdayguywhawdoes) Brightening the Covenant Chain: The Onondaga Land Rights Action and Neighbors of the Onondaga Nation in K. Bosselmann, R. Engel and P. Taylor *Governance for Sustainability: Issues, Challenges, Successes* (IUCN Environmental Policy and Law Paper No. 70, 2008).

Marsden, M. and T. A. Henare *Kaitiakitanga: A Definitive Introduction to the Holistic World View of the Maori* (1992, NZ Ministry for the Environment) 18 (emphasis added).

Ott, K., "A Modest Proposal about How to Proceed in Order to Solve the Problem of Inherent Moral Value in Nature", in L. Westra, K. Bosselmann and R .Westra (eds) *Reconciling Human Existence With Ecological Integrity* (Earthscan, 2008).

Sax, J., "Environmental Law Forty Years Later: Looking Back and Looking Ahead" in M. Jeffery (et al, eds) *IUCN Academy of Environmental Law Research Studies: Biodiversity, Conservation, Law and Livelihoods. Bridging the North South Divide* (Cambridge University Press, 2008).

Schrijver, N., *The Evolution of Sustainable Development in International Law: Inception, Meaning and Status* (Brill Academic Pub, 2008) 228.

Stone, C. D., "Should Trees Have Standing? Towards Legal Rights for Natural Objects" (1972) 45 California Law Review 450.

Taylor, P., "From environmental to ecological human rights: A new dynamic in international law" The Georgetown International Environmental Law Review, vol 10, issue 2, 309, 377.

Teubner, G., (et al, eds) *Environmental Law and Ecological Responsibility: the concept and practice of ecological self-organisation* (Wiley, 1994).

Weeramantry, Judge C. G., Foreword to M. C. Segger and A. Khalfan *Sustainable Development Law: Principles, Practices, and Prospects* (Oxford University Press, 2004).

Westra, L., *Environmental Proposal for Ethics: The Principle of Integrity* (Rowman and Littlefield, 1994).

Young, O., *International Governance: Protecting the Environment in a Stateless Society* (Cornell University Press, 1995).

Notes

[1] But see Jared Diamond's book on the collapse of ancient civilisations, *Collapse: How Societies Choose to Fail or Succeed* (Viking, 2005).

[2] See generally the Ecological Footprint Model at www.footprintnetwork.org (last accessed on 24 October 2009). It is a measurement tool that tracks human demands on the biological capacity of Earth. At present human demand for Earth's services exceeds what it can provide. This is termed overshoot.

[3] See United Nations Millennium Ecosystem Assessment at http://millenniumassessment.org ((last accessed on 24 October 2009), in particular the report: *Ecosystems and Human Well-being: General Synthesis* (Island Press, 2005).

[4] See generally, United Nations Millennium Ecosystem Assessment, *Ecosystems and Human Well-being: General Synthesis* (Island Press, 2005).

[5] The phrase 'right' relationship comes from P.G. Brown (et al) *Right Relationship: Building a Whole Earth Economy* (Berrett-koehler Pub, 2009).
[6] See, for example, the work of the following: J.B. Callicott, L. Westra, H. Holmes Roston III, J.A. Nash, B.G. Norton, D. Ehrenfeld, T.H. Berry, J.R. Engel.
[7] See for example, K. Bosselmann *When Two Worlds Collide* (RSVP Pub, 1995), and the Principle of Sustainability: Transforming Law and Governance (Ashgate, 2008), P Taylor *An Ecological Approach to Climate Change: Responding to the Challenges of Climate Change* (Routledge, 1998). See also, for example, the work of C. Cullinan, C. Stone, R. Eckesley, A. Dobson, G. Speth, P. Sand, M. Schroeter, G. Stutzin, E. Freyfogle and M.C. Wood.
[8] See generally G. Hardin 'The Tragedy of the Commons', Science 162, 1243–1248. and the notion that individual utility will take priority over protection of common resources.
[9] Sax primarily focused on property rights in this article, reflecting his long term interest in the public trust doctrine. However, there are many other areas of the law that environmental law has failed to penetrate or influence, to the point where they no longer incentivize ecological harm. These include tax law, contract law, company law and investment law, to name a few.
[10] *The Dictionary of Modern American Philosophers* (Thoemmes Continuum, 2005: 1247).
[11] The limited exception involves causes of action based on the doctrine of public nuisance – an unreasonable interference with a right common to the general public. Legal standing, an interest greater than that of the general public, is usually required.
[12] This is not to say that this approach is not without merit. Over time, regulatory measures can permeate the concept of rights and begin to change their character. This is, however a slow, incremental and indirect approach.
[13] Articles 1–6 (nature's rights) were approved on 7 July 2008. See www.celdf.org/Default.aspx?tabid=538 (last accessed on 4 June 2009).
[14] His metaphysical image of humanity, as part of nature and as having a special role in relation to nature, is not inconsistent with natural law. Note also the argument of Coyle and Morrow that the philosophical foundations of property rights include concern for intrinsic moral value and not solely concern for the reconciliation of conflicting utilitarian interests. They also make the point that it is not possible to turn back the clock to Locke and others, in our search for ideas about property. We must address modern utilitarianism and pre-occupation with human interests head on. See generally S. Coyle and K. Morrow *The Philosophical Foundations of Environmental Law: Property, Rights and Nature* (Hart Pub, 2004: 1–7).
[15] Philosophers such as Johannes Heinrichs have argued that we do not need a 'new' concept of humanity, but rather a more advanced or developed concept of humanity, see J. Heinrichs *Öko-Logic. Geistige Wege aus Klima and Umwelt Katastrophe* (2007, Steno Verlag).
[16] See generally L. Westra *Living in Integrity: A Global Ethic to Restore a Fragmented Earth* (Rowman and Littlefield, 1998), B. G. Norton *Toward Unity*

among Environmentalists (Oxford University Press, 1991) and *Searching for Sustainability: Interdisciplinary Essays in the Philosophy of Conservation Biology* (Cambridge University Press, 2003).

[17] Ott 2008: 40–59. Note also the work of Ron Engel, for example, "What Covenant Sustains Us" in L. Westra, K. Bosselmann and R .Westra (eds) *Reconciling Human Existence With Ecological Integrity* (Earthscan, 2008: 277).

[18] See above notes 6 and 7.

[19] *Ecosystems and Human Well-being: General Synthesis* (Island Press, 2005: 7).

[20] K. Bosselmann *Principle of Sustainability: Transforming Law and Governance* (Ashgate, 2008), R. Barnes *Property Rights and Natural Resources* (Hart Pub, 2009), D. E Fisher "Legal and Paralegal Rules for Biodiversity Conservation: A Sequence of Conceptual, Linguistic, and Legal Challenges" in M Jeffery (et al, eds) *IUCN Academy of Environmental Law Research Studies: Biodiversity, Conservation, Law and Livelihoods. Bridging the North South Divide* (Cambridge University Press, 2008: 94).

[21] IUCN Environmental Law Programme – Draft International Covenant on Environment and Development (3rd Ed: Updated Text) Environmental Policy and Law Paper, No 31, Rev 2, Foreword to 1st edition, xxi.

[22] IUCN Environmental Law Programme – Draft International Covenant on Environment and Development (3rd ed: Updated Text, 2004), Preamble 27.

[23] IUCN Environmental Law Programme – Draft International Covenant on Environment and Development (3rd ed: Updated Text, 2004), Preamble 27.

[24] Commentary to Fundamental Principles, IUCN Environmental Law Programme – Draft International Covenant on Environment and Development (3rd ed: Updated Text, 2004), 35 (hereinafter "Commentary").

[25] Commentary, Article 2, 35–36.

[26] IUCN Environmental Law Programme – Draft International Covenant on Environment and Development (3rd ed: Updated Text, 2004), Commentary, 33-39.

[27] This is re-enforced by interpretation of Article 11. On first reading is appears to be a traditional statement of state responsibility for transboundary harm. However, it must be read in light of developments in international law – including acceptance of the global environmental as an integrated whole and obligations to protect and preserve based on global environment being the CCH. Together these developments suggest that the state of a nation's territorial environment is no longer a matter solely within domestic jurisdiction of states. Commentary, 51, footnote 76. Note however, not all would agree that this position is currently accepted by the majority of states. See for example, K. Bosselmann *Principle of Sustainability: Transforming Law and Governance* (Ashgate, 2008: chapter 5).

[28] Commentary, 36–39.

[29] Commentary to Article 3, 37, at footnote 22. This is the most controversial aspect of CCH. There are some examples of this right of intervention emerging in the field of cultural heritage. See for example, F. Francioni "Beyond State Sovereignty: The Protection of Cultural Heritage as a Shared Interest of Humanity" *Michigan Journal Of International Law* (2004) 25(4)1209, 1221.There are few examples in law of *erga omnes* obligations giving rise to the rights of

popular action to enforce (*actio popularis*), in international law. See P. Taylor *An Ecological Approach to International Law* (Routledge, 1998) chapter 3.

[30] In the view of Oran Young(1995: 51–52): "a strong legal regime for climate change, particularly one that made humans liable for damages to physical and biological systems as well as for harm to each other, could lead "over time to a new system of ethics applicable to human-environment relations."

CHAPTER TWELVE

RECONCILING ENVIRONMENTAL HEALTH AND INVESTORS' RIGHTS IN INTERNATIONAL INVESTMENT LAW

VALENTINA VADI

Introduction

Environmental health is an increasing concern of both industrialised and developing countries. As environmental hazards seriously threaten health, the question is how to reconcile the need to promote environmental health with the need of economic development. As foreign direct investment is generally considered an essential means to promote development, this study will investigate the relationship between international investment law and the protection of environmental health. Are investment treaties compatible with states' obligations to protect public health, of which environmental health constitutes an important dimension?

At the substantive level, investment treaties provide an extensive protection to investor's rights in order to encourage foreign direct investment (FDI). Thus, a potential tension exists when a State adopts regulation interfering with foreign investments, as regulatory measures may be deemed to be a form of indirect expropriation. At the procedural level, investment treaties offer investors direct access to an international arbitral tribunal. Thus, foreign investors can directly challenge national measures aimed at protecting environmental health and seek compensation for the impact on their business of such regulation. This chapter will explore the conflict areas between investment treaty provisions and environmental health through an analysis of the recent arbitral jurisprudence. This survey shows that the regime established according to investment treaties does not strike an appropriate balance between the different interests concerned and that international law has not yet developed any machinery for the protection of environmental health through investment dispute settlement. The author will thus propose a series of policy options to reconcile the different interests concerned.

The Conceptual and Normative Scope of Environmental Health

Environmental health is a working definition which bridges the gap between environmental law and public health. There are two main reasons to deal with environmental protection from a public health perspective. First, such a perspective confers a degree of cogency and definition to environmental concerns. If the recognition of a distinct human right to environment remains controversial, there has been a growing recognition in public health law that the protection of the environment is a condition for the full enjoyment of health. Conceptualizing environmental health as a component of public health determines a paradigm shift that allows the translation of environmental concerns into the language of the state's concerns to protect public health. The protection of public health constitutes an essential function of state sovereignty. As one of the three constitutive elements of the state is its population – in addition to its territory and its organization – it is clear that the preservation of its population is an essential function of the state. Without such protection, the very existence of the state would be endangered. Protecting public health is a primary duty of States which arises from statutory, constitutional and international law. In sum, one may argue that the right/duty of the state to protect its population is a non-contested concept.

Second, as Professor Meron points out, "if the recognition of a distinct human right to environment remains controversial, there has been a growing recognition in international human rights case law that environmental factors are to be taken into account in the interpretation of traditional human rights" (Meron 2006: 451). The protection of the environment is a condition for the full enjoyment of a series of human rights.[1] This chapter assumes that the two conceptualizations – the right to health and public health – are not irreconcilable theories, but rather two sides of the same coin. While deeming health a public concern emphasizes the social contract between the state and its population (Fidler 2002: 156), conceptualizing it as a human right stresses the individual dimension of health. According to the human rights framework, not only do states have the obligation to refrain from acts which encroach upon people's health such as activities causing environmental pollution (*obligation to respect*), but they also have the obligation to take legislative action and other measures to protect people from health infringements by third parties (*obligation to protect*). Finally, States also have the *obligation to fulfil* environmental health, by providing a safe environment. Courts have applied human rights law to environmental concerns, considering environmental health as a component of a variety of human rights. In particular, courts have held that certain kinds of environmental impairment with severe harmful consequences for individuals may constitute a violation of other human rights. For instance, in the *López Ostra* case,[2] the ECtHR found a case of environmental damage and accompanying health problems to be a violation of Article 8 of the Convention on the protection of private and family life. In *Guerra v Italy*,[3] as Italian authorities had been too slow in closing down a fertilizer plant that caused toxic emissions, the ECtHR found that the State omission amounted to a violation of Article 8 of the Convention. Similarly, in *Fadeyeva v Russia*,[4] concerning the pollution caused by an iron smelter, the Court declared a violation of Article 8.

At the normative level, the linkage between environmental goods and health was originally – albeit implicitly – articulated by the founding document of international environmental law, the Stockholm Declaration on the Human Environment,[5] which recognized the interplay between an environment of quality and well-being. Similarly, the Ksentini Report,[6] concluded that environmental rights are a part of the existing human rights.[7] While these instruments have a mere hortatory value, they have paved the way for subsequent legal instruments.[8] Even the *General Comment No. 14* on the right to the highest attainable standard of health acknowledges the linkage between public health and the environment, affirming that the right to health extends to the underlying determinants of health such as… a healthy environment."[9]

Environmental Provisions in Investment Treaties

Environmental issues have been addressed by investment treaties only in recent decades, but have become a constant feature since the inception of the NAFTA. NAFTA presents several clauses concerning environmental measures. First, NAFTA Article 1114, entitled Environmental Measures, *inter alia* states that the Parties may adopt, maintain or enforce any measure otherwise consistent with Chapter 11 that they consider appropriate "to ensure that investment activity in its territory is undertaken in a manner sensitive to environmental concerns."

Second, at a more general level, NAFTA Article 104, in relation to Environmental and Conservation Agreements, gives priority to these treaties over the provisions in other parts of NAFTA, "provided that where a Party has a choice among equally effective and reasonably available means of complying with such obligations, the Party chooses the alternative that is the least inconsistent with the other provisions of this agreement."[10] Therefore, if the environmental measures are mandatory under one of the listed MEAs, they will be permissible under NAFTA. If they are not mandatory, but merely designed to implement one of the listed agreements, they will need to be as consistent with NAFTA as possible (Freedman 2003: 89–104)

Third, the NAFTA preamble commits the Parties to attain trade and investment goals in a manner consistent with environmental protection and conservation. Although preambular language is not deemed to be binding on the parties, it expresses the climate of an agreement. Fourth, the NAFTA was complemented by a side agreement, the North American Agreement on Environmental Cooperation (NAEEC) which is directed at fostering environmental cooperation among the Parties.[11]

Other subsequent investment treaties and the Energy Charter Treaty (ECT)[12] equally reflect environmental concerns in a variety of ways. They may include hortatory language in the preamble or separate provisions that emphasize the importance of environmental protection, environmental exceptions and so on and so forth (Fauchald 1996: 3–47). Even during the negotiations of the Multilateral Agreement on Investment, several proposals paralleled the NAFTA "not lowering standards" clause, furthered the affirmation of commitments to the MEAs, and made reference to the voluntary OECD guidelines in the preambular language.

There seems to be no irreconcilable conflict between investment treaties and environmental health objectives, at least at a theoretical–normative level. However, conflicts may arise in practice. Authors have stressed that the investment treaty clauses referring to environmental protection include "purely hortatory" language with unenforceable character (Kelsey 1999: 154–168). For instance, the breach of NAFTA Article 1114(1) would give rise to no more than consultations among parties, while "it may be questioned whether [Article 1114(2)] provides any meaningful relief for environmental regulations, or whether it is tautological, protecting only measures that are in any event 'consistent with this chapter'."(Freedman 2003: 94).

Arbitrating Disputes with Environmental Health Elements

As the number of disputes with environmental health elements continues to increase, it is important to reflect on the method for identifying and characterizing such disputes. In a preliminary way, as Professor Sands astutely highlights, no two parties will agree that a dispute is essentially "environmental" (Sands 2007: 313–325). For instance, in *Chevron v Ecuador*,[13] while Ecuador deems the dispute to have an overall environmental character, the investor deems it to be a case of denial of justice. The tribunal has upheld jurisdiction, dismissing, under the *prima facie* standard, an argument by Ecuador that the claimants are abusing the arbitration process in order to undermine ongoing environmental disputes in Ecuadorian courts.[14] The argument between the parties over the alleged environmental character of the dispute is a commonality in this kind of investment disputes. Therefore, according to Professor Sands, it would be more appropriate to talk about "disputes which have an environmental component", rather than to characterize a dispute as "environmental". Indeed, the mere definition of a dispute as an "environmental dispute" may have implications for a case (Sands 2007: 319). This neutral approach is based on the consideration that "environmental claims are rarely, if ever, raised in isolation of other international legal arguments" (Sands 2007: 315). The fact that environmental claims are raised together with other claims, does not erase their environmental quality, but simply emphasizes the need to ascertain such a quality in the merit phase of the proceeding.

In general terms, investment disputes with environmental health elements are characterized by the need to balance the legitimate interests of a state to protect or restore its environmental health and the legitimate interests of foreign investors to protect their property rights. Several issues arise in this context. While environmental concerns have been somehow integrated in investment treaties, often public health is not mentioned at all. Furthermore, environmental clauses remain rather vague and sometimes permit public health measures only if they are non-discriminatory. Some clauses even subordinate environmental measures to consistency with investment treaty provisions (Baughen 2006: 207–228). The very fact that the balancing process occurs in the context of investor–state arbitration could lead to the procedure being deemed biased in favour of the investors.

Turning our attention to the emerging environmental health case law, it is becoming clear that there is no such thing as a typical "environmental dispute". Investors may claim that certain forms of environmental regulation constitute an indirect expropriation or regulatory taking, and that compensation has to be paid. In other cases, investors may claim that a stabilization clause has not been respected due to environmental regulatory change or that certain environmental regulations amount to prohibited performance requirements. Other claims may concern the violation of the fair and equitable treatment or discrimination. Increasingly, environmental arbitrations have encompassed disputes regarding procedural rights (Rosenberg and Cheah 2001: 39–60).

The following analysis will scrutinize the way in which investment treaty guarantees have been interpreted in cases with environmental elements. While the expropriation provisions are the most commonly invoked in investment disputes with environmental elements, the expropriation claims are always accompanied by other claims such as discrimination, violation of the fair and equitable treatment and so on and so forth. Therefore, although the specific claims are hereby singled out for analysis, it is worth highlighting that they are often linked together and the respective arguments may overlap.

Indirect Expropriation

Several investment treaty arbitrations have dealt with the question as to whether regulation allegedly aimed to protect environmental health may be deemed to be an indirect expropriation. Scrutinizing the available awards, it emerges that there is no single notion of expropriation in customary international law and definitions vary in the context of different investment treaties. Comprehensibly, the arbitral tribunals have not adopted a consistent approach to the issue, but have elaborated different methodologies. In this context, two different trends may be highlighted: the *sole effects doctrine* and the *police powers doctrine*.

The sole effects doctrine in "environmental" disputes

The sole effects doctrine gives more weight to the effects of the regulatory measure on the investor than to the regulatory purpose of a given measure. The underlying assumption of this doctrine is that regulatory measures are generally adopted for the public good and may impose some burdens on private property. However, if the burden imposed on certain properties is overwhelmingly or almost completely to deprive them of any economic benefit, the owner is entitled to compensation. Several cases lend support to such a doctrine.

The *Metalclad* case[15] involved a successful claim for expropriation against the Mexican government by a US-based hazardous waste treatment facility. In 1993, the Mexican Federal government authorized Metalclad to operate a hazardous waste treatment station in the State of San Luis Potosi. However, the company was denied a municipal construction permit because of the complex hydrology of the region. Studies indicated that the site's soils were unstable and toxic waste could infiltrate the subsoil and carry contamination. After the municipal government denied Metalclad's request for a permit, the governor of San Luis Potosi declared the site part of an ecological zone. Metalclad sued the government of Mexico under NAFTA Chapter 11, alleging indirect expropriation. The arbitral tribunal held that the denial of the construction permit, as well as the construction of an ecological reserve, constituted acts tantamount to expropriation in violation of NAFTA Chapter 11. The tribunal adopted an extensive interpretation of indirect expropriation: Expropriation under NAFTA includes not only open, deliberate and acknowledged takings of property, such as transfer of title in favour of the host State, but also "covert or incidental interference with the use of property which has the effect of depriving the owner, in whole or significant part, of the use or ... economic benefit of property even if not necessarily to the obvious benefit of the host state."[16] The Metalclad case is considered to be emblematic of the sole effects doctrine, according to which the main criterion to establish indirect expropriation is the effect of the regulatory measure on the foreign property.

The subsequent TECMED case reaffirmed the sole effects doctrine.[17] Técnicas Medioambientales (TECMED), a company organized under Spanish law, filed a suit against the Government of the United Mexican States, arguing that Mexico's denial of a license renewal for the operation of a hazardous waste landfill amounted to an indirect expropriation of the foreign investor's property and to a violation of the fair and equitable treatment standard under the BIT concluded between the Kingdom of Spain and the United Mexican States. Since the decision was permanent and irreversible, the foreign investor could not exploit its assets, being deprived of any economic value. In turn, the respondent alleged that the denial of licence renewal was a legitimate regulatory measure issued in compliance with the state's police power in order to protect environmental health. As such, it would not amount to an expropriation under international law.[18]

The Tribunal primarily looked at the effects of the regulatory measure on the property as the fundamental criterion to decide whether indirect expropriation had taken place. As the effect of denying the licence was irreversible and the economic value of the assets was irremediably destroyed, the Tribunal found *de facto* indirect expropriation of the investment. The Tribunal stated that regulatory administrative actions are not *per se* excluded from the scope of the expropriatory measures even if they are beneficial to society as a whole. Also, the tribunal held that although states have the right to protect the environment, the proportionality between public and private interests should be taken into account.[19] Authors have highlighted that Tecmed represents a missed opportunity for Mexico to articulate important arguments concerning public health, (Suda 2006: 73–145) because the state merely claimed that the goal of the adopted measures was public health protection, but it did not articulate the claim. Therefore, if states more clearly articulated their claims, it would be possible for arbitral tribunals to better balance the different interests at stake.

The police powers doctrine in "environmental disputes"

The *police powers doctrine* refers to the inherent authority of a government to impose restrictions on private rights for the sake of public welfare, order, and security. The most common use of police powers is for the adoption and enforcement of environmental protection and public health regulations by relevant authorities. The doctrine derives from English common law principles mandating the limitation of private rights when needed for the preservation of the common good. Analogous principles exist in civil law countries: *sic utere tuo ut alterum non leadas* (use what is yours in a way that does not injure others) and *salus publica suprema lex esto* (public safety is the supreme law) (Reynolds and Kopel 2000: 511–537). The introduction of the police powers doctrine in investment treaty arbitration allows arbitrators with different legal backgrounds to refer to measures relating to public health and environmental protection.

In *S.D Myers v Canada*, the tribunal rejected the claim that the ban on the export of PCB constituted an expropriation because the ban was temporary and there was no appropriation or transfer of benefits to the host state. The tribunal affirmed that "regulatory conduct by public authorities is unlikely to be the subject of legitimate complaint under Article 1110 of NAFTA, although the Tribunal does not rule out that possibility".[20] In the *Methanex* case,[21] the arbitral tribunal expressly referred to the police powers doctrine. Methanex, a Canadian investor, initiated arbitration against the United States of America, claiming compensation resulting from losses caused by the ban on the use of a gasoline additive. As scientific evidence showed that MTBE (methyl tertiary-butyl ether) contaminated groundwater and was difficult and expensive to clean up, the

State of California enacted legislation to prevent the commercialization and use of MTBE. Methanex submitted that the Californian regulation was tantamount to expropriation as the US measures would not be meant to serve a public purpose, but rather to seize the company's market share to favour the domestic ethanol industry. Since no compensation was paid, Methanex argued that this violated due process of law and the minimum standard of treatment. However, the tribunal decided that there was no expropriation because it held that:

> "... as a matter of general international law, a *non-discriminatory* regulation for a *public purpose*, which is enacted in accordance with *due process* and, which affects, inter alios, a foreign investor or investment is not deemed expropriatory and compensable unless *specific commitments* had been given by the regulating government to the then putative foreign investor contemplating investment that the government would refrain from such regulation." [italics added][22]

The arbitral tribunal ascertained the non-discriminatory character of the measure and its public purpose, by looking at the procedure through which the national measure had been adopted. The award did not suggest that the study carried out by the University of California (the UC Report)[23] was scientifically correct. Instead, the award put emphasis on *due process*. Furthermore, since no specific commitments were ever given to Methanex, the tribunal held that the ban did not breach the legitimate expectations of Methanex.[24] Methanex had no reasonable expectation, as an investor, that it would be allowed to sell a product that was discovered to cause significant risk to environmental health. The *Methanex* tribunal adopted the police powers doctrine, putting emphasis on due process and the public health goal of the regulation.

Ascertaining Discrimination in "Environmental" Disputes

A crucial element in investment disputes with environmental elements is the ascertainment of non-discrimination. The non-discrimination principle is typically reflected in the investment treaty provisions of the national treatment (NT) and the most-favoured nation (MFN) treatment. These two standards do not guarantee a specific level of protection but are *relative* standards that require a host country to treat a foreign investor in the same way as a domestic investor or an investor from another country would be treated. The basic purpose of the MFN and NT clauses is to guarantee equality of competitive opportunities for foreign investors in the host state.

With regard to environmental regulation, the key question is whether foreign investments "are being regulated because the activity in question presents certain risks to the environment, or whether they are being regulated because they are foreign invest[ments]" or "simply by virtue of foreign ownership" (Mann and von Moltke 1999: 25). Nowadays it is undoubtedly difficult to spot openly discriminatory *language* in environmental or social regulations, although these characteristics have appeared in investment regulation and subsequent disputes. For instance, the recently instituted *Gallo* case is significant because it concerns *ad hoc* regulation that even in its title includes reference to the mine owned by the foreign investor. The case involves a project to transform a former open-pit mine in a landfill for non-hazardous municipal waste from Toronto. [25] The mine site presents a man-made lake, and the investor's proposal called for the deposit of waste in that lake, as well as the periodic withdrawal of contaminated water for treatment. Following a change in the provincial government, Mr. Gallo was not granted the renewal of an expired water removal permit and new *ad hoc* legislation prohibited waste disposal at the site on alleged environmental health grounds. As this move put Mr. Gallo out of business, the investor filed a claim, alleging various breaches of NAFTA Chapter 11 – including but not limited to discrimination – and seeking compensation for his alleged losses. As the dispute is still at a preliminary phase, it is not possible to predict how the case will evolve, or whether it will be declared admissible on jurisdictional grounds. However, the case is significant because it shows once again that environmental health measures may have a negative economic impact on foreign investments, and that they may appear to constitute *direct* discrimination.[26]

The discriminatory *effect* of a given regulatory measure is more difficult to assess. While environmental health policies may not openly discriminate against foreign investors, the use of apparently neutral criteria may affect this particular group of people. Arbitrators have tended to focus on the discriminatory effect of the given environmental measure, while the "likeness" test and the "legitimate objective" test have received scarce attention, if any. For instance, in the *Ethyl* case, Ethyl relied on trade law jurisprudence to argue that even equally formal treatment may result in less favorable treatment for a foreigner, and that the term "no less favorable" calls for effective equality in treatment.[27] As Canada had banned import of MMT, Ethyl argued that such a measure discriminated investors that imported MMT vis-à-vis domestic manufacturers of MMT or competitive products.[28] As the case was settled, no ruling on this argument was made.[29] In the subsequent *Myers* case, a US company, specialised in the remediation of PCB waste, alleged that Canada had violated NAFTA Chapter 11 by enacting an export ban on polychlorinated biphenol (PCB) waste. The ban denied the claimant the opportunity to undertake the PCB remediation business importing PCB from Canada to its remediation facility in the United States. The claimant argued that the ban had been applied in a discriminatory manner, in effect favouring Canadian operators who were not involved in transborder activities. The Tribunal dismissed the claim of expropriation, but found *inter alia* that Canada had violated the National Treatment provision. The tribunal extensively discussed the Canadian Minister for Environment's explicit instructions to reserve domestic waste processing industries for Canadians.[30] The evidence before the tribunal suggested that the ban arose out of economic protectionism, and had a *discriminatory intent* and *effect*. Very little attention was paid to the Basel Convention on the Control of Transboundary Movements of Hazardous Wastes and their Disposal (Basel Convention).[31] Indeed, Canada argued that even if the export ban favored domestic industry, as a signatory to the Basel Convention, the state was legitimately entitled, and in fact required, to ensure the availability of adequate in-country disposal facilities for PCB. However, the arbitral panel found impermissible discrimination against the claimant, since the export ban favored Canadian companies over non-Canadian companies.

Instead, in discrimination cases, a two-pronged test might be used. If discrimination is found in the first instance, adjudicators should enquire as to whether such discrimination might be justified by a legitimate objective. The litmus test would not be the mere assessment of *prima facie* discrimination, but a more structured and multilayered analysis. In the assessment of the likeness of circumstances, attention should be paid to the specific characteristics of each project. For instance, in the *Gallo* case, Canada was allegedly concerned with the potential contamination of drinking water.[32] Although the case has to be decided, it appears that both arguments presented by the parties – the alleged discrimination and the alleged justification of the same – will have to be taken into account by the arbitral tribunal to determine whether there was discrimination and if this was the case, whether this was justified by the circumstances of the case.

The Fair and Equitable Treatment Standard in Environmental Disputes

Arbitral tribunals have stressed the importance of procedural safeguards to ensure that private interests and public goods are properly balanced. An increasing emphasis is placed on administrative due process issues. In this context, the fair and equitable standard (henceforth FET) has become a key issue in disputes with environmental elements. This clause aims to provide a basic standard of protection that an investor may invoke, which is detached from the law of the host country. Therefore, although the exact meaning of the FET standard has to be determined case by case, it is considered to be an *absolute* standard that provides a fixed point of reference. A recent tendency considers the standard as embracing the notions of due process and denial of justice. According to this view, the FET standard includes the obligation not to deny justice in criminal, civil or administrative adjudicatory proceedings in accordance with the principles of due process embodied in the principal legal systems of the world. While historically, the FET standard was considered to be breached when the state conduct was of an egregious and shocking nature,[33] it has been applied in recent cases to conduct taken in good faith, when investors' expectations were frustrated by state action.

For instance, in *Metalclad*, the tribunal held that Mexico violated the FET standard, because the company was led to believe that the federal and state permits it secured allowed for the construction and operation of the landfill. Thus, the tribunal held that Mexico failed to provide "a transparent and predictable framework for Metalclad's business planning and investment"[34] and that the Mexican government had deprived Metalclad of its reasonable expectations. Although the Supreme Court of British Columbia held that the NAFTA panel erred when it imported the transparency provisions of NAFTA Chapter 18 into Chapter 11,[35] from a systemic perspective, authors have pointed out the importance of regulatory transparency and good faith in the conduct of the state (Weiler 2005: ch.19). In *Metalclad*, lack of transparency and coherence in state actions and the violation of the investor's legitimate expectations were weighted in the assessment as important indicators of indirect expropriation.

In the *TECMED* case, the Arbitral Tribunal found a breach of the FET standard as it deemed the licence to be denied on socio-political considerations, and not on compelling reasons of public interest. The Tribunal noted that none of the relevant authorities expressed concerns as to the dangers that the landfill may pose to public health and environmental protection. Moreover, the Tribunal pointed out that Mexican authorities had acted in a contradictory way, by, on the one hand, reassuring TECMED that they could operate the landfill until relocation to a different area was conducted, and, on the other hand, denying the renewal of the license.

Recent trends in the case law suggest that, on the one hand, the FET standard includes transparency, legitimate expectations and good faith

(Kreindler 2006: 7). On the other hand, authors submit that this standard ought to be relevant not only with regard to the state's conduct, but also with regard to the investor's behaviour in the conduct of business (Muchlinski 2006: 527–558). Crucially, according to the FET standard, an investor may be deemed to have, *inter alia*, the duty to refrain from unconscionable conduct. As Professor Lowe points out, "international law is developing so as to establish an expectation if not as yet a legal obligation on corporations, that they will exercise their powers and influence in such a manner as to promote the implementation of human rights" (Lowe 2004: 30).

Reconciling Environmental Health with Investor Rights

After having scrutinized the interplay between environmental health considerations and investor rights in investment treaty law and arbitration, the following sections will attempt to enucleate some legal tools that might prevent or help settling disputes with environmental health elements. First, Environmental Impact Assessments (EIAs) may represent a useful method of dispute avoidance (Francioni 2003: 229–243). While this tool already appears in many law instruments at the national, regional and international levels, investment treaties rarely, if ever, mention such a specific tool.[36] However, *de jure condendo*, the introduction of this specific mechanism may help reconciling the different interests at stake. Second, attention shall be paid to the process of balancing in investor – state disputes.

Environmental Impact Assessment

Environmental Impact Assessment (EIA) is an instrument of environmental governance that ensures that the environmental implications of decisions are taken into account before the decisions are made. Over 100 states and a number of international treaties require EIA in specified circumstances.[37] Certain advanced domestic systems and some international regimes require that the environmental impacts of proposed activities be made known not only to regulatory authorities but also to private stakeholders such as local communities. In this sense, EIA is an appropriate tool to integrate environmental health consideration in economic activities, and may be an instrument of dispute prevention (Francioni 2003: 235). However, recent investment treaty disputes with environmental health elements have questioned the very rationale of imposing EIA, and the way of conducting it.

In *Maffezini v Spain*,[38] an Argentine investor, established a joint venture in Spain for the production of chemicals. In 1992, the construction of the chemical plant had to be discontinued because of the investor's financial crisis. In the subsequent ICSID arbitration, Maffezini complained that the Spanish authorities had *inter alia* pressured the company to make the investment before the EIA process was finalized and before its implications were known. Therefore, according to the claimant, the Spanish authorities were responsible for the additional costs resulting from the EIA. The Tribunal stated that "the environmental impact assessment procedure is basic for adequate protection of the environment and the application of appropriate environmental measures. This is true not only under Spanish and EEC Law, but also increasingly so under international law".[39] Thus, the tribunal held that Spain had simply required compliance with its environmental laws in a manner consistent with its investment treaty commitments.[40]

In a pending NAFTA case, recently initiated against the Government of Canada, the Clayton family and their US corporation, Bilcon, object to the manner in which an environmental assessment has been undertaken.[41] The investors proposed to mine basalt in Nova Scotia and then ship it by tanker to their New Jersey site. The project attracted a large amount of public discussion in Nova Scotia, and was ultimately rejected following the EIA. In the Notice of Arbitration, the claimants acknowledge that an EIA was required for their project, but they claim that the process was unusually protracted, discretionary, and ultimately politically motivated. Because of failure of due process and the rule of law, the claimants allege discrimination and violation of Fair and Equitable Treatment.

As seen above, EIAs have come to the forefront of legal debate in investment disputes. In abstract terms, detecting the environmental consequences of the project before it is implemented, and ensuring that planned activities are compatible with sound environmental management may prevent the risks of environmental damage and ensure the reconciliation of private and public interests. However, the above examined disputes show that EIA does not exempt States from a general duty of good faith and transparency. In conclusion, the fact that to date EIA procedures have not prevented investment treaty disputes does not mean that they may not do so in the future. They clearly have such potential if they follow adequate procedural standards.

Balancing

If disputes with environmental health elements arise, arbitrators may have to balance the different interests at stake. It will be important for the states to show that their regulations are aimed at achieving legitimate public goals and follow due process of law. As an arbitral tribunal held, "[…] 'public interest' requires some genuine interest of the public. If mere reference to 'public interest' can magically put such interest in existence and therefore satisfy this requirement, then this requirement would be rendered meaningless since the Tribunal can imagine no situation where this requirement would not have been met."[42] Wälde and Kolo point out that acts of protectionism or mistreatment of foreign investors should not be "camouflaged in the much more palatable clothes of sacred environmental causes' and caution against 'not so holy alliances between protectionist interest and environmental idealism"(Wälde and Kolo 2001: 811–848).

If a case presented scientific uncertainty and a plurality of scientific opinions, authors suggest a two-fold method to deal with it (Orellana 2006: 48–72). First, adjudicators may put emphasis on the scientific method and on due process, particularly with regard to transparency and public participation. Accordingly, adjudicators should assess whether policy makers have adopted reasonable scientific method and due process in the risk assessment, that is, the determination of the levels of risk. Their enquiry would focus on whether regulations have been adopted in a transparent manner and whether they are subject to domestic remedies and/or judicial review. Second, they may adopt the *margin of appreciation doctrine* which was developed by the ECtHR, in order to secure some latitude for the host state to define and adopt societal public policies. The margin of appreciation doctrine would grant qualified deference to governmental regulations adopted on the basis of scientific evidence and would reduce the risk of arbitral tribunals second guessing democratic authorities (Orellana 2007: 671–789). The margin of appreciation doctrine would apply specifically to the risk management phase, where value judgments and other public considerations apply. The standard of review adopted by arbitral tribunals would be similar to that adopted by national administrative or constitutional courts. In this sense, such review can be seen as a "desirable constraint over the domestic political process" or as a "step towards good governance in international economic relations" (Wälde and Kolo 2001: 846).

The margin of appreciation doctrine would be particularly appropriate to the evaluation of governmental measures adopted to address public health issues in the light of the precautionary principle. The precautionary principle requires states to intervene and protect the public from exposure to harm where scientific investigation discovers a plausible public health risk. Decision makers may need to prevent harm before it occurs, as recognized by Principle 15 of the Rio Declaration: "Where there are threats of serious or irreversible damage, lack of full scientific certainty shall not be used as a reason for postponing cost-effective measures to prevent environmental degradation."Another element which may be taken into account to evaluate the legitimacy of a given regulatory measure is the existence of international standards, affirmed in instruments adopted by international bodies, such as the World Health Organization. If the adopted measures are based on international standards, they may be assumed to be legitimate.

The issue of whether or not some degree of compensation should be paid in case of lawful expropriation is not questioned. The problem arises in drawing the line between regulation and lawful expropriation, and determining the amount of the subsequent compensation. This debate has not reached a definitive conclusion and there is no settled practice. The fact that a certain regulation may have a negative impact on economic activities is undisputed. In certain cases, states have even provided compensation for their general regulations at a normative level. In other cases, the issue is to be dealt with in the investor–state dispute settlement mechanism. Examining the practice of the ECtHR in expropriation cases, it appears that the Court generally adopted the doctrine of 'fair balance' between the demands of the general community and individual fundamental rights. This general principle cannot be defined *a priori*, but has to be determined on a case by case basis. Whether the same approach may be used by arbitral tribunals is an open question.

Conclusions

This chapter has examined the particular intersection between environmental health and international investment law. Environmental health is a theoretical tool or working definition which highlights the linkage between the environment and public health. The regulatory competence of states to protect the common wealth has traditionally been part of state sovereignty, has somehow been reinforced by a web of multilateral environmental agreements which have reaffirmed the states' duty to protect the environment. Several investment treaties acknowledge such competence through co-ordination clauses and declaratory statements in their preambles. The synergy or mutual supportiveness between economic development and environmental considerations may be aptly expressed in the concept of sustainable development. Sustainable development may be conceived as a method or a process to integrate environmental considerations in developmental policies and market regulation but also as a unifying element of different principles of environmental protection that have emerged at both substantive and procedural levels (Francioni 2007: 43).

However, conflicts may arise between investment treaty guarantees and national regulations aimed at protecting environmental health. Foreign investors, who have direct access to international arbitral tribunals, may claim that such regulations negatively affect their investment, thereby amounting to indirect expropriation. Other claims may concern the alleged discriminatory effects of such laws and regulations. In this respect, on the one hand, states should co-operate in good faith to respect their international environmental obligations, and ensure that their investment treaty commitments do not frustrate their obligations under multilateral environmental agreements. On the other hand, foreign investors should comply with local environmental standards as long as these regulations reflect legitimate environmental concerns, respect due process of law, and have no discriminatory character. This chapter has shown the possible areas of conflict between environmental health concerns and investors' property rights and has scrutinized some legal tools that may help the reconciliation of these different values and goals within investment treaty law and arbitration.

References

Baughen, Simon, "Expropriation and Environmental Regulation: The Lessons of NAFTA Chapter Eleven", *Journal of Environmental Law* (2006).

Fauchald, Ole K. "International Investment Law and Environmental Protection", *Yearbook of International Environmental Law* 17 (2006).

Fidler, David, "A Globalized Theory of Public Health Law", *Journal of Law, Medicine and Ethics* 30 (2002).

Francioni, Francesco, "Dispute Avoidance in International Environmental Law" in Alexandre Kiss, Dinah Shelton and Kanami Ishibashi (eds) *Economic Globalization and Compliance with International Environmental Agreements* (The Hague: Kluwer Law International 2003).

—. "Sviluppo sostenibile e principi di diritto internazionale dell'ambiente" in *Il principio dello sviluppo sostenibile nel diritto internazionale ed europeo dell'ambiente*, ed. Paolo Fois (Editoriale Scientifica: Napoli 2007).

Freedman, Joseph, 'Implications of the NAFTA Investment Chapter for Environmental Regulation', in *Economic Globalization and Compliance with International Environmental Agreements*, ed. Alexandre Kiss, Dinah Shelton and Kanami Ishibashi (Kluwer Law International: The Hague 2003).

Kelsey, Jane, "International Economic Agreements and Environmental Justice", in *Environmental Justice and Market Mechanisms*, ed. Klaus Bosselmann and Benjamin J. Richardson (The Hague: Kluwer Law International, 1999).

Kreindler, Richard H., "Fair and Equitable Treatment: A Comparative International Law Approach", *Transnational Dispute Management* 3 (2006).

Lowe, Vaughan "Corporations as International Actors and International Law Makers", *Italian Yearbook of International Law* 14 (2004).

Mann, Howard and Konrad von Moltke, "NAFTA Chapter 11 and the Environment: Addressing the Impacts of the Investor–State Process on the Environment" (Winnipeg: IISD, 1999).

Meron, Theodor, *The Humanization of International Law* (Leiden: Martinus Nijhoff Publishers 2006).

Moore, J.B, *Digest of International Law*, vol VI, Section 1003 (1906).

Muchlinski, Peter, "'Caveat Investor?' The Relevance of the Conduct of the Investor under the Fair and Equitable Treatment Standard", *ICLQ* 55 (2006).

Orellana, Marco, "Science, Risk and Uncertainty: Public Health Measures and Investment Disciplines" in *New Aspects of International Investment Law*, ed. Philippe Kahn and Thomas Wälde (The Hague: Hague Academy of International Law 2007).

—. "The Role of Science in Investment Arbitrations Concerning Public Health and the Environment", *Yearbook of International Environmental Law* (2006).

Reynolds, Glenn H. and David B. Kopel, "The Evolving Police Power: Some Observations for a New Century", *Hastings Constitutional Law Quarterly* (2000).

Rosenberg, M. and M. Cheah, "Arbitrating Environmental Disputes", 16 *ICSID Review-FILJ* (2001).

Sands, Philippe, "Litigating Environmental Disputes: Courts, Tribunals and the Progressive Development of International Environmental Law" in *Law of the Sea, Environmental Law and Settlement of Disputes: Liber Amicorum Judge Thomas Mensah*, ed. Tafsir Ndiaye and

Rüdiger Wolfrum, (Leiden/Boston: Martinus Nijhoff Publishers, 2007).

Suda, Ryan, "The Effect of Bilateral Investment Treaties on Human Rights Enforcement and Realization", in *Transnational Corporations and Human Rights*, ed. Olivier De Schutter (Oxford: Hart Publishing 2006).

Von Moltke, Konrad *Discrimination and Non-Discrimination in Foreign Direct Investment Mining Issues* (Paris: OECD 2002).

Wälde, Thomas and Abba Kolo, "Environmental Regulation, Investment Protection and 'Regulatory Taking' in International Law", *International and Comparative Law Quarterly* (2001)

Weiler, Todd, "Good Faith and Regulatory Transparency: The Story of Metalclad v Mexico" in *International Investment Law and Arbitration: Leading Cases from ICSID, NAFTA, Bilateral Treaties and Customary International Law*, ed. Todd Weiler, (London: Cameron May 2005)

Notes

1 ICJ *Gabcikovo-Nagymaros Dam*, Judgment September 25 1997, ICJ Reports 1997. Judge Weeramantry Separate Opinion.

2 *López Ostra v Spain*, (Application no. 16798/90) judgment of 9 December 1994, Publications of the European Court of Human Rights, Series A, No. 303-C.

3 *Guerra and Others v Italy*, (116/1996/735/932) judgment of 19 February 1998, ECtHR, Reports of Judgments and Decision 1998-I, No 64.

4 *Fadeyeva v Russia* (Application no 55723) Judgment of 30 November 2005.

5 Stockholm Declaration on the Human Environment 16 June 1972 UN Doc A/Conf 48/14/rev. 1 11 ILM 1416 (1972).

6 UN Human Rights Committee Decision to Study the Possibility of a Human Right to Environment, UN Doc E/CN. 4/Sub2/1989/C23 (1989).

7 Draft Principles on Human Rights and the Environment UN Document E/CN.4/Sub 2/1994/9.

8 See for instance Article 11 of the San Salvador Protocol to the ACHR.

9 UN Committee on Economic, Social and Cultural Rights, *General Comment* n. 14, adopted on May 2000 (E/C.12/2000/4) at § 4.

10 NAFTA Article 104.1.

11 The North American Agreement on Environmental Cooperation (NAAEC), 32 ILM 1480 (1993).

12 Energy Charter Treaty, 34 ILM 374 (1995).

13 *Chevron Corp. (USA) and Texaco Petroleum Corporation (USA) v The Republic of Ecuador*, Interim Award of December 1, 2008, PCA, UNCITRAL Rules, available at
http://ita.law.uvic.ca/documents/Chevron-TexacovEcuadorInterimAward.pdf visited on 12 September 2009.

[14] *Chevron Corp. (USA) and Texaco Petroleum Corporation (USA) v The Republic of Ecuador*, Interim Award, § 77.
[15] *Metalclad Corporation v The United Mexican States*, Award 30 August 2000, Case No. ARB(AF)/97/1, 40 ILM 36.
[16] *Metalclad Corporation v The United Mexican States*, § 103.
[17] *Tecnicas Medioambientales TECMED S.A. v the United Mexican States*, Case No. ARB(AF)/00/2, 29 May 2003, (2004) 43 ILM 133.
[18] *Ibid.*, §125.
[19] *TECMED S.A. v the United Mexican States*, § 157.
[20] *Myers v Canada*, Partial Award, 13 November 2000, § 281.
[21] *Methanex Corporation v. United States of America*, Final Award, August 3 2005, 44 ILM 1345 (2005).
[22] *Methanex*, Part IV–Chapter D–Page 4.
[23] *Methanex* award, part III, chapter A, para 102(2).
[24] *Methanex*, Part IV–Chapter D–Page 5.
[25] *Vito G Gallo v Government of Canada*, Notice of Arbitration, 30 March 2007; *Vito G Gallo v Government of Canada*, Statement of Claims, 23 June 2008.
[26] *Vito Gallo v Government of Canada*, Procedural Order N° 1, 4 June 2008 §§ 37–38.
[27] *Ethyl Corporation v Canada*, Notice of Arbitration under the Arbitration Rules of the UNCITRAL and the NAFTA, April 14, 1997, § 15–17.
[28] *Ethyl Corporation v Canada*, Statement of Claim, October 2, 1997, § 33–36.
[29] The Canadian government agreed to settle Ethyl's claim, dropping the ban on MMT and paying Ethyl a reported US$ 13 million of legal expenses and lost profits.
[30] *S.D. Myers Inc v Government of Canada*, Award November 13, 2000 40 ILM 6, 1408–1492, § 183.
[31] Basel Convention on the Control of Transboundary Movements of Hazardous Wastes and their Disposal, 28 ILM 656 (1989).
[32] *Vito G Gallo v Government of Canada*, Statement of Defense, 15 September 2008, § 138.
[33] *Neer v United Mexican States*, Award, 4 RIAA 60 (1960).
[34] *Metalclad Corporation v The United Mexican States*, § 99.
[35] *The United Mexican States v Metalclad Corporation*, Supreme Court of British Columbia, 2001 BCSC 664, available at http://www.courts.gov.bc.ca/jdb-txt/SC/01/06/2001BCSC0664.htm
[36] For instance, EIA is mandated under Articles 2 and 10.7 of the NAAEC.
[37] Convention on Environmental Impact Assessment in a Transboundary Context (Espoo Convention), 30 ILM 800 (1991).
[38] *Emilio Augusto Maffezini v The Kingdom of Spain*, ICSID Case No ARB/97/7, Award of the Tribunal, 13 November 2000.
[39] *Ibid.*, §§ 67.
[40] *Ibid.*, § 71.

[41] *William Ralph Clayton, William Richard Clayton, Douglas Clayton, Daniel Clayton and Bilcon of Delaware Inc. v Government of Canada*, Notice of Arbitration, May 26, 2008.

[42] *ADC Affiliate Limited and ADC and ADMC Management Limited v Republic of Hungary*, ICSID Case ARB/03/16, Award 2 October 2006, § 432.

CHAPTER THIRTEEN

PUBLIC HEALTH IN THE FACE OF GLOBAL ECOLOGICAL AND CLIMATE CHANGE

COLIN L. SOSKOLNE
AND LAURA WESTRA

Introduction

Traditional public health over the centuries has focused on a few areas. It has ensured the health of communities through good sanitary practices, regulatory regimes that keep air and water clear of pollution and safe for inhalation and consumption purposes respectively, and soil quality that better ensures food safety and security. Also, since the advent of the smallpox vaccine by Edward_Jenner in 1796, various immunization programs against contagion from several other debilitating diseases have been the mainstay of public health.

Now, as humanity, both individually and collectively, tampers with the very fabric of life through the global expansion of the human enterprise, Nature's services that provide pure air, clean water, and uncontaminated soils are under siege. Additional changes include biodiversity loss, population growth with ever-widening disparities, over-consumptiveness and hazardous waste disposal, escalating demands on fossil fuels for energy and consumer products, and on technological gaps between the more advanced and less advanced economies, globally.

The effect of the changes that are under way is a net negative, with systemic impacts such as climate change, declines in air, water and soil quality, as well as food security issues around the world. If epidemiology is to usefully inform policy on matters of such global import, as it has traditionally done in the past on matters globally but bearing on more local public health concerns, then more support for the sub-specialty of eco-epidemiology is needed.

Eco-epidemiologists study the dynamics between global ecological changes and population health. By studying these dynamics, the public and policy-makers could be better informed about the need for policy interventions to serve ecological integrity, biocentric in its essence, thereby being better informed about the need to implement policies to sustain **all** life. In so doing, we would help humans to see their integral connection to the ecosystems in which we live, with consequences that will more ensure human well-being and the ongoing of civilization.

Pre- and post-global change practice of epidemiology

Those working in the field of epidemiology examine the distribution and determinants of illness and premature death at the community level. Epidemiology, derived from the Greek *epi* ("on") and *demos* ("populations"), is the science that is basic to the rational formulation of public health policy. Epidemiologists work to, among other things, provide information to policy-makers for interrupting the causes of disease, whether we are speaking about infectious and acute causes, or about causes related to people being chronically (or acutely) exposed to toxic contamination of any kind.

The goal of the discipline is to work to prevent harms to communities. Our job is to inform policy with a view to reducing harms by preventing disease and premature mortality at the community level. Exercising precautionary approaches is fundamental to what we do.

Ideally, epidemiology is proactive, maximizing opportunities to minimize harms to community health and well-being. The old maxim applies: "An ounce of prevention is worth a pound of cure." However, epidemiologists more often find themselves reactively responding to problems that could well have been prevented. In these latter circumstances, costs and harms are higher than what they would have been had early warnings been heeded.

Today, under the extensive range of global changes taking place, there is an even greater need to be proactive if we are to prevent harms on a massive scale. As we approach ecological system tipping points, we are less likely to avoid thresholds being exceeded at which point whole system collapses will not necessarily permit practical reactive responses. Therefore, how can we more effectively deliver on our professional obligation to prevent harms on a population scale under the current scenario of global ecological change?

It behooves those in the field of epidemiology, inasmuch as what we are trained to identify the causes of disease, to be able to identify the reasons for why early warnings so often go unheeded. The answer, in large part, lies in the effect of vested interests bringing pressure to bear on policy-makers. Information is presented by self-interested groups (e.g. tobacco companies and other corporate entities) that foment uncertainty and thus make the work of the policy-maker all the more difficult in the face of less rather than greater certainty.

The Nature of Epidemics and Pandemics

Epidemics occur when low background rates of specific diseases and premature deaths are replaced by increasing rates of these events. Pandemics occur when these events expand globally.

Major trends have been discernible in epidemiology over the past 100 years. Through the First World War, the focus was on epidemics of infectious/communicable diseases. With the advent of penicillin and other antibiotics since then, infectious diseases were considered controllable. Then, with people living longer, an inexorable rise was seen in the incidence of chronic, non-communicable diseases. These surfaced with tobacco, occupational diseases and environmental disasters that have resulted in epidemics of chronic conditions and premature death for those who survived beyond the initial exposure episode.

By 1981, with the advent of HIV/AIDS, the realization that communicable diseases were re-emerging came to the fore. It was only in the late 1990s, however, that the link was being made to the potential for catastrophic harms to populations on a rather massive scale from encroachment into the wild and with consequent declines in ecological integrity. Indeed, illness and premature death from global warming and desertification are already resulting in eco-refugees on a grand geographic scale, serving as a forerunner of multiple harms to come.

The Nature of Epidemics under Global Change

The recent consolidation of many indicators of ecological (dis)integrity suggests the imminent collapse of life as we know it. Collapse will take the form of not only premature death and excess morbidity, but also in the collapse of civilization through mass harms that are unfolding as we continue to disrupt life-supporting ecosystems through climate change and other stresses being placed on ecosystems.

Why should epidemiologists and public health researchers and practitioners be concerned with matters of a global nature? The reason is that we have the professional mandate to prevent harms to the health of communities; on a global scale, this mandate ought to be taken all the more seriously. Our mandate stems from ethical precepts based in the right to life. As such, and with looming global catastrophes, the Right to Life for both present and future generations (inter-generational equity/justice) is of concern to the eco-epidemiologist. This is especially so as epidemics turn into pandemics, and when risks to health and, indeed, to the survival of the species are so great as to extend from local communities to act on a global scale.

What is the real meaning of the right to health?

> "It has been shown that external agents have as great an influence on the frequency of illness as on its fatality, the obvious corollary is that man has much power to prevent as to cure disease … Yet medical men, the guardians of public health, never have their attention called to the prevention of sickness; it forms no part of their education … The public does not seek the shield of the medical arts … till the arrow of death already rankles in the veins …" (Gostin 2008: 18–19)

Perhaps a "vibrant conception of the common" (Gostin 2008: 6) is the most important aspect of public health. But the meaning of the "common good" is not obvious, although indeed health is an overarching collective right. Already in 1837, it was known that prevention was not at the forefront of the concerns of society or state governments. If, as Gostin argues, the physical protection of the citizens is the first concern of any government, then such protection should be extended beyond all other services expected from democratic governments, and it should start with preventing the exposures that are widely known in the 21st century to harm health and cause morbidity and premature mortality.

We consider the progression of public health in developed countries, first, from infectious and communicable diseases, to the non-communicable (chronic) diseases. Today, the poorest countries are facing the "epidemiological transition from infectious to non-communicable diseases" (Gostin 2008: 237). In earlier times, infectious diseases were the main concern of public health. But, just as many infectious diseases appeared to have been conquered (Gostin 2008: 237), other circumstances again arise (Garrett and Rosenstein 2005: 64–69):

> "Disease amplifiers are principally man-made and therefore controllable. Human beings congregate and travel, live in close proximity to animals, pollute the environment and rely on an overtaxed health system. This constant cycle of congregation, consumption, and movement allows

> infectious diseases to mutate and spread across populations and boundaries." (Gostin 2008: 234)

Gostin explains that "ecosystem degradation has multiple health effects" (Gostin 2008: 235), and he also explains the consequences of "Human/animal interchange" (Gostin 2008: 235), as he indicts the "overuse and inappropriate use of pharmaceuticals" by animal agriculture, but he does not go far enough.

Once the collective rights of all citizens are considered first, as they should be in public health law, and when the responsibilities and obligations of governments are taken seriously, it seems clear that the protection of the health of all should start *before* persons become ill and health care is required:

> Yet, just as infectious diseases move and change, so do the non-communicable diseases [NCDs]. The global rise in NCDs reflects significant transformations in diet habits, physical activity levels, and tobacco use worldwide. The process of industrialization, urbanization, economic development, and increasing food market globalization, has led to harmonization of behavior. What was once culturally attractive primarily in industrialized countries has gained popularity all over the world."

However, while many of these factors leading to ill-health represent deliberate (though often manipulated) individual choices, the major determining issues are beyond such choices, as they depend on the interface between state governance and multinationals' choices for "development".

A recent article published by The Lancet (Backman et al. 2008: 2047–2085) argues that the Universal Declaration of Human Rights (1948) "laid the foundations for the right to the highest attainable standards of Health", and concludes that "Right-to-health features are not just good management, justice, or humanitarianism, they are obligations under human rights law" (Backman et al. 2008: 2047).

In addition, Gostin cites the International Sanitary Regulations (ISR), adopted by the member states of the World Health Organization (WHO), pursuant to the WHO's Article 21 powers. Since 1969, these regulations were renamed International Health Regulations (IHR) (Gostin 2008: 246). In 2005, they were fundamentally revised to include many global pandemics, such as HIV/AIDS, SARS, avian flu, Marburg, and even bioterrorism (Gostin 2008: 246; Fidler 2005: 325–392; Forrest 2000: 153–179; Taylor 1997: 1327–1362). Article 1 of the IHR defines a public health risk as follows:

> "… a likelihood of an event that may adversely affect the health of human populations, with emphases on one which may spread internationally or may present a serious and direct danger."

But, it is not only infectious diseases that fit well within that definition; also, not only an "event", but ongoing practices that may and, indeed, do affect the health of human populations, as well as present "a serious and direct danger". This definition fits with the consequences of ecological degradation, including hazardous pollution, climate change and industrial activities that include those aimed at "development", particularly extractive and mining operations.

The public health effects that flow from these declines are based on solid and abundant evidence, as well as on the research of the WHO itself, the European Environmental Agency (EEA) and of scientists too numerous to name. Hence, at the very least, the current way that these activities are practiced, as well as their effects, should fall under the heading of evident and clear threats to public health. They should form part of the responsibility of states to oversee, correct and mitigate, or even to eliminate. Figure 1 shows only the "underlying determinants of health (e.g., water, sanitation, food, shelter and education)", without any reference to the negative aspects of globalized living that form the real "underlying determinants" of ill-health and abnormal development.

Figure 13-1. Right-to-health features of a health system underpinned by legal obligations based on general comment 14: preliminary working model

If we accept the claims advanced by *The Lancet's* article and the general sense of Gostin's seminal work, then the right to health appears to be a collective right "par excellence", or the clearest example of a collective right that must be considered primary and basic. It is a collective right not only to health care *after* the fact of various chemical and hazardous exposures, but also the right to health and normal human development before being exposed to the litany of harmful situations listed above. And, by extension, global effects from a degraded and unproductive environment, anthropogenically produced climate change, desertification leading to famine, and the like, must be considered similarly.

It is unfortunate that neither legal scholars, nor experts in public health declare the clear obligation of states, and of other non-state actors, to work to promote public health through prevention, first and foremost. First, preventive measures serve to reduce significantly or even eliminate the suffering of millions who either have not chosen the source of their health problems (such as cigarette smoking), or have not consented to the situations that engender those problems. Thus, second, it is far more equitable to reduce or prohibit all of the activities that cause the harms, than it is to attempt to redress the harms, once they have occurred.

Third, many of the harms, after they have been imposed on a population, are incompensable, and abnormal births, or children born only to acquire grave diseases both mental and physical, are clear examples (Grandjean and Landrigan 2006). Fourth, the cost of prevention to states and other responsible bodies is a fraction of the cost to attempt to redress the harms to public health after they have occurred, and that states themselves have either supported or helped to promote (e.g., by licensing unsafe operations, by not signing conventions intended to curb industrial harm, or by accepting support from corporations producing unsafe conditions and products).

The right to health and social justice

The uneven distribution in harms caused is of particular concern to epidemiologists and public health practitioners, especially if one believes in, or if one aspires to notions of fairness and justice in the world. For instance, it is not those causing the climate and ecological changes that are suffering the consequences arising from those changes.

Invocation of the ethical principle of environmental justice as well as a focus on the human right to an environment that can sustain both present and future generations are needed if we are to demonstrate to policy-makers the inequity and non-sustainability of current trends that are exacerbating climate and ecological system declines. In a rational policy environment, it is only by being able to demonstrate the connections between socio-economic activities and their attendant harms that policy is likely to be changed to create, on the one hand, incentives for behaviours (individually and collectively) that are less likely to contribute to ecological harms; and, on the other hand, disincentives for behaviours (individually and collectively) that are more likely to cause ecological harms. Hence the need to encourage the work of eco-epidemiology.

The World Health Organization on the social determinants of health

In 1986, the Ottawa Charter for Health Promotion was the first international document to recognize that health promotion is the process of enabling people to increase control over, and to improve their health. The Charter recognizes that

> to reach a state of complete physical, mental and social well-being, an individual or group must be able to identify and to realize aspirations, to satisfy needs, and to change or cope with the environment.
>
> Health is, therefore, seen as a resource for everyday life, not the objective of living.
>
> The fundamental conditions and resources for health are:
>
> - Peace
> - Shelter
> - Education
> - Food
> - Income
> - A stable eco-system
> - Sustainable resources
> - Social justice and equity

In the spirit of social justice, the Commission on Social Determinants of Health (i.e., the Commission) was set up by the World Health Organization (WHO) in 2005 to marshal the evidence on what can be done to promote health equity and to foster a global movement to achieve it. In 2008, the WHO produced a seminal report on the social determinants of health, recognizing that social justice is a matter of life and death. Social justice affects the way people live, their consequent chance of illness, and their risk of premature death.

The latter three bullet points above, as "conditions and resources for health", relate directly to concerns that emerge through declines in ecological integrity globally, particularly where impacts locally are driven by forces operating elsewhere. Climate change is but one example of drivers in rich countries causing direct harms in poorer countries.

Even without a focus on climate change and other drivers of global change, we watch in wonder at the disparities in health across the world. Life expectancy and good health continue to increase in some parts of the world, but not in others. And the gulf widens. A girl born today can expect to live for more than 80 years if she is born in some countries, but less than 45 years if she is born in others. Within countries there are dramatic differences in health that are closely linked with degrees of social disadvantage. "Differences of this magnitude, within and between countries, simply should never happen" according to the WHO report.

These inequities in health, avoidable health inequalities, arise because of the circumstances in which people grow, live, work, and age, and the systems put in place to deal with illness. The conditions in which people live and die are, in turn, shaped by political, social, and economic forces. These differences are only exacerbated under climate and other global changes. Increasingly, the health problems that rich and poor countries have to solve are converging around disparities.

The Commission calls on the WHO and all governments, in its 2008 report, to lead global action on the social determinants of health with the aim of achieving health equity.

> It is essential that governments, civil society, WHO, and other global organizations now come together in taking action to improve the lives of the world's citizens. Achieving health equity within a generation is achievable, it is the right thing to do, and now is the right time to do it.

The report calls on those with a passion for social justice, a respect for evidence, and a frustration that there appears to be far too little action on the social determinants of health to join forces recommending action now. And, with all of the global drivers of ever-widening disparities, this call is all the more important.

Where systematic differences in health are judged to be avoidable by reasonable action they are, quite simply, unfair. It is this that the report labels "health inequity". Putting right these inequities is, according to the report, a matter of social justice. Reducing health inequities is, for the Commission, an ethical imperative. Social injustice is killing people on a grand scale. On an eco-regional basis, the injustice of climate change is but one example.

The report of the Commission sets out key areas – of daily living conditions and of the underlying structural drivers that influence them – in which action is needed. It provides a global analysis and concrete examples to justify action regardless of the level of socio-economic development. In countries at all levels of income, health and illness follow a social gradient: the lower the socioeconomic position, the worse the health. Three principles of action are recommended in the report of the Commission:

1. Improve the conditions of daily life – the circumstances in which people are born, grow, live, work, and age.
2. Tackle the inequitable distribution of power, money, and resources – the structural drivers of those conditions of daily life – globally, nationally, and locally.
3. Measure the problem, evaluate action, expand the knowledge base, develop a workforce that is trained in the social determinants of health, and raise public awareness about the social determinants of health.

Moving the EcoHealth agenda forward

Eco-epidemiology is emerging. Starting with the Journal "Ecosystem Health", which then morphed into the Journal "Global Change and Human Health", and then into "EcoHealth", the field is evolving. Indeed, research undertaken by members of the recently established International Association of Ecology and Health (IAEH) is critical for making the linkages between the dynamics of ecosystem change and population health.

So promising are the signs of interest in EcoHealth that, in December 2008, an international conference convened some 700 delegates in Merida, Mexico. This conference produced a draft Call to Action, included as an appendix to this chapter. This is a response to the changing world. It signals what public health should be in this new world and what epidemiology should tell us if its mission to prevent health harms is to be encouraged.

References

Aron JL, Patz JA. (Eds) 2001. *Ecosystem Change and Public Health: A Global Perspective*. Baltimore, The Johns Hopkins University Press.

Backman, Gunilla et al. 2008. "Health Systems and the Right to Health: An Assessment of 194 Countries", *The Lancet*, 2047–2085

Butler C. 200[illegible]. Inequality, global change and the sustainability of civilization. Global Change & Human Health 2000;1(2):156-172.

Butler C. D. 2007. "Globalization, population, ecology and conflict." *Health Promot J Austr*, 18(2): 87–91.

Butler C. 2008. "Sustainable health for all by the year 2100?" *Int J Public Health*, 53: [illegible]23–224.

Butler C. D. 2008. "Environmental change, injustice and sustainability" *J of Bioethica Inquiry* 5(1): 11–19.

CSDH (2008). *Closing the Gap in a Generation: Health Equity through Action on the Social Determinants Of Health*. Final Report of the Commission on Social Determinants of Health. Geneva, World Health Organization. Accessed on November 24, 2009 at: www.who.int/social_determinants/thecommission/finalreport/en/index.html

Daily G. C. (Ed.) 1997. *Nature's Services: Societal Dependence on Natural Ecosystems*. Washington, DC: Island Press.

Diamond J. 20[illegible]5. *Collapse: How Societies Choose to Fail or Succeed*., London: Allen Lane.

European Environment Agency. 2001. *Late lessons from early warnings: the precautionary principle 1896 – 2000*. Luxembourg, Office for Official Publications of the European Commission,.

Fidler DP. 20[illegible]5. "From International Sanitary Conventions to Global Health", *Chinese Journal of International Law,* 4

Forrest M. 200[illegible] "Using the Power of the World Health Organization: The international Health Regulations and the Future of the International Health Law", *Columbia Journal of Law and Social Problems*, 33

Garrett, Laurie and Rosenstein, Scoot. 2005. "Missed Opportunities: Governance of Global Infectious Diseases", *Harvard Int. Law Rev* vol. 27

Gostin L. O. 2007. *Meeting basic survival needs of the world's least healthy people: toward a Framework Convention on global health*.. Georgetown University. http://lsr.nellco.org/georgetown/ois/papers/1

Gostin, Lawrence O. 2008. *Public Health Law,* 2nd ed., University of California Press, Berkeley

Grandjean, Philippe and Landrigan, Philip. 2006. "Developmental Neurotoxicity of Industrially Chemicals," *The Lancet*, Nov.06.
Heinzerling L. 2007. Climate change, human health, and the Post-Cautionary Principle. Georgetown University. http://lsr.nellco.org/georgetown/ois/papers/4.
Last J. M. 1998. *Public Health and Human Ecology* (2nd Edition). Stamford, Connecticut: Appleton & Lange.
Lovelock J. 2009. *The Vanishing Face of Gaia: A Final Warning*. London: Allen Lane.
McMichael T. 2001. *Human Frontiers, Environments and Disease: Past Patterns, Uncertain Futures*. Cambridge, Cambridge University Press.
McMichael A. J., Powles J., Butler C. D., Uauy R. "Food, agriculture, energy, climate change and health". *The Lancet*; 370: 1253–1263.
McMichael A. J., Beaglehole R. 2000. "The changing global context of public health". *The Lancet*; 356:495–499.
McMichael A. J., Butler C. D., Folke C. 2003. "New Visions for Addressing Sustainability". *Science*; 302(5652):1919–1920.
Michaels D. 2008. *Doubt is Their Product: How Industry's Assault on Science Threatens Your Health*. New York: Oxford University Press,.
Millennium Ecosystem Assessment. *Ecosystems and Human Well-Being. Volume 3: Policy Responses*. (Eds. Kanchan Chopra, Rik Leemans, Pushpam Kumar, Henk Simons). Chapter 16: Consequences and Options for Human Health. Coordinating Lead Authors: Carlos Corvalan, Simon Hales, Alistair Woodward. Lead Authors: Diarmid Campbell-Lendrum, Kristie Ebi, Fernando De Avila Pires, Colin L. Soskolne. Contributing Authors: Colin Butler, Andrew Githeko, Elisabet Lindgren, Margot Parkes. Island Press, Washington DC: 467–486. Also accessible at: www.maweb.org.
Ottawa Charter for Health Promotion. First International Conference on Health Promotion. Ottawa, 21 November 1986 – WHO/HPR/HEP/95.1 Accessed on November 24, 2009 at www.who.int/hpr/NPH/docs/ottawa_charter_hp.pdf
Smith R. 2005. "Medical journals are an extension of the marketing arm of pharmaceutical companies". *PLoS Medicine*, www.plosmedicine.org.;2(5), e138: 0364–0366
Soskolne C. L. and R. Bertollini. 1999. *Global Ecological Integrity and 'Sustainable Development': Cornerstones of Public Health: A Discussion Document*. World Health Organization, European Centre for Environment and Health, Rome Division, Italy, 74 pages. Also published at www.euro.who.int/document/gch/ecorep5.pdf
Soskolne C. L. 2004. "On the even greater need for precaution under

global change". (Reprinted from The European Journal of Oncology Library, Vol 2, 2003). *International Journal of Occupational Medicine and Environmental Health*;17(1):69–76. PMID: 15212208

Soskolne C. L., Westra L., Kotzé L. J., Mackey B., Rees W. E., Westra R. (Editors) 2008. Sustaining Life on Earth: Environmental and Human Health through Global Governance. Lexington Books, a Division of Rowman & Littlefield Publishers, Inc., Lanham, Maryland, USA.

Soskolne C. L. 2008. "Eco-epidemiology: On the need to measure health effects from global change". Chapter 7 in *Reconciling Human Existence with Ecological Integrity: Science, Ethics, Economics and Law*, edited by Laura Westra, Klaus Bosselmann and Richard Westra. Earthscan, London, UK; Sterling, VA; pp. 109–123.

Soskolne C. L., D. Lilienfeld, B. Black. 1994. „Epidemiology in Legal Proceedings in the United States". In: *The Identification and Control of Environmental and Occupational Diseases* (Eds. M. A. Mehlman and A. Upton). Advances in Modern Environmental Toxicology,:Part 1 (Volume XXII) 101–117. Princeton Scientific Publishing Co., Princeton, NJ.

Stern N. 2006. *The Economics of Climate Change.* Cambridge: Cambridge University Press.

Strong M. 2000. *Where on Earth Are We Going?* Toronto: Alfred A Knopf.

Taylor, Allyn L. 1997. "Controlling the Global Spread of Infectious Diseases: Toward a Reinforced Role for the International Health Regulations". *Houston Law Review,* 33.

Westra L. 2009. "Biological Integrity, Ecological Integrity: The Right to Life and the Right to Health in Law", *Transnational Law and Contemporary Problems*, Vol. 18, No.1, Winter 2009.

Appendix

DRAFT CALL FOR ACTION

TO THOSE DEDICATED TO RECONNECTING HUMANS TO THE ECOSYSTEMS OF WHICH THEY ARE A PART AND THEREBY TO RESTORING AND MAINTAINING THE HEALTH OF ALL LIFE FORMS

Colin L. Soskolne
Chair of Synthesis Plenary Session
International EcoHealth Forum 2008
Mérida, Mexico, December 1–5, 2008

THE ORIGINS OF THIS CALL FOR ACTION:

The ***International EcoHealth Forum 2008*** brought together researchers, policy-makers and civil society representatives to share experiences, evidence and theories at the intersection of ecology and health, locally, regionally and globally. The goal of the Forum was to identify options for the sustainability of ecosystems and human health.

Details about the Forum can be found at: www.ecohealth2008.org

The purpose of this **CALL FOR ACTION** is to rally *EcoHealth Forum 2008* participants and others to strive for changes in research, practice and policy that will better protect both ecosystem and human health, now and for generations to come.

CALL FOR ACTION

Overwhelmingly, interdisciplinary evidence from the *International EcoHealth Forum 2008* was compelling on the need for action to avert the rapid rate at which ecological decline is occurring around the world. Improved and sustained human health depends on healthier ecosystems and on sustainable ways of living. Indeed, in order to achieve health as a sustainable state of self-regulating harmony between humans and their physical, biological and social environments that enables them to co-exist indefinitely, urgent action is needed. We – responsible and concerned professionals, practitioners and civil society representatives – can and must help society, wherever in the world we live, to change course, from one of ecosystem destruction, to one of global sustainability – locally, regionally and globally.

We, therefore, promote this **CALL FOR ACTION** by calling on our professional organizations, and on civil society and governments, to act with urgency on the following 14 points from our *International EcoHealth Forum 2008*:

1. We must work to engage with both community-based grassroots efforts and in the formulation of national, international and private-sector policies to move the world onto a sustainable path to secure a healthy future. Existing national laws and international agreements favouring sustainability and health must be upheld and enforced.
2. We must work to reconnect humans to the natural ecosystems of which we are a part and to which our futures are inextricably linked. Where existing research supports action, we encourage action; where more research is needed to reduce uncertainty and/or to eliminate gaps in knowledge, more high quality research should be undertaken.
3. We must adjust our lifestyles and expectations, both individually and collectively, to transform current paradigms destined to collapse to sustainable ways of life. All around the world there is a need to recognize the importance of ecosystem sustainability for human health and well-being.
4. Efforts to meet the 2015 **Millennium Development Goals** must be increased if the goals are to be met.
5. Our approach to research and practice must integrate the social and environmental dimensions of health with ecosystems, including the elimination of social inequalities and health disparities, regionally and globally.
6. We must expand awareness across generations, disciplines and sectors about the importance of shared values and ethics, foundational to a path of sustainability.
7. We must work to inform and influence policy at the highest levels. We do so by engaging with the actors ultimately able to make changes in policy and practice based on the evidence, be they in government, in the private sector, or in the community.
8. We must work to show that people need to better manage themselves rather than work to manage ecosystems.
9. We must work together and more effectively with local knowledge sources and a broad range of other stakeholders in adopting

integrated, transdisciplinary approaches to face the complex problems of today.

10. Researchers among us must win the community's trust if they are to remain relevant and have lasting impact.
11. We must work to reduce fragmentation in Governance.
12. We must work to establish stronger inter-sectoral public health policies that include those communities traditionally marginalised or excluded from decision-making processes.
13. Delegates at future EcoHealth conferences must continue to strive to ensure the participation of a broad range of stakeholders; from funders and employers, to affected communities, indigenous peoples, policy-makers, and international development agencies, and from NGOs to business interests.
14. We respect those among us who have worked for years and even decades to bring attention to our collective ecological plight. It is on their shoulders that we stand today. We recognize the forces that have undermined their efforts and brought us to the current state of ecological decline. We therefore commit to joining with their legacy to protect the global public health interest above any other narrow short-term interests.

CHAPTER FOURTEEN

RE-BUILDING BABEL: COLLECTIVE THINKING, DEMOCRACY, ENVIRONMENT AND HEALTH

VALERIE A. BROWN

Abstract

A democracy designed to address the needs of all individuals in a small local population does not readily transfer to 6 billion people and the entire planet. The difficulties are well illustrated in attempts to combine the many interests in the interrelationships between human health and global ecological integrity. The situation is comparable to the story of the city of Babel. Once the citizens were no longer a single people with a shared language, they could not understand one another and the city collapsed.

Two pathways for re-interpreting issues of democracy and global ecological integrity offer a solution. The first is to accept that each issue is a wicked problem, that is, one that falls outside the current thinking of the society that produced it. A fresh approach is required. The second is to use the experiential learning cycle as the basis for collective learning among the interest groups that make the decisions, namely, key individuals, community, specialists, organizations and holistic thinkers. Decision-making then becomes the pursuit of a collective inquiry. Preserving the contributions of each of the interest groups while interpreting the whole replaces the past reliance on a single language. The conclusion is that Babel can be re-built as a democracy, once we learn how to approach the issues collectively.

Context

As other chapters in this book argue, it is time to re-examine the practice of democracy. Such a re-examination is by no means unusual.

Democracy has been mutating ever since the idea first appeared in ancient Greece. Given the basic paradox that democracy is the rule of the governed (Plato 1991), we need to ask how that ideal is translated into practice under the conditions of life of today. Issues central to a democratic state include how to achieve equitable access to health and a sustainable environment. In each of these fields, the marginalized and disadvantage are at far higher risk than the rest of their society.

There is therefore need to take seriously the goals for a democratic state as expressed in Lincoln's Gettysburg Address and the Constitution of France: government of the people, by the people, for (all) the people (my insertion). Decisions on these crucial issues are now being made against a backdrop of a deteriorating social–environmental system and a narrow compartmentalization of knowledge (IPCC 2007; Funtowicz and Ravetz 1993; Funtowitz and Ravetz 1993).

At the beginning of Western civilization, democracy rested on individuals' interpretations of the issues (Plato 1962). Democratic communities were small enough to be familiar with all their members, while certain classes of people – slaves, non-property owners and non-residents – were excluded. Fast forward, and we expect to include whole populations in numbers undreamt of then, coupled with an unprecedented technological power for our decisions to alter the planet. It is being argued that our responsibilities include not only all human beings, but all living things (Berry 1999).

This level of power and responsibility were still in the future when our current ruling version of democracy emerged almost simultaneously in Britain, United States and Europe. Forerunners in each country, Edmund Burke in 1757 (Burke 1999), Thomas Paine in 1776 (Paine 1776), and Francoise Voltaire in 1759 (Voltaire 1998), sought to achieve a balance between individual autonomy and equity among all citizens on the one hand; and between local community well-being and efficient social organization on the other. All three writers sought to resolve these tensions through the medium of citizens who hear, according to Paine "the simple voice of nature and of reason" (Paine 1776, p24).

In this 21st century the ideals for democracy remain much the same, at least on paper. However, there is now need to consider how the simple voice of nature and reason can make itself heard throughout a population six times the size of that of the 18th century and on behalf of the whole planet. Much is made of the difficulty of achieving the 18th century version of democracy within different national cultures, even different civilizations (Falk 1999).

This chapter examines the implications for democratic decision-making at a more grounded level. It explores the ways that local decisions on health and environment are influenced by the divided interests that contribute to the Western knowledge base. The chapter offers a collective learning framework that reconnects the decision-making interests as they work towards a goal of healthy people and a healthy planet.

The Issues

The key interests required to collaborate in moving towards a humane and ecologically sustainable future are widely agreed to be those of community, specialists, and government (World Bank 1998–1999), UNEP (Habitat (United Nations Centre for Human Settlements) 1997), (World Social Forum 2004). Action research with over 300 communities in Australia and Asia added two more. The knowledge constructed by individuals and knowledge derived from a holistic perspective proved to make essential contributions to decisions on community governance and well-being (Brown 2005, 2008; Keen, A. et al. 2005). This leads to the conclusion, long recognized in social change theory and in community development strategies, that the essential set of contributors to democratic decisions on important issues is a combination of key individuals, members of affected communities, relevant specialists, influential organizations and holistic thinkers.

To harness these interests to a common purpose is not as simple as bringing the different contributions together and hope that they can agree. Their degree of separation is much greater than that. Individual experience and local knowledge have become all but excluded from major decisions. Holistic thinking is actively discouraged. As Kuhn (1970) predicts, changes in approaches to specialised knowledge are initially rejected. It took over a decade for acceptance of Lovelock's Gaia hypothesis of the interconnections that create the living planet (Lovelock 1979).

During the 20th century, individual voices have increasingly been buried in mass opinion polls (as, for instance, in the national opinion polls that masked the wisdom of the early protests against the Vietnam and Iraq wars). In both civil and civic decision-making, the role of citizen is being replaced by that of consumer. Purchasing power has become as important a political influence as the vote. One example is the continued industry marketing of obesity-building foodstuffs in children's television times, in the face of widespread protests by nutritional and health experts, and the public. Community experience has been given little or no credibility in the social marketing campaigns that have become a regular tool of government. The expensive Australian Government websites advertising food and petrol prices proved to be of little local benefit in practice.

Specialists who provide the major amount of advice to government and to industry offer competing recommendations, which can then be safely ignored in governance decisions. This has been the case with the choice of responses to global warming. Organizations, both government and industry, have developed a style of management based on meeting internal objectives, with little consideration of the public good. Infamous examples have been continuation of asbestos in building materials and lead in petrol long after the health risks were known. The use of intuition and imagination on how the world could be has been increasingly discouraged in favour of objective-based planning. The knowledge contributions needed for working collaboratively to meet democratic ideals can fairly be compared with the people from the Tower of Babel. The knowledge bases of the individuals, communities, specialists, organizations and holistic thinkers needed to effect whole-of-community well-being are widely scattered and losing their capacity to understand one another:

> "And the LORD came down to see the city and the tower, which the sons of men had built. And the LORD said, 'Behold, they are one people, and they have all one language; and this is only the beginning of what they will do; and nothing that they propose to do will now be impossible for them.
>
> Come, let us go down, and there confuse their language, that they may not understand one another's speech' So the LORD scattered them abroad from there over the face of all the earth, and they left off building the city". (Genesis 11: 1–9)

This chapter argues that, like the people of Babel, the great democratic inspiration of the 18th century writers has become dispersed, captive to the fragmentation of the knowledge base on which it was built. Decisions have become responses to special interest pleading, rather than to considerations of individual equity and community well-being. Results of this cacophony of voices are the log-jams in conflicting responses to issues such as responses to climate change and the uses of nuclear energy (Diesendorf and Hamilton 1997).

The separation of essentially interconnected concerns is particularly notable in matters of health and of environment. In all Western countries these are administered from different departments. In most, the departments of Health and of Environment are internally divided in turn, into agencies of health and of environment, plus finance, social services, planning and more. The fragmentation becomes absurd. Academic departments follow a similar trail, and are even further sub-divided into the fiercely defended specializations of the social and biophysical sciences. Decisions based on the fragmented organizations and specializations are marked by the absence of the receivers of their services, as individuals and as communities.

Further, each form of contribution to decisions is scorned by the others. Individual knowledge is rejected as biased, community knowledge as anecdotal, specialised knowledge as jargon, organizational knowledge as self-interested and holistic knowledge as airy-fairy and impractical. The grain of truth in each accusation serves to builds the defenses even higher. The surprise is that any coherent democratic decisions are made at all.

In summing up the conditions affecting democratic decision-making on matters of environment and on health, fragmentation and compartmentalization of knowledge rank high. Decision-makers from all categories of decision-making tend to believe that they are (and the only people who are) well equipped to define problems and engineer solutions. Allowing other interests to participate in such processes is considered by each interest group to lead only to compromise, delay and incompetence. It is not surprising that the inhabitants of Babel left off building, not only the tower, but the city itself, once they could no longer understand one another.

Changing the Agenda

In pursuit of a just and sustainable future we cannot afford to leave people scattered and unable to communicate. The Babel tower can be re-built to suit the times. It is possible to argue that the re-building is already in train. Another agenda is emergent: in all the key decision-making arenas. A compensatory emphasis on establishing connections rather than the maintaining divisions is beginning to become apparent. Transparent summits, open forums, unaligned think tanks, brainstorming round tables, centres for dialogue and the like have been gradually taking the place of some of the bilateral agreements and back-room bargaining. Previous modes of divided decision-making have not vanished; many forms of collective decision-making have gained in strength.

In Australia at least, almost no major initiative in health and in environment is without a community consultation requirement (Commission for Preventative Health, 2008; Caring for Country, 2009). How this is honoured is another matter, but it is always there, to an extent unknown in twentieth century decision-making.

The world of information technology has driven the most spectacular change. During the twentieth century, the development of search engines that could retrieve information from anywhere at a speed seemed almost miraculous. In 1992 a new search engine named Google appeared, with searches based on relationships between stored data that transformed our social and professional lives. Not only could information be retrieved; it could be linked to other information previously stored in separate files. New knowledge was generated from the newly established relationships. The success of the global response to SARS (Severe Acute Respiratory Syndrome) was dependant on real time linkages among and within diverse groups of specialists, agencies, and communities. A decade later, the idea of building an information system on connections rather than divisions has generated the social networking sites of Facebook, Twitter and Myspace.

In "hard" science, the shift from concentrating on parts to exploring wholes has been just as great as the shift in information technology. Research into limiting factors and baseline differences has opened sufficiently to include the field of chaos theory, in which everything is connected to everything else. The seminal paper on how the beat of a butterfly's wing could cause a hurricane in Texas marked a departure point (Lorenz 1963). Chaos theory came into mainstream consideration with its description of fractals, resilient systems and strange attractors (Briggs and Peat 1998). Bohr's concept of complementarity came into its own, namely that the one system could have different and valid explanations, as in his example of the particle and wave theories of light (Bohr 1955). It became possible to talk about parallel realities without resort to metaphysics.

Key advances contributing to decisions on health and on environment, such as mapping the human genome and predictions of climate change, were possible only through realizing the potential of cooperation among international teams of administrators, specializations, and talented individuals, supported by government funds and community support. The control of the threatened pandemic of AIDS was avoided in Western countries by the collaboration between the at-risk community and professional services (Brass and Gold 1985). Recovery of over-exploited soils is being made possible by an alliance between increasing public priorities for organic food, farmers interested in sustainable agriculture, a shift in the orientation of expert advice from agronomists, and government development funding (Mollison 1997).

In making lasting and effective decisions on interactions between health and environment, where do we now go for guidance? Two quite separate developments can help. Each requires the participants to accept a collaborative approach to complex issues. Each has proved well-founded and effective over time. The concepts of wicked problems and of collective learning form the basis of a new agenda, going beyond the Enlightenment search for objectivity and certainty and the one right answer to accepting contributions of creative thinking and uncertainty.

There has been a considerable growth in interest in Rittel and Webbers' formulation of "wicked problems", complex issues that resist current methods of problems solving, and cannot be resolved without changes in the society that produced them (Rittel and Webber 1973). Examples of wicked problems of concern locally and globally are energy use; urban violence and drug dependence. Rittel and Webber describe wicked problems as too complex to be precisely defined; having no permanent solutions since each solution brings further change; not right or wrong, true or false, since they combine fact and value; and not sitting conveniently in any one discipline or department, making subsequent action difficult (Rittel and Webber 1973). Responses to wicked problems have included the development of a framework for open, critical and imaginative inquiry (Brown, Harris and Russell, 2010).

Open and imaginative inquiry requires acceptance that there are diverse ways of viewing the world and multiple forms of construction of knowledge. Applied to wicked problems involving health and environment, Table 14-1 demonstrates the potential for confusion and argument. However, once one accepts that here are parallel realities, there is potential for combining rich sources of ideas and actions.

Linking the Knowledges

In Table 14-1 the participants in sustainability decision-making take up different positions on the same topic, depending on their field of interest in the topic, and their role at the time. The positions they take are not necessarily compatible with one another. Nor are they necessarily predictable. The choices of an individual may differ from those of their community of place or their community of interest. They may differ from or they may follow the direction indicated by their professional training. The individual's place in an organization will influence to what extent they adopt the values of the organization or give priority to their personal or professional decisions. Taken as isolated decisions, the possibilities are endless.

Following Table 14-1, first row, an individual may easily find they are on a collision course between their personal well-being, "my vitality" (e.g. using fossil fuels for recreational purposes); their concern for the environment, "my place" (reduction of fossil fuel emissions); and their democratic rights as a citizen, "my vote" (they may not agree with the position of either political party on controlling emissions).

The second row of Table 14-1 draws attention to the tension between current threats to a community's well-being, their past experience of managing their region, and the democratic goals of autonomy and equity. The third row describes how specialists take part in debates on whether the solution lies in more human action, or in leaving environmental systems to their pre-human state. Until very recently, legal systems have not included environmental justice in their deliberations so there has been no referee in that direction (Westra, 2008).

Table 14-1. Five sources of knowledge on health, environment and democracy

Source of knowledge	Health	Environment	Democracy
Individual	My vitality	My place	My vote
Community	Our communal well-being	Our shared history	Equity and Autonomy
Specialized	Physiology, Psychology, Immunology, genetics, Epidemiology, etc	Ecological systems, Biological sciences, Environmental management, etc.	Law, statistics, voting patterns
Organizational	Health services, hospitals, community health centres	Environmental Protection Agencies, resource industries	Parliaments, Electoral Offices, Administrations
Holistic	Life	Landscape	Rules for living

In the fourth row, organizational knowledge, organizations from health and from environment compete for power and resources, through lobbying democratic structures which are in their turn based on a legal system of opposing arguments. Thus the divisions are reinforced and opportunities for collective agreement reduced. The last row demonstrates a final reason why the three crucial components of life on the planet become matters for debate, not collective action. The core aims of life, landscape and rules for living are approached as conflicting rather than complementary.

Re-visiting Table 14-1 as a whole reveals that the democratic process alone cannot resolve the tensions between the evidence drawn from health and from environment. This is of serious concern since the changes in global environmental systems are threatening human life support systems. Human health and the health of the global environment are intricately connected.

There is another way to approach these dilemmas, other than painstakingly collating responses to the myriad factors listed in Table 14-1. Taking a collective approach in line with an overall social shift in that direction, the task moves from preoccupation with the parts to exploring the system that creates a whole greater than the sum of those parts. First, to establish the parts.

Evidence from a six-year study of health and environment decision-making supported by the arguments of philosophers of knowledge finds that each of the contributing interests to whole-of-community change works from a different knowledge base (Brown 2008). Each knowledge base has its own key terms, source of evidence and tests for truth. Once these are clarified for individual, community, specialist, organizational and holistic contributions the connections between them can be established, and the sum of their contributions can be evaluated.

Individual knowledge

Each individual mind is the primary site of the construction of knowledge, albeit mediated by the society in which it is developed (Berger and Luckmann 1971). Polanyi identifies the difference between individuals' explicit and tacit knowledges, that is, what you know you know, and what you continually draw on without knowing that you know Polanyi (Polanyi 1958). Added to this is the important difference between “knowing that you don't know”, and “not knowing that you don't know”. In classical science, these finer points are excluded from being examined in an inquiry. Only the individuals' externally validated observations are considered to contribute to valid knowledge (Chalmers 1994). Yet an individual’s reflection on their experiences is crucial to any understanding of the dynamic connections between environment and health.

Community knowledge

A community's knowledge is developed within the community, and can only validly be accessed from within that community. While there can be interest-based, and place-based communities, the concern here is with the latter, in an environmental context. In any inquiry, local communities can be considered as collection points for and interpreters of a diverse body of local knowledge, the knowledge being legitimated through shared events and long-term shared memories. Anthropologist Clifford Geertz (Geertz 1983) describes the knowledge of a local community as gained through "citizens not just using their eyes and ears, but using them collectively, judiciously and reflectively to understand their own locality".

It is important to note that the review of community knowledge does not subsume indigenous knowledge. Indigenous knowledge (traditional knowledge) and community knowledge are often regarded as equivalent. This can be misleading. In most traditions, first people's knowledge is created by a synthesis of identity, community, landscape, social governance through traditional stories, fusing the five separate constructions of knowledge in Western thought.

Specialised knowledge

Each specialization forms a distinct community of practice, with their own research models and Kuhnian paradigm (Kuhn 1970). This is equally true of practitioners of medicine, law, ecology, or engineering, each of whom exhibit all the characteristics of a community of practice (Wenger 1998). The rigour and validity of a specialised inquiry rests on well-defined questions, critical doubt, empirical observations, and the capacity to generalize the findings (Popper 1975; Chalmers 1994). The result will often be delivered in a specialised language that increases accuracy but excludes the contributions made by other forms of knowledge.

Organizational knowledge

Government, industry and the major non-government agencies have increasingly adopted a managerial approach. All forms of organization now tend to function under a similar framework of strategic decision-making: planning, designing, applying and reviewing. The language used to describe results, namely, cost/benefits, objectives, timelines, inputs and outcomes, is limited to matters of concern to the organizational knowledge culture (Giddens 1990; Ralston Saul 1992). The validity of organizational knowledge is tested against its consistency with the organization's objectives, in a circular self-justifying process. While this approach increases the efficiency of the organization it leaves little room for incorporation of broader goals. Many organizations would describe themselves as working for public good goals such as social well-being and socio-environmental sustainability, but these goals are not often stated and even less often evaluated.

Holistic knowledge

Holistic knowledge is achieved through an understanding of the whole. However, a holistic understanding is pursued in two contrasting ways. One school of thought seeks to document the parts of the whole as units in a hard (technical) or a soft (social) system (Forrester 1969; Churchman 1979). The findings of such an inquiry are usually represented as a grid, a hierarchy of detailed lists, or a flow chart. The second interpretation of holistic, as seeking the essence of the matter is the one applied in this chapter. The validity of this construction of knowledge is established by the acceptance of its validity by all the contributors: a shared "*aha!*" moment. .Picasso's Guernica, the fall of the Berlin Wall and the finding of the gap in the world's ozone layer all brought a shared understanding across a wide range of people.

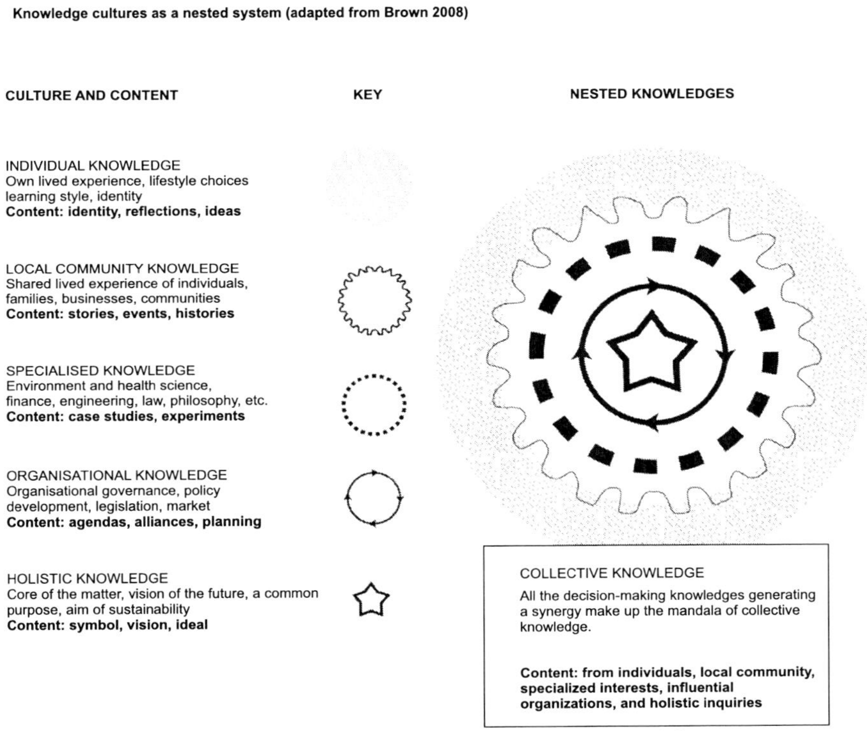

Figure 14-1. A collective knowledge mandala

With a deeper understanding of the origins of the knowledges that contribute to health and environment decision-making, it is possible to reconsider the relationships between them. We have already discussed how, in the normal conduct of decision-making, the different knowledge bases contribute separately, often as conflicts of interest (Table 14-1). The goal is to change that relationship to one of collective learning in which each contribution is valued both for itself, and for their collaborative whole.

In standard practice, in decision-making on sensitive issues the knowledges form a hierarchy. Specialised and organizational knowledge vie for supremacy. Community and individual knowledge are recognized in principle, but are seldom consulted. Holistic knowledge is by-passed almost entirely. It is possible to change that relationship to joining in the movement to collective thinking outlined in the section above. One such relationship is a nested set, in which each knowledge builds on the other

(Figure 14-1). Another relationship is a network, each with their own position, and at the same time forming a collaborative system (Figure 14-2).

Once the knowledge contributions have established themselves as a collective, rather than as isolates, then another question arises, can they maintain this relationship throughout the decision-making process?

A framework for collaboration between the knowledges on complex issues has built on the foundations of Kolb's Experiential Learning Cycle (Kolb 1984). Kolb and his colleagues conducted decades of inquiries, resulting in their formulation of four stages in adult human learning: developing principles, describing the facts; designing to make use of potential, and taking action. Kolb's work confirmed that long-term learning did not take place unless all four stages had been completed, and the learning was put into practice.

The decision-making process on the complex issue of human health and environmental futures can be approached as collective learning. Accepting the Rittel and Webber (1973) view of wicked problems as open ended, uncertain and with no final solution, any decision-making on such problems becomes an on-going critical inquiry. To turn this into collective inquiry will involve the different interests coming together at each stage of the learning cycle, and the cycle becoming a spiral of cumulative experience and learning (Figure 14-2).

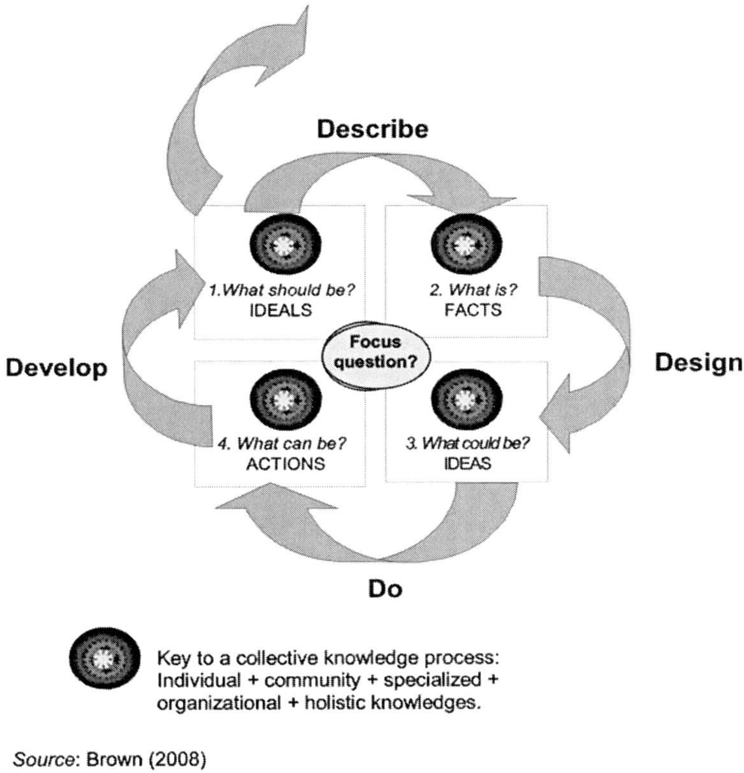

Figure 14-2. Collective learning spiral
Source: Brown (2008) after Kolb (1984)

An example of how this process has been applied in practice comes from a project undertaken in Canberra in 2007. A collective social learning spiral was initiated in a whole-of-community change project, a Sustainability and Health Action Strategy for the city of Canberra in the Australian Capital Territory. The strategy adopted a "snowball" process: key individuals from each of the interest groups of community, specialists and government took part in a collective learning workshop. The participants learnt from one another as they followed each of Kolb's four stages of adult learning, in the form of four questions: "what should be?" (sharing their ideals); "what is?" (giving their different versions of the facts), "what could be?" (contributing their ideas); and "what can be?" (accepting collaborative follow-up actions).The aim was not to achieve consensus, nor was it to set priorities. The aim was to arrive at a synergy from the combination of responses at each stage. The design of the process was for this to be a pilot, and to spread the thinking outwards to other projects across the city (Brown 1997).

The first event was an invited workshop to determine *what should be*, the shared aims that would drive their projects. Their collective vision was that ACT would be "an active, collaborative, networking, more peaceable community, using its wisdom and knowledge capital to build a just, sustainable and healthy city and society". These goals reflected aspects of the city as the country's administrative and political core with a strong university presence. Different participants espoused different parts of the vision, but all could find their place in the collective vision.

The next step was to decide *what is* that is, what are the parameters of the project. Canberra is the national political capital, where parliament meets and lobby groups and foreign embassies have established themselves. It is also a university town, with two universities, including the well-endowed national university. With a population of 350,000 in a natural bush setting, Canberra identifies itself as "a bush capital". Standards for health and conservation of the natural environment are higher than in comparable developed-country cities. On the other hand, the existence of poverty, especially among the small indigenous population, is all the more marked and unnecessary in a well-to-do community. The urban sprawl was spreading into the bushland setting, much as in other Western cities. "What is" was constructed by each of the knowledge groups contributing their version of these "facts". As demonstrated in Table 14-1, these contributions can be expected to be very different from one another.

"What could be" was part vision and part criteria for the different projects to meet. In an open, brainstorming session, workshop members agreed that a collective environment and health project could:

- be feasible if it is practical, financial, and has the necessary resources to ensure it is could be completed;
- support ecological sustainability and advance community well-being;
- make a real impact on the mainstream community by projects stretching thinking and going beyond today's norms;
- rest on collaboration and engagement of the community, especially those who are disadvantaged under present conditions;

- build individual and social capacity through action learning, problem solving and behavioural change; and
- enable a connectedness among all forms of knowledge and interests, including relationships with all species

At a follow-up workshop to design "*what can be*" that met the criteria, 50 people were ready to join in the program. The snowball effect and the creative synergy were beginning to gather strength. Twelve projects were developed as open-ended possibilities in a design workshop, and later fine-tuned to fit the conditions the group had identified under "*what is*". One project was Community Gardens. Here a neighbourhood adopted a disused recreation area, and developed it into fertile ground using Permaculture principles. The focus was on community conviviality, nutritious food and no waste. Another project was an indigenous design for an indigenous Wellness Centre that met cultural as well as physical health needs. A third project was named the Art of Moving, a publicity campaign for the people of Canberra to combining healthy exercise with reducing high-energy using forms of transport. This project used drama, sculpture, artwork and dance in buses, schools, public streets and community health services around the city.

One year later, a workshop brought the projects together to review what had been able to be put into practice: *what can be*. Of the 12 projects, 10 had proceeded to action. Seven of these had received significant funding and several had spun off still further projects. A social learning outcome from the projects was the prototype of collective thinking and action, combining the diversity, skills and learning of the five knowledge cultures.

Experience has shown that there are many pathways by which the conditions for collective thinking the project developed under "what could be" can be met. Where they are put into effect, there is a marked change from present practice to individual, community, specialised and organizational interests including each other in whole-of-community change initiatives.

It is worth revisiting the lesson carried by the biblical story of Babel:

> "Behold, they are one people, and they have all one language; and this is only the beginning of what they will do; and nothing that they propose to do will now be impossible for them"

Re-building Babel does not mean going backwards to one language and one belief system. It means moving forward to a world which accepts that there are many knowledge cultures, each constructing a different reality, and all capable of understanding, learning from and collaborating with the others. The question remains, is there time to establish collective decision-making so as to re-build ecological integrity that provides the life-support systems which ensure human health?

Acknowledgments

I would like to thank my colleagues in the Local Sustainability Project at the Australian National University which provides the empirical basis for this chapter; the Global Ecological Integrity Group for the discussions that led to the conclusions, and Erica Fisher and John Harris who read and critiqued the draft.

References

Berger, P. and T. Luckmann (1971). *The Social Construction of Reality: A Treatise in the Sociology of Knowledge*. Harmondsworth, Penguin.

Berry, T. (1999). *The Great Work: Our Way Into the Future*. New York, Bell Tower.

Bohr, N. (1955). *The Unity of Knowledge*. New York, Doubleday and Co.

Brass, A. and J. Gold (1985). *Aids and Australia: What everyone should know*. Kensington, Bay Books

Briggs, J. and D. Peat (1998). *Seven life lessons of chaos: spiritual wisdom from the science of change*. New York, Harper Collins Publishers.

Brown, V. (2008). *Leonardo's Vision: A guide to collective thinking and action*. Rotterdamm, SENSE.

Brown, V. A. (1997). *Managing for Local Sustainability: Policy, Problem-Solving, Practice and Place*. Canberra, National Office of Local Government.

— (2005). Knowing: Linking the knowledge cultures of sustainability and health.in *Sustainability and Health: supporting ecological integrity in public health*. V. A. Brown, J. Grootjans, J. Ritchie, M. Townsend and G. Verrinder. Sydney, London, Allen and Unwin, Earthscan: 131–152.

Burke, E. (1999). *A philosophical enquiry into the sublime and beautiful and other pre-revolutionary writings*. UK, Penguin.

Chalmers, A. F. (1994). *What is this thing called science? An assessment of the nature and status of science and its methods*. St Lucia, University of Queensland Press.

Churchman, C. W. (1979). *The systems approach and its enemies*. New York, Basic Books.

Diesendorf, M. and C. Hamilton, Eds. (1997). *Human Ecology, Human Economy: Ideas for an ecologically sustainable future,*. Sydney, Allen and Unwin.

Falk, R. (1999). *Predatory globalization: a critique*. Cambridge, Polity Press.

Forrester, J. (1969). *Urban dynamics*. Cambridge, Massachusetts, MIT Press.

Funtowicz, S. O. and J. Ravetz (1993). "Science for the post-normal age." *Futures* **25**(7): 739–755.

Geertz, G. (1983). *Local Knowledge. Further Essays in Interprative Anthropology*. New York, Basic Books.

Giddens, A. (1990). *The Consequences of Modernity*. Cambridge, Polity Press.

Habitat (United Nations Centre for Human Settlements) (1997). *Environmental Planning and Management Source Book: Implementing the Urban Environment Agenda*. Nairobi, UNEP.

IPCC (2007). Summary for Policymakers. Geneva, Intergovernmental Panel on Climate Change Working Group 1.

Keen, M., B. V. A., et al. (2005). *Social Learning in Environmental Management*. London, Earthscan.

Kolb, D. A. (1984). *Experiential Learning: Experience as the Source of Learning and Development*. New Jersey, Prentice-Hall.

Lovelock, J. (1979). *Gaia: A New Look at Life on Earth*. Oxford, Oxford University Press.

Mollison, B (1997) Introduction to Permaculture, Sisters Creek, Nrth Queensland, Tagari.

Paine, T. (1776) "Common Sense. Of the origin and design of government in general."

Plato (1962). *On Socrates' The Republic*. Cambridge, Mass, Harvard University Press.

—. (1991). The republic of Plato. A. Bloom. NewYork, Basic Books.

Polanyi, M. (1958). *Personal Knowledge*. Chicago, The University of Chicago Press.

Popper, K. (1975). *The Logic of Scientific Discovery*. New York, Bantam.

Ralston Saul, J. (1992). *Volataire's bastards: the dictatorship of reason in the West*. Toronto, Viking.

Rittel, H. and M. Webber (1973). "Dilemmas in a General Theory of Planning." *Policy Sciences* **4**: 155–169.

Voltaire, F. (1998). Candide or optimism. *Candide and other stories*. London, Oxford University Press.

Wenger, E. (1998). *Communities of Practice: Learning, Meaning, and Identity*. Cambridge, Cambridge University Press.

Westra, L., (2008). *Environmental Justice and the Rights of Indigenous Peoples: International and Domestic Legal Perspectives* London, Earthscan

World Bank (1998–1999). *World Development Report 1998/99: Knowledge for Development*. Washington, World Bank.

World Social Forum (2004). *Forum Statement* World Social Forum, Johannesburg.

PART IV –

WATER, FOOD AND CLIMATE CHANGE IN INTERNATIONAL LAW

INTRODUCTION

KLAUS BOSSELMANN

Part IV shows in more detail how ineffective recent environmental governance has been. When even the basics of human existence – the water we drink, the food we eat and the climate that sustains us – are seriously at risk, something fundamental must have gone wrong. Despite their different subjects, all chapters have their origins in the observation of profoundly misconceived concepts in current legal regimes. And each chapter touches upon ethical foundations that need changing, not merely regulatory instruments at national and international levels.

Continuing the theme of the previous parts, this part is concerned with applying ecological integrity thinking to law and governance, in this case food, water and climate change. Rather than treating food, water or greenhouse gases as pure commodities that can be subjected to any kind of legal regime or management scheme, we need to think of ecological interconnections first. What we eat and how it is being produced, has implications for the wider food chain and the integrity of related ecosystems such as agricultural environments. Water is an integral part the planets ecological system and inextricably linked, for example, with the global climate system. And climate change, in turn, affects agricultural and water supply systems in the most dramatic way.

How then to regulate and manage these interconnections more effectively is the overarching concern of the following chapters. One dilemma of current environmental regulations has been that they are either too general (hence not sufficiently prescriptive) or too specific (hence not sufficiently comprehensive). Typically, environmental legislation is designed around specific subjects; for example, food security. This is in itself problematic as no single concern can be dealt with irrespective of wider social and ecological implications. But more alarmingly, most subject-focused laws follow an economic rationality that captures issues of access, cost efficiency, demand and supply, but little more. The integrity of the associated ecosystems is not included, in many instances not even a general sustainability benchmark.

To overcome the traditional piecemeal approach, we must, first of all, realize the ecological functions of soils, plants, animals ("food"), water or climate. The fact that they perform functions for human consumption is secondary and, while important, not the sole concern. The chapters on food (Gottwald), animals (Cassuto), water (McIntyre, Karageorgou) and climate (Karageorgou, Brown) show how much of pure commodities these subjects have become; their ecological functions hardly feature in existing laws.

The second requirement is, therefore, to draw ethical conclusions from given ecological realities. It is logical, therefore, that the authors raise fundamental ethical issues, sometimes implicitly (e.g. McIntyre), sometimes expressively (e.g. Gottwald, Brown). Only then comes good regulatory technique and coverage, the third requirement.

The message throughout this part, and in fact the entire book, is that any normative framework for sustainability governance must include ethical and legal dimensions. Without considerate ethical principles, legislation only reinforces business as usual. Without law, ethical principles remain toothless.

Franz-Theo Gottwald's analysis of genetically modified (GM) food includes the entire life cycle of GM crops (in humans, plants, animals and the environment). They are all affected and all need to be considered for regulatory regimes. Ethical problems, therefore, arise in an ecological context as much as in the areas of production and consumption. Building upon, and critically evaluating, the ethical assessment model of Busch et al., Gottwald shows how ethical assessment can be built into law-making procedures. Among the most important prerequisites are the independence of ethical assessment bodies and the rigorous application of the precautionary principle.

David Cassuto examines another sector of industrial food production, i.e. animals, to show how current legislation has been modeled around efficiency rather than ethics. Unlike most studies on animal rights and laws, Cassuto is not concerned with animal welfare, but with the broader question of how humans interact with animals. This allows for more focus on the wider ethical issues involved and how the law has responded to changing socio-economic conditions. The transition from agriculture to modern agribusiness has resulted into efficiency as the all dominating benchmark, thereby reducing complexities of human–animal relationships to a mere concern for animal welfare. Ironically, the efficiency approach has led to ever increasing inefficiencies that can only be ignored if social and ecological costs continue to be externalized. True efficiency would require to overcome the perceived dichotomy between efficiency and ethics and to unravel the process that has made animals into almost complete commodities.

The following two chapters center around water management. Owen McIntyre questions an often assumed theorem, namely that a human right to water and private investment in water are opposed to each other. Private sector involvement in water management has steadily increased, although McIntyre notes that it may have peaked now. The international debate in support of a human right to water, on the other hand, has not (yet?) led to its formal recognition. It is timely, therefore, to more closely follow the development of international investment law to see how corporate responsibilities can be improved. According to McIntyre, human rights considerations are increasingly likely to be raised before investment dispute tribunals. This way, failures of states to consider human rights with respect to water management can be argued as violations of international law.

Like McIntyre, Vicky Karageorgou looks for ways to break through (assumed) dichotomies, in her case, the separation of water management from climate change measures. While climate change adaption strategies have been developed for some time, water management regimes are largely devoid of such concerns. As climate change significantly impacts on water cycles and availability, water law needs to incorporate adaptation strategies to match. Karageorgou discusses legal developments at international and regional (European) and national levels to call for a "paradigm shift" in water law and environmental law making. This would involve ongoing assessments of regulatory impacts and the shift to the "non-equilibrium paradigm" and "adaptive management", both consistent with the concept to ecological integrity. Such a shift would have to include national and transboundary regimes alike.

Don Brown, finally, examines the ethical issues raised by the much debated emission trading regimes. The standard assumption that these regime are both cost effective and carbon reducing may or may not be true, the problem with the wide acceptance of cap and trade regimes is, however, the lack of "ethical spotting". Brown aims for identifying ethical issues, not necessarily solving them. The ethical blind spots include the justification of caps, the treatment of the climate – a global commons – as property rights, the environmental effectiveness, and issues of distributive and procedural justice. The complexities of cap and trade regimes are far greater than their presentation in the public arena suggests. Essentially, we need to debate the efficiency of climate change measures and not their economic feasibility. The former includes ethical considerations, the latter excludes them.

CHAPTER FIFTEEN

GENETICALLY MODIFIED FOOD: ETHICAL IMPLICATIONS ALONG THE FOOD CHAIN

FRANZ-THEO GOTTWALD

Genetically Modified Food: Actual Benefits, Empty Promises and Incalculable Risks

By manipulating the genetic make-up of plants and animals, gene material from one organism is artificially inserted into another species. This is supposed to give genetically modified (GM) organisms new capabilities.

There are many promises, the gentech-industry made during the last decades. However, none of the supposed pioneering scientific findings really turned out to be transferable to reality. First of all, the industry promised to produce genetically modified (GM) seeds which would result in better crop quality, higher resistance against certain pests and diseases and higher yields. Hence, the farmers would benefit from GM seeds, requiring fewer pesticides, herbicides and fertilizers, resulting in better economic income. As a positive side-effect, the environment would benefit from less use of harmful chemical substances in agricultural production.

Reality has shown that the opposite has come true. After decades of research there are no GM food crops that live up to all this hype. In the United States, farmers have been sueing producers of GM seeds for poor harvests for years (Bové and Dufour 2001, p 145). Newspapers and NGOs consistently report about Indian farmers who have committed suicide; the poor quality of GM cotton and new, so far unknown pests, in combination with debts due to overpriced seeds take away their means of existence. Farmers who were supposed to reap the benefits of GM technology instead run the risk of facing financial ruin.

GM plants have shown to be very susceptible, and prone to various other diseases and pests. One example for the misleading promises of the industry is MON810 maize, patented by the controversial American *Monsanto* company. This maize is genetically engineered to produce a modified insecticide (Cry1Ab) that naturally occurs in the soil bacterium *Bacillus thuringiensis* (Bt). The production of this toxin is supposed to protect maize plants from European corn borer larvae (ECB, Ostrinia nubilalis). However, the maize made headlines due to poor performances, concealed facts about potential health risks and doubtful usefulness.

That GM pollen should be a source of concern is confirmed by several studies. One study, carried out by Professor John Losey of Cornell University examined the effects of GM maize pollen on the larvae of Monarch butterflies. These were fed on milkweed leaves with either GM pollen, in which case 44 percent died within four days, or untransformed (non-GM) Bt-maize pollen, in which case none of the larvae died. The experiment demonstrates the toxicity of this GM product to at least one animal species (Losey et al. 1999). Another biological system that suffers from GM crops are bees. French researchers have discovered that some varieties of the transgenic oilseed rape *canola* can harm bees by destroying their natural ability to recognize flower smells. This development is alarming for agriculture and food industry worldwide, since bees are the most effective pollinators (Nikiforuk 1997). The damage to the world economy, caused by declining pollinating insects, might amount to 190 to 310 billion Euros per annum (Gallai et al. 2008).

In contrast to statements of the industry the use of herbicides increases rapidly when GM seeds are sown (Friends of the Earth 2008). The causes for this development are obvious: growing of plants that are engineered to withstand herbicides leads to increased spraying with these chemicals as the GM crops will not be damaged by them. Hence, it is not only the technology itself which should cause concern – its implications such as excessive use of total herbicides also must be considered. One example is the total herbicide Roundup® by *Monsanto.* Although Roundup®, particularly the active ingredient glyphosate, have been declared not carcinogenic by several authorities, there exists axiomatic scientific evidence, that the ingredients of the herbicide not only could damage human health (Walsh et al. 2000; Richard et al. 2005), but also can be hazardous for a variety of organisms and the eco-system itself (e.g. Giesy et al. 2000; Daruich et al. 2001; Relaya 2005). A recent French study proved that Roundup® might seriously damage human cells. In the test series, Roundup® formulations caused total cell death of human cells within 24 hours (Benachour and Séralini 2008). In addition to health risks and harm to the environment, many farmers increasingly get problems with resistant weeds. In 2006, for instance, Palmer Amaranth was responsible for total losses of Bt-cotton in North Carolina and Georgia (Schattenblick Infopool 2006).

Another critical point with regard to GM plants is the myth of co-existence. Seeds of GM crops have been proven to spread far, resulting in uncontrollable risks for the environment and contamination scandals (in which non-GM crops become polluted with GM material). A study of the Exeter School of Bioscience showed a huge variation in the amount of cross-pollination between GM and non-GM crops of maize, oilseed rape, rice and sugar beet. Levels vary according to whether the GM field is upwind or downwind of the non-GM field given the direction of prevailing wind over the flowering period of the crop (Hoyle and Cresswell 2007). The consequence of uncontrolled cross-pollination is not only an unpredictable risk for the environment, but also the loss of freedom to chose between GM and non-GM products – for farmers and consumers. Both must have the opportunity to decide for or against genetically modified food. At the moment, this freedom of choice cannot be guaranteed.

The telling argument brought forward by the industry is the potential contribution of GM technology to fight world hunger. In fact, the opposite is true. Since GM crops have been shown to be equally or less productive than conventional plants, the technology is not suitable for producing higher quantities. Second, GM seed is expensive, more precisely: it costs about three times more than traditionally bred seed. Poor farmers, smallholders and subsistence farmers, who are particularly threatened by hunger and malnutrition, can hardly afford GM seeds. But the most important reason, why GM crops are not the answer in combating world hunger is: hunger and malnutrition are not caused by general food shortages. It is caused by a variety of factors, i.e. an unfair market system, corrupt governments, lacking infrastructure and – most important for this discussion – the unacceptable lifestyle and food habits of the industrialised nations.

In order to satisfy the enormous hunger for meat in the industrial and, increasingly, emerging countries, gigantic quantities of valuable crops are fed to farm animals. Meat consumption and its concomitant animal feed production has become a worldwide problem, especially in juxtaposition to the problem of world hunger. Worldwide, 923 million people go hungry (Welthungerhilfe 2008). And yet, world populations theoretically should be superfed. But despite a total worldwide grain harvest of about 1,7 billion tons, hunger reigns supreme as almost half of global grain yields end up being fed to farm animals.

GM Animal Feed

Cows, pigs and hens in Europe are already largely dependent on GM feeds. Their feed usually contains ingredients made from GM plants or additives and enzymes that are produced with genetically modified microorganisms. Due to the high demand of animal feed, large quantities of feed must be imported. Soy meal is the single most important animal feed in Europe, accounting for more than half of protein-rich animal feed. Other important feeds imported include maize, rapeseed, cottonseed, oats, rye and wheat. Many of these imported feed stuffs are genetically modified. Table 15-1 below summarizes the proportions of GM varieties of the total harvest for four crops.

Table 15-1. Quantities (ha) of GM soya, maize, cotton and oilseed rape grown worldwide in 2007 as a proportion of the total harvests

Global cultivation	Total	GM varieties
Soya bean	86 million	58.6 million (68%)
Maize	140 million	35.2 million (25%)
Cotton	34 million	15 million (44%)
Oilseed rape	23 million	5.5 million (24%)
Total for the above 4 crops	283 million	114.3 million (40%)

Source: GMO Compass (2007)

EU countries import large quantities of GM feed from Argentina, Brazil and the United States. For instance, 40 million tons of soy are imported annually into the European Union. About 90 percent of these feed stuffs contain considerable amounts of genetically modified ingredients. In contrast to soy beans, the demand for maize mostly can be met by EU production.

Besides GM plants, animal feed also contains GM ingredients such as vitamin B2 and B12, biotin, amino acids, enzymes, beta carotene and colouring.

GM feed does not even offer an advantage to the feed industry, since GM maize and soy do not improve yields or quality, and neither are cheaper than conventional crops (Friends of the Earth International 2008). However, GM feed for animal production raises many ethical questions. These concern potential health risks for animals and humans, insufficient regulations on labelling, questions about fairness and food justice in the developing countries which provide the feed, as well as general questions about meat consumption and its concomitant social and environmental effects.

There are some strong indications that GM feed might be harmful to the health of animals and it cannot be excluded that human health might also be adversely affected by the consumption of GM-fed animals or their products. This is confirmed by many interviews with farmers who describe animals refusing to eat GM feed or, if forced to do so for lack of an alternative, losing weight, having fertility problems or otherwise failing to thrive (Soil Association 2002).

Although these concerns exist, consumers are not able to chose between GM animal products and non-GM animal products. GM feed must be labelled, but as mentioned before, the end products like milk, eggs, and meat do not require labelling. Backtracking is only possible via analysis of feed ingredients. These deficits in consumer protection and freedom of choice are ethically problematic, particularly when it comes to the fundamental right to food.

Meanwhile, most of animal feed in the US is genetically modified. Since BSE, there is an enormous need for protein-rich feed in order to fatten animals. Hence, the cultivation of GM soy beans has increased dramatically. Worldwide, the acreage for GM soy beans has risen from five million hectares in 1997 to almost 60 million hectares in 2007 (Transgen 2008a). In the US and Argentina more than 90 percent of soy is genetically modified. Market leaders in cultivation of GM soy are the United States, Argentina and Brazil. Interestingly, Brazil is a leading exporter of agricultural goods, while an estimated 50 percent of the population suffers deficiency symptoms due to poor nutrition. This raises the question, whether it is ethically justifiable to import huge amounts of valuable crops from developing or threshold countries for animal feed, while in these countries people suffer from hunger and malnutrition.

The example of animal feed shows, that ethical problems not only exist when talking about GM feed and food, but also call for a holistic and binding ethical evaluation model in order to trade off potential risks against potential benefits of this technology, and determine its ethical implications and limitations.

Consumers and Farmers: How to Handle Uncertainty

A broad majority of consumers in Europe does not want genetically modified food. Supporters of genetic engineering claim that the rejection of GM food is based on irrational fears (Potrykus 2008) and a general scepticism against complex scientific inventions. However, surveys show that other fields of application, for instance in medicine, are far more accepted by society. Consumers seem to feel deep, intuitional discomfort when it comes to genetically modified food. In ethical debates, the role of intuition and "gut instinct" is controversial. A certain issue cannot be condemned due to a vague and spontaneous reaction of dislike. Nevertheless, these feelings must be taken seriously, because they are, among other reasons, the result of a failed participatory and informational policy. The whole debate on GM food takes place without adequate participation of the population. While the industry has enormous influence on politics and legislative processes (e.g. Vollborn and Georgescu 2006), the consumers were not informed properly about potential risks for human health and the environment. Furthermore, the approval procedures by the EFSA (European Food Safety Authority) and the risk management of the Commission and the Council have been criticised by NGOs (e.g. Greenpeace 2006, 2008a, 2008b; Müller 2008), as well as by EU member states for many years now. Critics urge for real and effective application of the Precautionary Principle, which will be discussed in detail later on. They claim that potential risks and safety aspects have not been adequately considered in approval procedures, field studies and GM crop cultivation. This atmosphere of uncertainty has stoked fears about the safety of GM food and the trustworthiness of industry and politics. In addition, consumers have not had the opportunity to experience the advantages of GM technology – in many cases because there obviously seem to be none.

Also, there is a large informational deficit. Most consumers do not understand the complicated scientific processes behind GM technology. Although there are many information platforms on the World Wide Web, it is worth asking who supports these websites. The platform TransGen.de, for instance, which claims to be comprehensive, neutral and independent, is, among others, subsidised by Bayer CropScience, BASF, Dow Agro Sciences, Monsanto Agrar, Du Pont/Pioneer Hi-Bred International and Syngenta Agro.

Another problem regarding GM food is the non-transparent situation along the food chain. In fact, we have been eating genetically modified foods for more than a decade now. About 75 percent of processed food that is produced in the US contains some GM ingredients. Almost everything that contains soy or corn, including crackers, cereals, or cooking oils, also contains GMOs. However, only few consumers know about that fact. Many believe that they can choose between GM and non-GM foods, recognising GM products via labelling. But there still are no consistent standards for labelling. The situation is particularly precarious when it comes to animal products. Meat, milk, eggs and other food products from animals fed with GM feed do not have to be labelled. As a consequence, consumers do not know whether the food is genetically modified or not. The reason for this legal loophole is that animals fed with GM feed are not themselves seen as genetically modified organisms; GM feed ostensibly is not traceable in the end products. However, this issue is disputed controversially since some studies have detected traces of GM feed in milk and other animal food products (e.g. Schubbert et al. 1994; Doerfler and Schubbert 1999; Einspanier et al. 2001; Klotz et al. 2002; Phipps et al. 2003).

But not only consumers distrust this alleged revolutionary technology. Opposition against GM seeds on the part of producers is increasing. Many farmers have joined regional initiatives, so-called "Gentechnikfreie Regionen" (GM-free regions). Farmers in these networks agree on agricultural production without the use of GMOs. In 2008, the area under GM crops within the EU has slightly declined to 108.000 hectares (Transgen 2008b). One reason for this development is the ban of Bt-maize in France. However, in some countries agricultural production largely depends on GM seeds. In the US, GM crops are cultivated on 60 million hectares (Transgen 2008b). American maize, soy and cotton plants are genetically modified in most instances (80–98 per cent).

Global seed industry is controlled by a handful of huge international companies. Five corporations (Dow, DuPont, Syngenta, Aventis and Monsanto) controlled three quarters of the patents issued on GM technology in the last decade. Monsanto owns or licenses 90 percent of the global GM seed planted. The whole technology is under patent protections.

Regulations like the EU Directive 2008/62/EG support these alarming developments. As a result, fewer and fewer seeds are unrestricted and available for farmers. Furthermore, site-specific, traditional crops are eliminated and not obtainable anymore. Farmers are forced to buy GM seeds and associated products such as fertilizers, pesticides and herbicides. Once established on their fields, it is almost impossible to return to conventional seeds – wherever GM seed germinates, farmers must pay for licensing. From an ethical perspective, this can be seen as an intrusion into freedom of profession and estate, infringing upon the autonomy of farmers.

Ethical Considerations

Integrity of creation

When considering genetic engineering of organisms from a religious or ethical point of view, the fundamental question arises, if humans generally are allowed to intervene in a living being in such a drastic way. However, many arguments are diffuse and hard to legitimate, for instance that humans must not "play God" (Vogt 2004). On the other side, many arguments advocating GM technology are diffuse as well. For instance, many scientists claim that genetic engineering of plants is just a further development of traditional breeding methods. From an ethical point of view (and common sense, too), it is not. Traditional growing methods are based on interbreeding; plants are selected for certain traits, in order to make them more resistant or adaptive to local conditions. Farmers themselves decide which traits are most useful and beneficial; they are autonomous in their entrepreneurial decisions. Plant breeding in a traditional way is based on evolutionary principles. New breeds form slowly, specific traits evolve gradually. Traditional breeding methods are based on natural processes; they are well established, expedient and reasonable. Not everything is possible. Even if mutations are induced artificially, plants still remain intact in their genetic makeup.

Furthermore, property and patent rights regarding GMOs arouse ethical debates. Living organisms and their components must not be regarded as patentable inventions. They are part of creation and not result of innovative business designs (Vogt 2005).

Plants are living beings, striving at well-being and the avoidance of adverse effects. We still know very little about plants, their mechanisms of survival and their abilities. Looking at the world from a "gene perspective" fails to recognize the complex dimensions and interdependency of creation. There is no "life program" made of genes. The genetic code only plays an important role as information carrier within a far more complex system (Koechlin 2007). Plants can be modelled from a genetic perspective, but they are more than the simple sum of their genes. As living organisms, they have inherent rights which cannot be taken away and which should remain unaffected by human intrusion. This *Integrity of Creation* must serve as ethical barrier – profit does not justify everything feasible.

Precautionary principle

Genetic engineering does not leave room for mistakes. Results of flaws in this technology cannot be recalled and fixed, but become the negative inheritance to countless future generations. This is why the Precautionary Principle must be applied in a very careful and diligent way.

The Precautionary Principle is a moral and political principle which states that if an action or policy might cause severe or irreversible harm to the public or to the environment, in the absence of a scientific consensus that harm would not ensue, the burden of proof falls on those who would advocate taking the action (Raffensperger and Tickner 1999). This implies a responsibility to intervene and protect the public from exposure to harm where scientific investigation discovers a plausible risk in the course of having screened for other suspected causes. The protections that mitigate potential risks can be relaxed only if further scientific findings emerge that more robustly support an alternative explanation.

The Precautionary Principle has been endorsed internationally on many occasions. In 1992, at the Earth Summit Meeting at Rio, world leaders agreed on Agenda 21, which recommended the widespread application of the Precautionary Principle in the following terms:

> "In order to protect the environment, the precautionary approach shall be widely applied by States according to their capabilities. Where there are threats of serious or irreversible damage, lack of full scientific certainty shall not be used as a reason for postponing cost-effective measures to prevent environmental degradation" (Rio Declaration 1992).

With regard to GM technology, the Precautionary Principle is not applied adequately, particularly because the field of application is of tremendous importance – our daily food. Hence, the fundamental right to food is affected by GM food. Since the direct and indirect implications of GM technology for food security, food safety, human and animal health and for the environment are not foreseeable, this technology has to be handled very carefully.

However, lobbying of a powerful industry, enormous market potentials, exaggerated trust in this new technology, competitive strategies for technological progress – all these factors have led to broad neglect of the principle. The most debated problem when adopting the Precautionary Principle in GM technology is that the principle must not lead to total blockage of new technologies based on speculative, irrational fears; the application of the principle cannot imply the proof of total harmlessness of GM technology. Speculative worst-case scenarios are, in contrast to hypothetical worst-case scenarios, where the causal effects for a potential catastrophe are known (for instance the use of nuclear energy), based on speculative risks. These scenarios result in a total ban of any new technology, since every innovation might comprise unknown risks (Busch et al. 2002).

However, these arguments simplify the problem and do not contribute to a solution. With regard to GM food, one has to consider the far-reaching consequences of minor disturbances in the food chain. Furthermore, long-term consequences must be considered carefully – particularly with regard to future generations.

Ethical approaches

GM technology in the food sector raises many problems which have ethical implications as well: potential (health) risks for humans, animals and the environment, integrity rights of producers, consumers and GM organisms themselves, co-existence, contamination of non-GM fields, monopolism within seed industry, labelling, informational and participatory deficits and neglect of the precautionary principle in risk evaluation and approval procedures. However, the list of ethical approaches in order to deal with these problems is long as well, as shown in Figure 15-1 below.

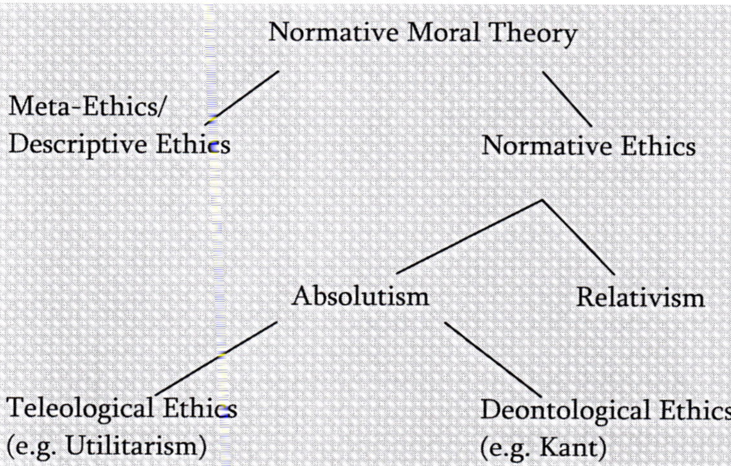

Figure 15-1. Simplified illustration according to Honderich, 1995: 940f.)

There are different paradigms within absolutistic ethics which are applicable to ethical questions of GM technology: deontological anthropocentrism, physiocentrism, biocentrism, pathocentrism. However, none of these approaches seems to be capable of fully considering, deliberating and evaluating all related problems.

There are many endeavours to establish ethical models in order to assess the costs and benefits of GM technology. Following, the practicability and adequacy of the Ethical Assessment Model by Busch et al. (2002) is analysed using a case study on GM animal feed.

The Ethical Assessment Model by Busch et al.

The model

The model is based on the work of an interdisciplinary study group. The authors claim that their model is designed to rationalise the public debate on genetic engineering, since it is applicable to all kinds of case studies.

First, a decision is made whether the existing case can be compared to other, previously evaluated cases. If no comparable cases exist, the method is evaluated with regard to its potential ecological, economical and social advancements. The model differentiates between research projects and applications; for applications (e.g. introduction on the market, field trials, etc.), limitations for the technology are more rigorous than in scientific use.

The model substantiates the person-centred approach with the principle of sustainability. The authors use the three dimensions of sustainability, as defined at the Conference in Rio in 1992. For operationalising the principle of sustainability, the authors apply a simplified version of the Pareto criterion. Considering periods of time and assuming that the current ecological, economic and social situation is sufficiently balanced, the model indicates the following: if the application of a GM technology is appropriate to improve at least one of the factors above (economic, ecological or social circumstances) without degrading another, the application of the technology is not only ethically acceptable but ethically desirable (= Pareto-superior sustainability). If one category is enhanced by the application of GM technology, but another situation is diminished, the Pareton criterion cannot be applied (ethically problematic).

The authors demonstrate the different considerations on the basis of a paradigmatic decision tree, the core of the model (Figure 15-2).

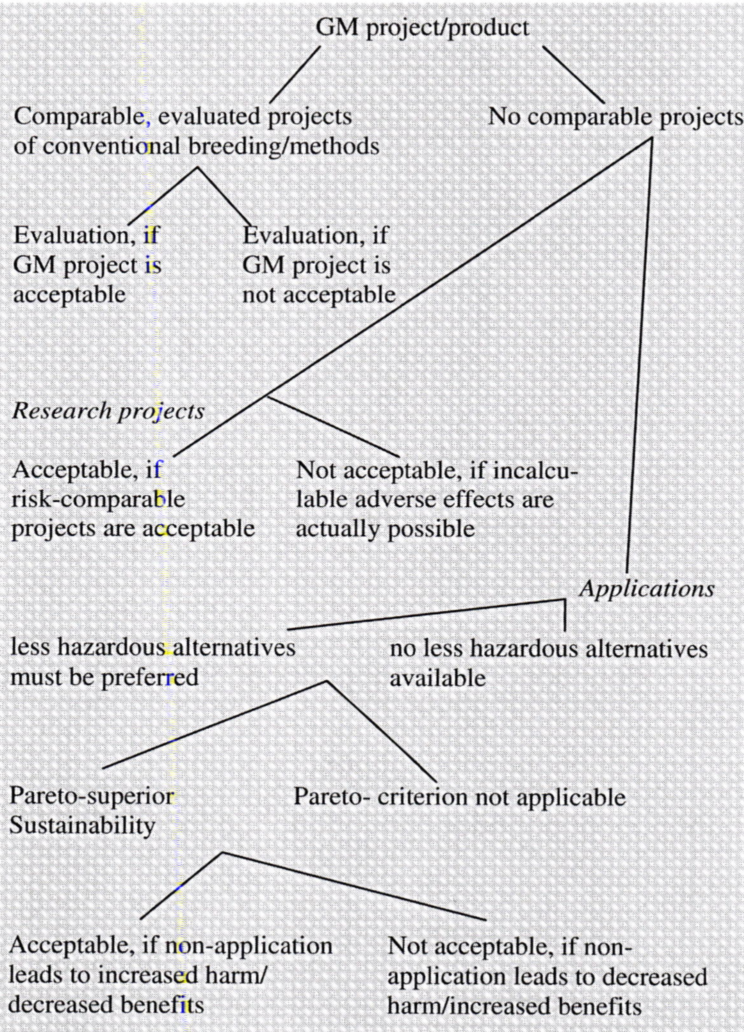

Figure 15-2. Decision Tree according to Bush et al. (2002, p. 62)

Application of the model

Following, the model is applied to a case study, in order to ethically assess the implementation of methionine-soy.

Methionine is a physiologically important amino acid. Since fodder plants, in this case soy, contain relatively low amounts of methionine, it is chemically produced and commonly added to animal feed. It is now possible to genetically modify soy plants by transferring a gene of Para nuts which encodes the proteine 2S albumin. This proteine contains methionine in relatively high amounts. It is theoretically possible to increase methionine concentration in soy using conventional breeding methods, e.g. by mutagenesis via ionising radiation. However, due to the existing GM technique, no alternative breeding method has been established yet.

Following the decision tree of Busch et al. (2002), the project is classified as an "application" with "no comparable projects":

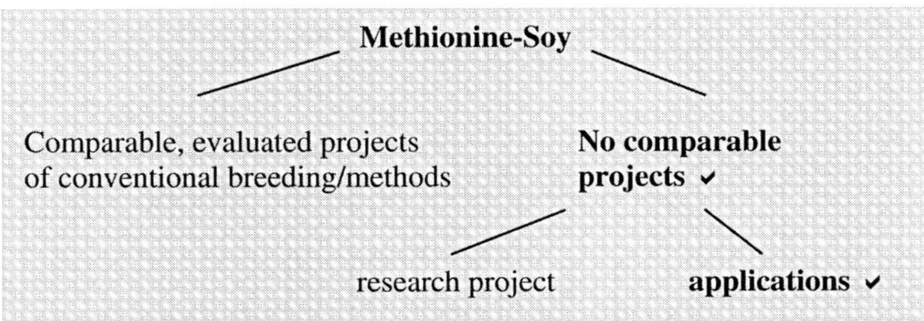

Figure 15-3. Application of the Decision tree by Busch et al.

In the next step, Busch et al. argue that there is no less hazardous alternative. However, one could argue that an alternative, well-tried technology (conventional breeding) or the *further development* of these conventional methods should be generally preferred.

Implications of GM soy on the three sustainability factors (ecology, economy, social aspects) must be traded off against each other. Following Busch et al. (2002), there are no obvious ecological risks. From an economic point of view, marketing of methionine-soy could have positive consequences for the industry as well as for suppliers and farmers.

However, this soy contains genes from nuts which are known to cause allergic reactions in some people. Hence, it was tested at a very early stage of development. It was found that by transferring the 2S-albumin-gene, its allergenic potential was transferred likewise. Due to the high importance of soy within food production and its widespread use, marketing of the methionine-soy could have had severe consequences on people suffering from allergies. Therefore, the Pareto criterion cannot be applied in this case: although methionine-soy could have positive economic effects, it might have severe social effects (=allergic reactions).

Busch et al. (2002) draw the conclusion, that potential health risks must be ethically higher valued than potential economic advantages. Therefore, methionine-soy is not ethically acceptable (Figure 15-4).

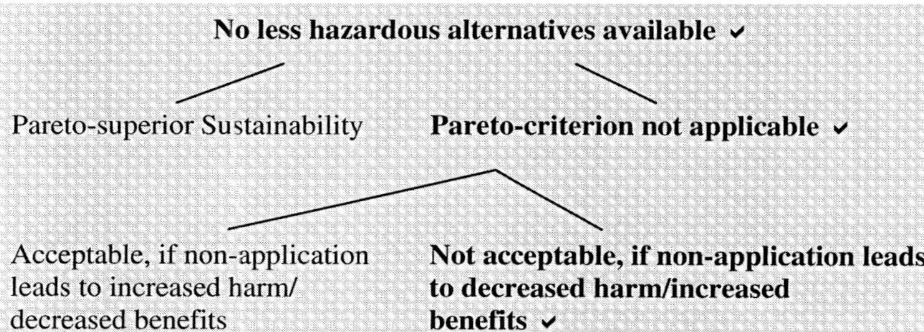

Figure 15-4. Is methionine-soy ethically acceptable? Decision tree according to Busch et al. (2002)

Criticisms and modifications

This case is clear-cut, since the risk of allergic reactions is very obvious and does not constitute an ethical dilemma itself. What if the methionine-producing protein came from another organism which is less allergenic than a nut? Genetically modified foods are reputed to cause allergic reactions. Food allergies are caused by allergens, commonly segments of proteins (allergic epitopes). If genes producing proteins that cause allergic reactions are engineered into corn, for example, an individual who is allergic to that particular protein may experience an allergic reaction. Although the industry is obliged to report suspect proteins in altered food, unknown allergens could potentially slip through the system.

Even the first step, the distinction between comparable and non-comparable projects, lacks a profound debate on the topic. The authors avoid elaborate justification for equating conventional, already approved projects and GM projects. They simply reject arguments like worst-case scenarios or the argument of potential abuse or malpractice, without assessing the technology in its entireness. The authors argue that every technology theoretically could be abused. However, GM crops and food impact the entire world population at their most vulnerable point. Furthermore, monopolistic structures within the seed industry are highly susceptible to aggressive lobbying strategies, misuse of power and the development of controversial technologies e.g. terminator technologies.

Also, Busch et al. do not take risks into account which indirectly affect or might affect ecological, economic or social issues. In GM technologies, it is not enough to assess the obvious; unbiased and independent scientific evidence (if available) must be evaluated, taking all direct and indirect implications into account. When analysing the case studies in Busch et al. (2002), the authors often fail to consider these potential side effects.

Although weighing risks and benefits is necessary, it is neither easy nor the sole concern in considering the ethics of GM technology in agriculture and food industry. When assessing potential risks and benefits, experiences and reports of concerned farmers must be taken into account. Also, based on past experiences and results, the question needs to be answered who benefits from a particular GM project and how these benefits are shared. One example is *Golden Rice*, the showcase project of the GM industry. The rice is genetically modified in order to produce beta-carotene which is supposed to be metabolised to vitamin A. Thereby, widespread vitamin A deficiencies could be avoided. However, the dilemma of dietary micronutrient deficiencies originally developed due to efforts of the seed industry to market their products. Through agricultural industrialisation, indigenous plants, well adapted to the nutritional requirements of the population, were eliminated. Furthermore, vitamin A is liposoluble and can only be absorbed by the body if fatty acids are ingested. In at-risk countries, supply with fat is equally insufficient as other aliments. Hence, it is questionable if GM technology is the adequate solution in order to combat malnutrition. In addition to all uncertainties and potential risks, GM crops are only appropriate to contribute to a better nutritional situation in developing countries, if seeds are cheap, accessible and not patented. None of these conditions are currently extant. This example shows how important it is to appraise the background behind the technology and its fields of application.

Ethical analysis of the pros and cons of GM crops is a highly complex issue that requires stringent, independent research. Legislators and executive authorities are required to safeguard independent research and to rigorously apply the Precautionary Principle. Decision-makers too often rely on the findings of organisations like the ISAAA (International Service for the Acquisition of Agri-Biotech Applications), which are funded by the biotech industry and have a clear interest in promoting certain products.

Furthermore, when weighing costs and benefits, potential economic benefits must not be valued higher than social and/or environmental issues; alternatives, such as traditional methods or technologies from organic agriculture must be taken more seriously than in the past.

Until now, GM food has not been accepted by the European population, and it has not kept one of the manifold promises made by the industry. Hence, any ethical assessment should be based on careful consideration of all direct and indirect implications this technology might have – including worst-case scenarios: global food security might be at stake.

References

Benachour, N. and Seralini, G.-E. (2008) Glyphosate Formulations Induce Apoptosis and Necrosis in Human Umbilical, Embryonic, and Placental Cells. In: Chemical Research in Toxicology (im Druck). Download under: http://pubs.acs.org/doi/abs/10.1021/tx800218n.

Bove, J. and Dufour, F. (2001) Die Welt ist keine Ware. Bauern gegen Agromultis. Rotpunktverlag, Zürich.

Busch, R.J.; Haniel, A.; Knoepffler, N.; Wenzel, G. (2002) Grüne Gentechnik – Ein Bewertungsmodell. Herbert Utz Verlag, München.

Daruich, J.; Zirulnik, F; Gimenez, M.S. (2001) Effect of the herbicide glyphosate on enzymatic activity in pregnant rats and their fetuses. In: *Environmental Research* 85(3) 226–231.

Doerfler, W. and Schubbert, R. (1999) Fremde DNA in Säugersystemen. In: *Rundgespräche der Kommission für Ökologie Band 16, Lebensmittel und Gentechnik*. Verlag Dr. Friedrich Pfeil, München, pp. 53–60.

Einspanier R.; Klotz, A.; Kraft, J.; Aulrich, K.; Poser, R.; Schwagele, F.; Jahreis, G.; Flachowsky, G. (2001) The fate of forage plant DNA in farm animals: a collaborative case-study investigating cattle and chicken fed recombinant plant material. In: *European Food Research and Technology* 212:129–134.

Friends of the Earth International (2008) Who benefits from GM-crops? The rise in pesticide use. Friends of the Earth International, Amsterdam.

Gallai, N; Salles, J.-M.; Settele, J.; Vaissière, B.E. (2008) Economic valuation of the vulnerability of world agriculture confronted with pollinator decline. In: *Ecological Economics* 68(3) 810–821.

Giesy, J.P.; Solomon, K.R.; Dobson, S. (2000) Ecotoxicological Risk Assessment for Roundup Herbicide. In: *Reviews of Environmental Contamination and Toxicology* 167: 35–120.

GMO Compass (2007) Animal Feed and Genetic Engineering – Feeding the EU's Livestock. http://www.gmo-compass.org/eng/grocery_shopping/processed_foods/153.animal_feed_genetic_engineering.html, Zugriff am 17.12.2008.

Greenpeace (2004) The European Food Safety Authority (EFSA) Failing Consumers and the Environment. Download under: http://www.greenpeace.at/uploads/media/EFSA_critique__April_2004.pdf, Zugriff am 15.12.2008.

Greenpeace (2008a) Efsa soll Zulassungsverfahren für Genpflanzen überprüfen. http://www.greenpeace.de/themen/gentechnik/nachrichten/artikel/efsa_soll_zulassungsverfahren_fuer_gen_pflanzen_ueberpruefen/, Zugriff am 15.12.2008.

Greenpeace (2008b) Heftige Greenpeace-Kritik an EU-Behörde EFSA. http://www.greenpeace.at/news_gen_081102.html, Zugriff am 15.12.2008.

Honderich, T. (ed.) (1995) The Oxford Companion to Philosophy. Oxford University Press, Oxford.

Hoyle, M. and Cresswell, J.E. (2007) The effect of wind direction on cross-pollination in wind-pollinated GM-crops. In: *Ecological Applications* 17:1234–1243.

Klotz A.; Mayer, J.; Einspanier, R. (2002) Degradation and possible carry over of feed DNA monitored in pigs and poultry. In: *European Food Research and Technology* 214: 271–275.

Koechlin, F. (2007) Zellgeflüster. Streifzüge durch wissenschaftliches Neuland. Lenos, Basel.

Losey, J.E.; Rayor, L.S.; Carter, M.E. (1999) Transgenic pollen harms Monarch larvae. In: *Nature* 399: 214.

Müller, W. (2008) Die EFSA führt die Europäische Kommission und die Öffentlichkeit über genetisch veränderte Organismen (GVO) in die Irre. Download under: http://www.keine-gentechnik.de/fileadmin/files/Infodienst/Dokumente/08_05_15_werner_mueller_dt.pdf, Zugriff am 15.12.2008.

Nikiforuk, A. (1998) The Bad Seed, as quoted in *The Ecologist* 28(5) 273.

Phipps R.H.; Deaville, E.R.; Maddison, B.C. (2003) Detection of transgenic and endogenous plant DNA in rumen fluid, duodenal digesta, milk, blood and feces of lactating dairy cows. In: *Journal of Dairy Science* 86: 4070–4078.

Potrykus, I. (2008) GMO-technology and malnutrition – public sector responsibility and failure. In: Busch, R.J. and Prütz, G. (2008) Biotechnologie in gesellschaftlicher Deutung. Herbert Utz Verlag, München, pp. 157–168.

Raffensperger C. and Tickner, J. (eds.) (1999) Protecting Public Health and the Environment: Implementing the Precautionary Principle. Island Press, Washington DC.

Relyea, R.A. (2005) The impact of insecticides and herbicides on the biodiversity and productivity of aquatic communities. In: *Ecological Applications* 15:618–627.

Richard, S.; Moslemi, S.; Sipahutar, H.; Benachour, N.; Seralini, G.-E. (2005) Differential Effects of Glyphosate and Roundup on Human Placental Cells and Aromatase. In: *Environmental Health Perspectives* 113(6) 716–720.

Rio Declaration (1992) Rio Declaration on environment and development. Download unter:
http://www.un.org/documents/ga/conf151/aconf15126-1annex1.htm.

Schattenblick Infopool (2006) GENTECHNIK/247: Chemiecocktail Glyphosat fördert resistente Unkräuter (SB).
http://www.schattenblick.de/infopool/umwelt/redakt/umge-247.html, Zugriff am 13.12.2008.

Schubbert, R.; Lettmann, C.; Doerfler, W. (1994) Ingested foreign DNA survives transiently in the gastrointestinal tract and enters the bloodstream of Mice. In: *Molecular and General Genetics*, 242: 495–504.

Soil Association (2002) Seeds of Doubt – North American farmers' experiences of GM crops. Soil Association, Bristol.

Transgen (2008a) Gentechnisch veränderte Pflanzen – Anbauflächen weltweit.
http://www.transgen.de/anbau/eu_international/201.doku.html, Zugriff am 16.12.2008.

Transgen (2008b) Anbau gv-Pflanzen: Rasanter Anstieg weltweit, vorsichtiger Beginn in Europa.
http://www.transgen.de/anbau/eu_international,
Zugriff am 16.12.2008.

Vogt, M. (2004) GenEthik: Grüne Gentechnik in ethischer Sicht. Vortrag im Rahmen des Veranstaltung „Gemachte Natur? Chancen,Risiken und ethische Bewertung von Grüner Gentechnik". Universität Bayreuth, 19.11.2004. Download unter: http://www.uni-bayreuth.de/forum-kirche-universitaet/tagungen/gruene-gentechnik/GenEthik.pdf, Zugriff am 18.12.2008.

—. (2005) GenEthik: Ethische Orientierungen im Konflikt um Grüne Gentechnik. In: Schriftenreihe der Katholischen Landvolkbewegung zu zeitgemäßen Themen 5. Internationaler Landvolkdienst ILD, Rhöndorf/Rhein.

Vollborn, M. and Georgescu, V.D. (2006) Die Joghurt-Lüge – Die unappetitlichen Geschäfte der Lebensmittelindustrie. Campus, Frankfurt/M.

Walsh, L.P.; Mc Cormick, C.; Martin, C; Stocco, D.M. (2000) Roundup inhibits steroidogenesis by disrupting steroidogenic acute regulatory (StAR) protein expression. In: *Environmental Health Perspective* 108: 769–776.

Welthungerhilfe (2008) Aufgaben und Ziele der Welthungerhilfe. http://www.welthungerhilfe.de/aufgaben_und_ziele.html, Zugriff am 16.12.2008.

CHAPTER SIXTEEN

OWNING WHAT YOU EAT: THE DISCOURSE OF FOOD

DAVID N. CASSUTO[1]

Introduction

Discussions of animal treatment within the global food industry often devolve into debates about animal rights. Such detours needlessly distract from an ongoing social and environmental catastrophe. This essay attempts to reframe the global food debate in a manner that more directly acknowledges our obligations to and the needs of the billions of animals enslaved within the industrial food apparatus.

Industrial agriculture has refashioned animal husbandry into a mechanized process that ignores historic methods of human/nonhuman animal interaction (methods that evolved over millennia) as well as ethical mores. These industrial methods – cloaked in the mantle of efficiency – have become deeply entrenched despite clear evidence of their unsustainability and unworkability. This intractability results from a systemic flaw inherent in the role of efficiency in society. Not only is efficiency an amoral concept devoid of any normative component, but those who lionize it also routinely exclude externalities from their calculus. This makes any cost-based risk equation potentially unsound and misleading.

Consequently, using efficiency as an ethical barometer is flawed both hermeneutically and practically. It should never have acquired a normative aspect and it should never have been defined to exclude externalities. The upshot of this double mistake is that the prevailing mode of human/animal interaction is unsustainable (inefficient) and ethically bankrupt. Reframing that interaction will require refashioning the legal system that enables it.

Part II of this essay examines the role of communication in the formation of law and social norms and the implications of that role for animal law and ethics. Part III contextualizes animal law within contemporary risk society. Part IV looks at how efficiency has transformed from an economic concept into a normative guideline and discusses how that transformation has affected animals and agriculture. It tracks the rise of industrial agriculture and ties it to this fundamental misunderstanding of the concept of efficiency. The essay concludes with some thoughts on how to reformulate contemporary notions of efficiency and ethics to account for the idealism that should be a necessary component of communication and, consequently, of law.

Communication and Law

Law governs interactions between and among members of society. It codifies shared goals that reflect an ideal vision of a just society.[2] This vision of justice arises through communication. For communication to be coherent there must be a shared belief in the possibility of consensus and mutual understanding. Participants must evince a willingness to come to an agreement about the truth/correctness of the matter under discussion.[3] This shared commitment to agree forms the foundation of discourse, which in turn forms the foundation of norms that then become codified into law.

Consensus-driven communication requires a common language. Laws governing human interaction (property, contract, criminal, torts, etc.) all fit within the discursive framework of shared goals and commitment to the perpetuation of society. Animal law, however, does not. (Non-human) animals do not share a language with humans. They do not participate in human discourse nor do they share the goals of human society.

Without a common normative vision, there is no consensus from which to create laws. Consequently, animals are not merely an unwilling participant in the law-making process; they *do not participate at all.* It therefore makes no sense to talk about animal law as such; it is more properly described as a set of laws governing how humans interact with the animals.

This distinction is more than merely semantic. Human interaction with animals lies within human control. However, the other side of that process – animal interaction with humans – resides entirely outside of human control. Given that the animal perspective is both varied and unknowable and that humans are social organisms who interact with other species, it is understandable and necessary that humans would create a set of rules to guide those interactions. But because those interactions lack any shared commitment to consensus, attempts to impart meaning to the process are necessarily counterfactual. Herein lies what Aristotle might have described as the tragic nature of animal law. The impossibility of communication coupled with the immutable need for communication creates a crisis borne of conflicting truths that undermine meaningful interaction.

Despite this seemingly unsolvable dilemma, there is some cause for hope. *All* communication is arguably counterfactual; that does not make it pointless. When humans communicate they do not truly understand each other; they merely share the goal of achieving that understanding. As Habermas argues, it is not the existence of truth, but rather a shared commitment to its possibility that makes communication (and therefore society) possible.[4] Similarly, with respect to animals, true, meaningful communication is impossible. However, if attempts at communication were made in good faith and predicated on a willingness to exclude ideology and self-interest, then the human component of the human/animal interaction would not differ fundamentally from other forms of discourse.

The impossibility of an ideal state need not doom a society founded on a commitment to its attainment. Consequently, the tragedy of animal law does not lie in its aspirational nature. It lies rather in the way that human nature undermines those aspirations. This is clearly visible in the principles of risk management.

The Risks of the Risk Society

How risk society came into being

Ulrich Beck explains that: "Just as modernization dissolved the structure of feudal society in the nineteenth century and produced the industrial society, modernization today is dissolving industrial society and another modernity is coming into being."[5] In industrial society, wealth production overshadowed risk production because (among other reasons), the risks accompanying the ascendancy of industry were poorly understood and because the remnants of feudal society imputed a sense of preordained destiny both to social status and to the workings of environment.[6] Furthermore, the omnipresent struggle against scarcity engendered a willingness to endure detrimental side effects.

Beck notes, for example, that in the early 1800s, the Thames was so polluted that people who fell in suffocated instead of drowning – the result of inhaling the poisonous gases carpeting the river.[7] These and other risks arose as a consequence of modernization and were easily apparent. Today (at least in the developed world), scarcity no longer drives production. Society now produces at such a rate that many of its hazards are associated with *over*-production. Furthermore, the accompanying risks of post-industrial society are less visible (e.g., toxins in the food web, ozone depletion, climate change, etc.). These shifts have precipitated a fundamental reordering of society. Managing risk has become at least as important as wealth amassment.

Risk management in post-industrial society is reflexive; it is the "systematic way of dealing with hazards and insecurities induced and introduced by modernization itself."[8] Because risks are often not readily perceptible, the task of identifying risk falls most often to science, propelling scientists into the role of neutral and benevolent expert. Yet, identifying risk is not a neutral act. It has profound societal implications that transcend science. Risk identification determines what constitutes harm (an inherently subjective determination) and assesses whether that harm rises to a level requiring mitigation. In this manner, science, when wielded by the powerful, becomes the organizing principle around which society constructs its response to danger.

Through its role as risk creator/assessor, science becomes the source of what Mary Douglas calls "taboo thinking," which uses the threat of danger to create and uphold community values.[9] Shared danger bonds society through the shared goal of mutual survival. Because modern threats are invisible, "experts" who inform the public of the existence of the threat and the proper response wield a powerful tool of mass coercion. This aggregation of power in the hands of a select few would be troubling enough by itself. The situation is exacerbated by the fact that the risks that science is tasked with identifying and mitigating are themselves the *creations* of science.

Creation here refers not just to the social construction of risk (determining whether a given behavior constitutes a threat) but also to the behavior itself. For example, once science identifies mass confinement of animals as a potential cause of pandemic influenza – it then must determine the proper reaction to that threat by determining whether the risks from continued confinement outweigh the benefits of (and to) industrial agriculture. This is the process of defining risk and responding to it. Embedded within this identification/mitigation heuristic lies the fact that the confinement methodology is itself is a creation of science.

The multi-tiered role of science in risk creation means that it is tasked with the impossible chore of neutral, critical self-evaluation. It engages in risk behavior, defines the risk created by that behavior, evaluates the level of threat produced, and then advises society on how to respond.[10] Allocating all these tasks to experts amounts to a wholesale abdication of responsibility by the lay public and by law-making bodies.

While the public has become more aware of the existence of threats, it has also become increasingly unable to recognize them and complacent in its incompetence. It has effectively forfeited its "cognitive sovereignty."[11] Such behavior amounts to complicity in a social structure constructed not around consensus and common belief but rather around the goals and conclusions of an elite sub-group.[12] Horkheimer and Adorno's warning of a post-enlightenment world wherein reason metamorphosed into technology seems profoundly and disturbingly apt.[13]

Despite its seemingly privileged status, the societal position of science remains unenviable. Risk management decisions that science must make regarding its own behavior and creations have potentially draconian implications. As the number of risks increase and their consequences become more severe, the margin for error shrinks to the point where admitting mistake becomes a profoundly anti-social act.

> To admit today that one had been mistaken in setting the acceptable values for the safety of pesticides – which actually would be a normal case in science – amounts to the unleashing of a *political* (or economic) catastrophe, and must be prevented for that reason alone. The destructive forces scientists deal with in all fields today impose on them the inhuman law of *infallibility*. Not only is it one of the most human of all qualities to break this law, but the law itself stands in clear contradiction to science's ideals of progress and critique ...[14]

Science relies on experimentation and hypothesis. Uncertainty is inseparable from scientific method. Yet in a risk society, if science admits error, the implications can be catastrophic both for society and for science's privileged role within it. Consequently, science must act as the legitimating body for the regulatory apparatus that straitjackets it. It is trapped in a self-perpetuating cycle of taboo thinking.

This is the cycle of contemporary risk management. In order for society to function efficiently, risks must be assayed and behavioral guidelines issued with the goal of attaining a smooth-functioning, efficient society. In this manner, efficiency attains the status of norm, with science as its arbiter. Smooth functioning requires minimizing exposure to risk. And science determines risk and also the proper techniques for mitigation.

Were science not used to cause so many societal risks, thereby rendering it unable to objectively analyze potential threats, the process of identifying and mitigating risk would be a quintessentially scientific task. Simply theorizing about how to reduce threat levels, presents little conflict with scientific method. However, when science is called upon to determine an acceptable level of risk, it takes on a normative function that becomes self-undermining. The unworkability of this arrangement becomes starkly apparent in the context of agriculture.

Animals at risk

In the early years of the last century and prior, farms were hotbeds of cultivated biodiversity. Farmers often raised a dozen or more species of fruits and vegetables, including corn, alfalfa and hay to support the pigs, cattle, chicken and horses that populated the farm. Less than a hundred years later, the animals are gone, as is just about everything else save one or two crops – usually corn or soybeans.

The metamorphosis of small, diversified farms into large, monocultural, agribusinesses has many causes. However, as Michael Pollan explains, the biggest factor behind the transformation was cheap, abundant corn made possible both by the advent of nitrogen-based fertilizer and a generous (if misguided) program of agricultural subsidies.[15] The availability of synthetic fertilizer meant that rotating crops became unnecessary and the nation's growing demand for corn could be met with an even larger output by the nation's farms.

Under normal economic conditions, corn's superabundance would have glutted the market and caused prices to fall. However, New Deal farm programs set a target price for the corn and allowed farmers to take out loans, using their surplus corn as collateral. When prices recovered, farmers sold the corn and paid back the loans. If prices failed to recover, they kept the loan money in exchange for the government's keeping the corn. The government could afford to wait until demand ticked up to put it on the market. In either case, surplus corn stayed off the market until demand revived.

Everything changed for agriculture in the United States during the 1970s. Earl Butz, President Nixon's Secretary of Agriculture, introduced a new system of price supports guaranteeing farmers a set price for their corn no matter the market price. Butz exhorted farmers to "get big or get out" and to regard themselves as "agribusinessmen" rather than farmers.[16] A predetermined price per bushel of corn meant that growers had no incentive to decrease production when demand slacked. Instead, they were spurred to grow as much as possible and dump it into the market, which in turn caused prices to crater still more.

As prices fell, successive farm bills enacted by Congress lowered the guaranteed price paid for corn, causing farmers to have to grow still more to eke out a profit. Consequently, the market became perennially glutted with corn, small growers all but disappeared, and the need to utilize the ever-growing surplus became ever more urgent. Growers began feeding the corn to animals, including cattle, whose digestive systems cannot tolerate it without prophylactic antibiotics and other medications. The feed itself was cheap but the consequences of the cattle ingesting that feed were not. From this tangled attempt to make efficient use of what should never have been grown, the factory farm emerged.

The story of other animals' journeys from farm to Concentrated Animal Feedlot Operation (CAFO) is similar, albeit different in certain key respects. For example, animal agriculture for pigs and chickens (not cattle) is highly vertically integrated. Growers do not own the animals and have no input into the manner in which the animals are fed or housed. They must simply obey the dictates of their corporate overlords. The growers also have little leverage with respect to the price they are paid for their labor. They cannot command prices sufficient to cover environmental degradation and waste disposal. As a result, costs get externalized and passed along to the general public. Those costs arenot reflected either in the cost of production or in the retail price of the product.[17] Instead, they become hidden costs, which, along with corn subsidies, become woven into the national tax burden.

Efficiency

Efficiency as norm

In order to turn agriculture into agribusiness, growers had to embrace large commercial enterprises as superior and preferable to small-scale farming. Farmers who bristled at this mission, they faced replacement by others more sympathetic to the corporate goal. That goal was simple: make agriculture more efficient. The missing option – one that to this day remains unexplored – involves interrogating the utility of efficiency as a bounding principle in agriculture.

In economic terms, efficiency means getting the best possible return on an investment. Any resources spent should lead to a greater yield. In this sense, efficiency is a fundamental principle of a market economy. However, agriculture – and specifically animal agriculture – is not economics. While economics drives many facets of agriculture, that does not make the two equivalent any more than gasoline is equivalent to the cars that it powers.

Agriculture is grounded in ecology. And, while ecology's definition has evolved over the years, it has always centered on the relationship between living things and their surroundings. It has been variously defined as: the study of the interaction between organisms and their environment, the study of the distribution and abundance of organisms, and the scientific study of the processes influencing the distribution and abundance of organisms, the interactions among organisms, and the interactions between organisms and the transformation and flux of energy and matter. For none of these definitions, however, does efficiency serve as the organizing principle.

When organisms interact with their environment, redistribute themselves, or when the environment is in flux, one sees an almost wanton expenditure of energy. Whether it be the volume of water cascading into the ocean from a river after a rain or the playfulness of a polar bear cub frolicking on the ice, energy and resources get expended in a seemingly heedless environmental potlatch. The river's swollen discharge will help keep the estuary's salinity constant during a future dry time and the bear cub will use precious calories but that play will help it learn coordination and how to interact. Yet, at no time is there any regimented, perfect equilibrium sought, or any real method of calculating efficient behavior.

Agriculture relies on human/animal interactions, which are subsumed within the ecological web even while forming part of the human economy. Ecology has no defined bounding principles while the latter cleaves to efficiency. The relationship between ecological unpredictability and bottom line analytics has existed for centuries but the twentieth century witnessed a shift. Market dynamics became dominant; ecosystemic expenditures fell prey to accounting principles and animal behavior became almost irrelevant to animal "science." As a result, the barnyard has become the stockyard, the farmyard a CAFO, and the manure pile a sewage lagoon.

When market efficiency displaced ecology as the foundation of agriculture, another crucial component was lost as well. Ethics once provided the moral principles underlying human/animal interactions. While one could argue (persuasively) that those principles were never sufficiently sensitive to the needs and welfare of the nonhuman animals, it lies beyond cavil that the treatment and care of animals in the days prior to industrial agriculture differed markedly from their treatment now. Farmers used to house and feed animals in a manner designed to keep them comfortable and to develop relationships with each other and with humans. These relationships did not necessarily maximize yield but were rather based on a set of normative guidelines even as the ultimate reality of the animals' commodity status inevitably imbued that bond with a sense of unreality.

One sees vestiges of this bifurcated relationship in agricultural education organizations like 4-H, where children are given an animal to care, raise and nurture. Often, the children grow to love these animals, even as they know they are raising them for slaughter. The culmination of their efforts usually happens at the county fair where the animals are auctioned for slaughter, leaving the children grief-stricken and heartsick even as parents and teachers congratulate them on a job well done.

This complicated and problematic relationship between the children and their animals is emblematic of the tensions underlying traditional approaches to animal husbandry. It was impossible to escape the animals' commodity status even as ethics demanded their decent treatment. The advent of industrial agriculture simplified eliminated any tension within the human/animal relationship by completely commodifying the animals and removing ethics from the equation. Instead, the focus became exclusively on profit generation. Animals went from partially commodified beings whose value could be measured both in individual terms and as units of exchange value to simple merchandise whose value derived from decreased costs of care and increased value at alienation.

An object's market value increases with marketability leading to an increased desire to alienate it. The greater an object's value, the greater the temptation to divest oneself of it. Thus, the realities of the marketplace make it both difficult and counterproductive to form a relationship with a commodity, a fact that further facilitates objectification. In the case of animals, it also enables mistreatment.

Animals' commodification confers an exchange value that, in the case of "meat" animals, is realized through death and dismemberment. For animal producers (milk cows, breeding sows, etc.), value emerges through maximizing productivity while minimizing costs. In neither instance does the animals' quality of life enter the equation. Rather, in both cases, the economic incentive (which is, after all, what drives exchange value) lies with minimizing expenses associated with the thing while maximizing its yield.[18] It is easy to see how this logic leads to factory farms designed to maximize economic efficiency regardless of the impact on animals. It also catalyzed the industry's vertical integration. As a result, a few large conglomerates now control the animal production process "from squeal to meal."[19]

Classification as a thing also exempts objects from any moral calculus. Exchange value is amoral; it is a function of demand. Owners maximize value through increasing marketability. Moral arguments as to the market value of things carry little practical weight because the two systems (moral and market) do not share a common mode of valuation.[20]

This basic incompatibility between relationships predicated on ethics and those predicated on commoditization means that the role of agriculture could never be other than problematic.[21] However, the ascent of the factory farm was not inevitable. It could have been avoided and will eventually be undone primarily for two reasons.

First, the factory farm system is not efficient. It ignores externalities that render the cost/benefit calculus profoundly against its continuation and relies on a massive subsidies and a (government abetted) propaganda campaign to sustain itself in the face of increasing public opprobrium. Second, circling back to Habermas's thesis – communication (and society) does not require an ideal state in order to exist.[22] They rather require a shared commitment to its attainment. Consequently, the incompatibility of agriculture, ethics and a market economy need not lead inevitably to catastrophe. The discord can rather serve as the basis for a discourse grounded in aspiration.

Agribusiness is bad business

As noted above and as has been well-covered elsewhere, industrial agriculture's dependence on cheap corn, which is a product of unsustainable government subsidies, means that the price of food does not reflect the actual cost to produce it. Nor does it reflect the loss of biodiversity or the increase in the national carbon footprint that comes from increased reliance on petroleum-based fertilizers (crop rotation once served to replace depleted nitrogen in the soil but that practice has all but disappeared with the ascent of monoculture) and mass production of methane-emitting animals. Furthermore, industrial agriculture is the single largest producer of sewage waste in the United States. The waste collects in sewage lagoons, makes its way into the groundwater, surface water, and soil. This leads to massive and sometimes irremediable contamination problems, dead zones in the ocean, pernicious algae blooms, and other environmental problems. In addition, when the lagoons rupture or spill, it causes contamination and loss of life in rivers and surrounding ecosystems.

When one further factors in the increase in antibiotic resistant bacteria as well as the respiratory infections that result from inhaling the poisonous fumes from the facilities, the costs of maintaining large-scale animal confinement facilities becomes far from economical. Recent events have also shown the lurking dangers of deadly influenza, for which confinement agriculture provides an ideal incubator. The threat from an influenza pandemic is almost incalculable.

Given all this, even if efficiency provided a normative basis for the treatment of nonhumans, industrial agriculture would remain woefully inefficient, and therefore unethical on that basis alone. But, as already discussed, norms derive from ethics, which do not cleave to any economic model. They are a set of principles designed to govern human interaction with each other and with their environment. In that sense, ethics represent a form of human ecology.

Because ethics govern behavior and behavior implicates interactions with the human and nonhuman environments, ethics have an ecological component. And, since ecology is the study of flux (ecosystems constantly evolve; stasis – a lack of energy flow in or out of the system means that the system is inert), ethics must constantly evolve to reflect that change. Ethics are fluid because both the environment and human relationships must always remain in flux. Aldo Leopold's famous Land Ethic aptly captures the character of the human commitment to environment as well as the shifting parameters of that commitment.[23] Nonetheless, ethics embody a dedication to the attainment of a just society and in that sense remain static.

While the tension between ethical behavior and an agricultural system based on commoditization of nonhumans was unavoidable, it has also long been present. The shift from uneasy coexistence to monolithic dominance by the market-based approach to animal management was triggered by the emergence of the risk society. Both with respect to physical dangers – where threats became more dispersed and beyond the ken of the lay person – and in the realm of the economy, where the complexity of commerce has increased exponentially in the past half-century, the public has come to rely more and more on experts to interpret the data and advise on the correct course of behavior.

With the state supporting large agribusiness, the views of the "experts" employed by those businesses increasingly came to dominate the discourse. Furthermore, the government experts worked within a regulatory regime that supported that market-based approach. As industrial agriculture become integrally woven into the fabric of the economy, the risks inherent in that approach (pollution, economic collapse, systemic animal abuse, etc.) became hidden costs embedded in a downward spiraling feedback loop. The risks grew worse as the risk behavior became more systemic (industrial agriculture has become a worldwide phenomenon) and those increased risks meant that the dangers associated with their exposure also increased.

Science as gatekeeper

The same "experts" who invent and propagate the methods that generate societal risks must also assess the danger from those methods and, if necessary, expose it and suggest mitigating measures. Those obligations put scientists and economists in the unenviable position of assessing whether the risks created by their behavior are worthwhile. They must determine how much danger and damage from the sanctioned behavior is acceptable. And it is precisely here that risk assessment leaves the realm of science and enters the realm of ethics. It also creates an unsustainable conflict of interest. Unfortunately, this is precisely the situation in which we find ourselves today with respect to industrial agriculture.

The inhumane treatment of animals has been rationalized as the inevitable byproduct of efficient agriculture. As a result, industrial agriculture flourished and became interwoven with the global economy. Now, as the flaws in the industrial model become increasingly clear, so too do the risks inherent in moving away from that model. Society has grown to depend on and expect cheap, mass produced meat, which involves enormous amounts of corn, which in turn requires government subsidies, inhumane, confinement agriculture and antibiotics. Yet, abandoning that model will cause significant economic upheaval and social unrest, as well as undermining faith in those entrusted with safeguarding society. In sum, protecting society from the danger will involve endangering that same society. This scenario calls for a risk analysis that those who created the situation are ill-equipped to make.

Conclusion

Untangling this Gordian knot of dependence involves acknowledging the essential incompatibility of certain foundational ethical precepts (i.e. humane interaction with nonhumans) with a market-based approach to agriculture. This intractable conflict could lead to systemic paralysis or collapse. However, it need not, so long as the aspirational nature of society is recognized and embraced.

As Habermas observes, communication (and society) is built on the counter-factual premise that the conditions for ideal discourse (a just society) exist and that true consensus (free of any externally imposed constraints) is possible. Similarly, social ideals and the laws that codify them are counter-factual as well. They are aspirational – representing a shared vision of a society where externally imposed behavioral standards would not be necessary. Norms (and laws) presuppose a shared commitment to consensus and ethical behavior even as their very existence testifies to the absence of that consensus.

In this sense, human society is predicated on contradiction. On the one hand, this might be fodder for a legitimation crisis. On the other hand, it is profoundly reassuring in that this foundational contradiction within society does not mean its inevitable demise nor does it mean that nihilism will envelop us all. On a practical level, however, the problem of industrial agriculture remains unresolved.

Recalibrating our relationship to animals within the ecology of the market will require several steps. The first involves recognizing the dichotomy between efficiency and ethics. Efficiency has no normative component and ethics have no necessary relationship with efficiency. In the context of agriculture, acknowledging this schism will necessitate a through reexamination of efficiency's role in determining acceptable treatment of nonhumans. This must be coupled with a recognition that a deliberate indifference to life does not align with any extant ethical vision.

The second step requires acknowledging that the current vision of agricultural efficiency is based on flawed assumptions and bad science. It has wrought ecological havoc and undermined the global economy. In addition, as factory farming spreads, so too do worldwide health problems, including obesity, diabetes, and heart disease.

Industrial agriculture has grown ascendant because its drawbacks have been deliberately obscured. It has used subsidies to obscure a lack of profitability, sited its facilities far from population centers to keep the animals from prying eyes, and successfully lobbied for the exemption of its methods from animal cruelty laws, All of these strategies allowed the public to embrace ignorance and to assume that their safety was entrusted to a sagacious and risk averse set of experts. But as Aldo Leopold observed, "too much safety seems to yield only danger in the long run."[24] Ignorance has led to a dangerous loss of agency. People are no longer in touch with their food. Even as the animals they consume have devolved almost entirely into commodities, those who consume them have divorced themselves from the process that turned the animals into food. Ignorance has obscured responsibility. Therein lies the last and most important step in the unmaking of the factory farm. We must learn to own what we eat.

Notes

[1] A version of this paper was also presented both at the Conference on The Animal Within the Sphere of Human Needs sponsored by the University of Québec at Montreal, International Research Group in Animal Law, Montreal, Canada in May 2009. I am indebted to Joe Edgar and Stephen Iannacone for their research assistance with this project.

[2] See Jürgen Habermas, *Legitimation Crisis* (Thomas McCarthy trans., Polity Press, 2007; originally published 1976). Note that this vision is necessarily counterfactual.

[3] See Thomas McCarthy, 'Introduction' to Jürgen Habermas, *Legitimation Crisis*, at xvi (Thomas McCarthy trans., Polity Press 2007; originally published 1976).

[4] E.g. *id.*

[5] Ulrich Beck, *Risk Society* 10 (Mark Ritter trans., Sage Publications, 2007; originally published 1986).

[6] *Id.* at 12-14.

[7] *Id.* at 21.

[8] *Id.*

[9] Mary Douglas, *Risk and Blame* 4 (Routledge Press 1992).

[10] As Beck observes, eventually, "one comes up against the law that so long as risks are not recognized scientifically, *they do not exist*–at least not legally, medically, technologically, or socially, and they are thus not prevented, treated, or compensated for." Beck, *supra* note 5 at 71 (emphasis in original).

[11] *Id.* at 53.

[12] See Jenny Steele, *Risks and Legal Theory* 33 (John Gardner ed., Hart Publishing 2004) ("[T]he growing *recognition* that we are ignorant, and especially that we are ignorant of the way that our actions will interact with other factors, means that we are increasingly 'culpable' … Ignorance loses its innocence as awareness of limits to knowledge grows.") (emphasis in original).

[13] See, e.g. Lash, et al., *Risk, Environment and Modernity* 3 (Sage Publications, 2000; originally published 1996).

[14] Beck, *supra* note 5 at 54.

[15] Michael Pollan, *The Omnivore's Dilemma: A Natural History of Four Meals* 54-64 (Penguin, 2006)

[16] *Id.* at 52.

[17] *Putting Meat on the Table: Industrial Farm Animal Production in America* (A Report of the Pew Commission on Industrial Farm Animal Production) p6.

[18] See Gary Francione, *Animals, Property, And The Law* 29 (1995); Robyn Mallon, 'The Deplorable Standard of Living Faced by Farm Animals in America's Meat Industry and How to Improve Conditions by Eliminating the Corporate Farm', 9 *Mich. State Univ. J. of Med & L.*389, 399 (2005) (explaining that even when a law protecting farm animals exists "[t]he law is violated due to the traditional corporate attitude of wanting to maximize profit and minimize expenses.").

[19] Smithfield Foods, the world's largest vertically integrated producer of pig products, proudly embraces both the term and the concept. This quotation, from

Lewis Little, president of Smithfield's largest subsidiary, appears in the company's promotional material. See "Understanding Smithfield," available at: staging.smithfieldfoods.com/Understand/Family/smithfieldFerme.asp

[20] See Francione, *Animals, Property and the Law*, *supra* note 18 at 27 ("[O]ur allocation to humans of rights in the bodies of animals reflects the notion that it is more efficient to relegate animals to property status … than it is to value animals for themselves and to accord them dignity and respect. Moral concern for animals is not a cost-justified policy.").

[21] The commoditization discussion is drawn from (and covered in greater detail in): David N. Cassuto, *Bred Meat – The Cultural Foundation of Factory Farms,* 70-WTR LAW & CONTEMP. PROBS. 59 (Winter 2007)

[22] Jürgen Habermas, *Moral Consciousness and Communicative Action* 133-36 (Christian Lenhardt and Shierry Weber Nicholson trans. MIT Press 1990).

[23] "A thing is right when it tends to preserve the integrity, stability, and beauty of the biotic community. It is wrong when it tends otherwise." Aldo Leopold, *A Sand County Almanac* 262 (Oxford University Press 1949)

[24] *Id.* at 141

CHAPTER SEVENTEEN

PRIVATE INVESTMENT IN WATER AND SANITATION SERVICES: RIGHTS-BASED APPROACHES AND INTERNATIONAL INVESTMENT LAW – A POSSIBLE WAY FORWARD?

OWEN MCINTYRE

Introduction

It is often assumed by those opposed to the involvement of the private sector in the provision of water and sanitation services, particularly by those concerned about private sector involvement in developing countries, that recognition of the human right to water ought to operate to hinder such involvement (Smets 2005: 177–183; Morgan 2006: 215). However, while practical application of the right to water should help to ensure that certain aspects of social protection, relating particularly to affordability, equality of access, restrictions on disconnection, public participation, and quality of supply, are respected and protected, for example by means of their inclusion in contracts concluded between public authorities and private companies for the provision of water-related services, the normative requirements emerging from this right do not *per se* restrict the privatisation of water services. In fact, the only statement of international declarative law impacting directly on private sector participation in this area is that included in the 2001 Bonn Recommendations, adopted by 118 States at the Bonn International Conference on Freshwater, to the effect that 'Private sector participation should not be imposed on developing countries as a conditionality for funding'.[1] Indeed, the emergence of the human right to water may serve to facilitate private investment in water services by identifying broadly accepted standards of service and social safeguards, which private investors can factor into their investment decisions and which make it difficult for public authorities to alter the regulatory environment arbitrarily.

The inclusion of water and sanitation targets among the Millennium Development Goals (MDGs) has highlighted the need to mobilise private investment in order to have any realistic chance of meeting such targets. For example, it is quite clear that a number of key UN agencies envisage a significant role for private sector water providers in making progress towards the MDG target.[2] It is telling that the World Panel on Financing Water Infrastructure, which was convened by the UN Secretary General, chaired by Michel Camdessus, a former President of the World Bank, and made up of representatives of the international financial institutions, banks and businesses involved in funding private investments in the water sector, adopted a report providing unqualified support for the existence of right to water.(Winpenny 2003; Smets 2005: 175; van Hofwegen 2006) In addition, the CEOs of at least two leading water multinationals, Suez and Véolia, have expressly recognised the universal nature of the right of access to water.(Smets 2005: 175–176)

Though the level of private sector water service provision remains relatively small in percentage and numerical terms, estimated in 2001 at no more than 5 per cent globally covering no more than 300 million individuals,[3] Morgan points out that

> "A remarkable 7,300 per cent increase on 1974–1990 private-sector investment levels in water services occurred between 1990 and 1997' and that 'intergovernmental activities in relation to water since then have intensified and continue to incorporate the private sector as a key partner in their vision." (Morgan 2006: 215–246)

Though Morgan also notes that the rate of increase in private sector involvement may have peaked, Schreiber, writing in 2008 and citing sources from 2006 and 2007,(Prasad 2006: 669; Mann 2007: 122) notes 'a vast increase in the amount of water privatization contracts between states and investors resulting in roughly 10 percent of global water consumers today receiving their water from private enterprise.'(Schreiber 2008: 431–478) Of course, private sector involvement in the provision of water and sanitation services may take a range of forms depending, for example, on whether assets, operations, or both move from the public to the private sector, on how much ownership or control is transferred to the private sector, on what types of functions the private company performs, and on how long the arrangement is due to last.(Williams 2007: 493; Petrova 2006: 584) The form of privatisation employed in any particular instance will have obvious implications for the nature and extent of State regulation required so as to meet social objectives of the human right to water, for the private operator's incentive to invest in infrastructure so as to improve or extend the services provided, for its ability to recoup capital invested and, consequently, for the availability of financing.

Of course, States retain primary responsibility for ensuring that the actions of private service operators do not compromise realisation of the human right of access to water and sanitation. Under the obligation 'to protect', States are obliged to ensure equal and affordable access to water and sanitation services, which might require that they establish and maintain an appropriate regulatory system for private sector providers based on, *inter alia*, independent monitoring, genuine public participation, and effective penalties for non-compliance.[4] Under the obligation 'to fulfil', States are required to consider the requirements of equity and affordability, which might involve legislating for appropriate pricing policies or the introduction of a system of income supplements for socially disadvantaged groups.[5] In addition, under the obligation 'to respect', General Comment 15 includes a requirement for States to take steps to prevent its citizens or companies based within its jurisdiction violating the right of access to water in other States.[6] This requirement might be addressed in a number of ways, such as by providing financial and technical support for building the necessary capacity for the public sector in other States in which its companies operate to perform the regulatory tasks required.

However, it is increasingly important to understand the mechanisms by means of which emerging human rights requirements relating to access to water and sanitation services might apply to and influence the behaviour of private sector providers, especially those operating in developing countries. There is a wide range of such mechanisms, which includes standards for water services provision developed by international bodies, various policies and codes of conduct adopted voluntarily by key transnational operators, an ongoing UN initiative to clarify the options for ensuring that transnational corporations respect human rights requirements, and the increasing significance of human rights factors in the practice and jurisprudence of international investment law.

Corporate Obligations

Despite recent interest in the idea of extending the application of key international human rights norms so as to apply directly to corporations, (Ratner 2001: 443) which could make the requirements of the human right to water central to arrangements for the privatisation of water and sanitation services, it is clear that such norms could only as yet dictate corporate behaviour where individual corporations have voluntarily agreed to abide by codes of conduct which explicitly or implicitly require compliance with international human rights norms.(Murphy 2005: 389) The UN Global Compact is a high-profile initiative which aims to encourage the adoption of such instruments and provides a set of 10 core principles which, while not referring explicitly to water rights or entitlements, does provide that companies should comply with international human rights norms, which might be argued to include a right to water and sanitation.(Williams 2007: 488–491) Indeed, Williams points out that the corporate code of conduct of at least one large transnational water services company expressly alludes to the UN Global Compact, thereby indirectly accepting international human rights obligations at the corporate level. [7] Though such acceptance remains voluntary and is not accompanied by any formal oversight or enforcement mechanisms, voluntary corporate policies or codes of conduct are increasingly likely to be linked to various forms of governmental assistance for corporations operating abroad. In a 2009 report, the Special Representative of the Secretary-General on the issue of human rights and transnational corporations and other business enterprises (SRSG) notes that

> "a growing number of States are adopting corporate social responsibility (CSR) policies. … In some cases, access to official assistance, such as export credit or investment insurance, may be linked to companies having a CSR policy, participating in the United Nations Global Compact, or confirming their awareness of the OECD Guidelines."[8]

The OECD Guidelines for Multinational Enterprises represent another legally non-binding initiative which might operate to support application of key elements of the human right to water to private companies. They consist of recommendations providing voluntary principles and standards for responsible conduct for multinational corporations operating in or from States which adhere to the OECD Declaration and cover business conduct in such areas of relevance to the implementation of the human right to water as human rights, environment, information disclosure, combating bribery, and consumer interests.[9] Though voluntary, it is significant that the Guidelines benefit from a formal monitoring apparatus as each of the 40 adhering States are required to set up a National Contact Point (NCP) responsible for the promotion of the Guidelines at the national level and for handling all enquiries and matters related to the Guidelines in that specific State, including the investigation of complaints made about a company operating in, or headquartered in that State.

There is a clear trend in the declarative practice of States towards extending responsibility for respecting human rights to private companies involved in the provision of private services. For example, the Draft Declaration on Guaranteed Access to Essential Services,(Morgan 2004: 181–182) proposed by France in the context of the 2002 World Summit on Sustainable Development held in Johannesburg, relates to essential services indispensible for a dignified life, expressly including drinking water and sanitation,[10] and would apply equally to both public and private sector providers. The Draft Declaration regards as central to the practical implementation of the right to access essential services the provision of an adequate and appropriate regulatory framework and, to this end, it appends to the right 'key principles of a code of sustainable management that would regulate the provision of such access'.(Morgan 2004: 181) However, this approach emphasises the role of the State in regulating private service providers, rather than extending international human rights obligations *per se* so as to apply directly to private business entities.

In respect of the development of an appropriate regulatory framework for private sector water service providers, which is informed by international human rights law and widely accepted by States and leading private operators, the adoption in 2007 of three sets of guidelines by Technical Committee 224 of the International Standards Organisation (ISO) represents a significant step.[11] The guidelines set standards for service activities relating to the provision of drinking water supply and sewerage services, which apply equally to both public and private actors, both as service provider and user. With 35 Participating Countries and 17 Observer Countries involved in their development and adoption, in liaison with a range of interested international organisations, including the World Health Organisation, World Bank and International Water Association, and with leading industry interests, the ISO guidelines must at least enjoy the indeterminate status of 'soft law' under international law and are likely to prove influential in determining an acceptable level of service provision where a dispute arises with a private sector provider, such as in the context of investment arbitration proceedings.

Of potentially far greater significance in this regard is the ongoing UN initiative on the issue of human rights and transnational corporations and other business entities which is developing practical recommendations for operationalising a framework for ensuring that private corporations respect human rights. As part of this initiative, the UN Sub-Commission on the Promotion and Protection of Human Rights had approved in August 2003 the ill-fated draft Norms on the Responsibilities of Transnational Corporations and Other Business Entities with Regard to Human

Rights,(Weissbrodt and Kruger 2003: 901)[12] which had sought to compel transnational corporations to comply with certain human rights obligations, traditionally applying only to State actors. ,(Weissbrodt and Kruger 2003: 901)[13] Recognising the very significant role played by non-State, corporate actors,(Knox 2008: 19) the Draft Norms sought to augment rather than replace State responsibilities with corporate obligations.[14] The document went on to list the key economic, social and cultural rights by which it purported to bind such corporations, including 'in particular the rights to development, adequate food and drinking water, the highest attainable standard of physical and mental health, adequate housing'.[15] Thus, the UN Draft Norms promised to do much to ensure that private sector water service providers respect the right of access to water services and, thereby, to contribute significantly to the general realisation of the right. However, the legal status, and thus the true normative value, of the Draft Norms was always in question.[16] At the request of the Human Rights Commission, the Economic and Social Council of the United Nations affirmed in 2004 that the Draft Norms 'have no legal standing' and thus that they represented nothing more than aspirational standards.(Williams 2007: 490–491) Indeed, in 2005 John Ruggie was appointed as Special Representative of the Secretary-General on the issue of human rights and transnational corporations and other business enterprises (SRSG) with a broad mandate to re-examine this entire issue, and he concluded in a 2006 Interim Report that 'the flaws of the Norms make that effort a distraction from rather than a basis for moving the Special Representative's mandate forward'.[17] Indeed, in April 2009 he further concluded that recent confusion in respect of business and human rights is due to the Draft Norms, which 'so co-mingled the respective responsibilities of States and companies that it was difficult if not impossible to disentangle the two'.[18] Therefore, it seems safe to assume that the UN Draft Norms are well and truly a dead initiative in their current form.

In 2008, the SRSG proposed a new approach for understanding the issue of human rights and transnational corporations, based on the "protect, respect and remedy" policy framework. The framework is described as resting on three complementary pillars:

> "the State duty to protect against human rights abuses by third parties, including business, through appropriate policies, regulation, and adjudication; the corporate responsibility to respect human rights, which in essence means to act with due diligence to avoid infringing on the rights of others; and greater access by victims to effective remedy, judicial and non-judicial."

Though much of the detail of this policy framework continues to be elaborated through the ongoing work of the SRSG, this approach appears already to enjoy considerable support from the Human Rights Council[19] and among States, leading business entities and civil society.[20]

In respect of the State's duty to protect, the SRSG's 2009 Report states that this requires each State to ensure the protection of rights 'against other social actors, including business, who impede or negate those rights' and that it 'applies to all recognized rights that private parties are capable of impairing, and to all types of business enterprises'.[21] The SRSG makes it clear that this duty relates to a standard of conduct rather than a standard of result and explains that States 'may be considered in breach of their obligations where they fail to take appropriate steps to prevent it [abuse] and to investigate, punish and redress it when it occurs'.[22] In respect of the corporate responsibility to respect, the SRSG's Report states that, in addition to the need to comply with all applicable national law, companies must at a minimum have regard to the one 'social norm' which 'has acquired near-universal recognition by all stakeholders, namely the corporate responsibility to respect human rights, or, put simply, to not infringe on the rights of others'.[23] He goes on to explain that in order to satisfy the corporate responsibility to respect, which 'exists independently of State duties and variations in national law',[24] and encompasses 'all internationally recognized human rights',[25] '[W]hat is required is an ongoing process of human rights due diligence, whereby companies become aware of, prevent, and mitigate adverse human rights impacts'.[26] Also, as water privatisations will often involve commitments on the part of companies to invest in water infrastructure, which usually involves huge capital costs, it is significant that the SRSG views the due diligence obligation to respect human rights as applying to banks and other lenders.[27] Such lenders or investors would obviously include international financial institutions (IFIs) or multilateral development banks (MDBs) supporting private sector investment in water. In respect of the provision of greater access by victims of human rights violations to effective remedy, the SRSG's Report outlines the role of the State obligation 'to take appropriate steps to investigate, punish and redress corporate-related abuse of the rights of individuals within their territory and/or jurisdiction',[28] (as distinct from the individual right to remedy recognised in a number of international and regional human rights conventions), and highlights the emphasis placed by some international and regional human rights bodies, in elaborating upon the nature of this State obligation, on 'ensuring that effective remedial processes exist for abuses by private companies carrying out "State functions".'[29] The Report also outlines the role of non-judicial mechanisms, such as grievance mechanisms established at the company, national or international levels, and proposes a set of principles which should underpin all such mechanisms.[30]

Therefore, while undoubtedly immensely constructive, and much more in keeping with the traditionally accepted precepts of international law,[31] the "protect, respect and remedy" framework currently being pursued by the SRSG is rather softer in terms of the nature of the human rights obligations that it seeks to impose on private business entities than the ill-fated UN Draft Norms. However, due it's more legally cautious and considered approach and its greater emphasis on the practical and extra-legal extension of such obligations, it is likely to have a more profound impact on the day-to-day operations of private sector water and sanitation service providers. For example, the SRSG's commitment to exploring the significance of international human rights obligations for the negotiation and application of investment and trade agreements is likely to prove significant for water services companies operating in developing States.

A number of commentators, most notably Knox, highlight the hazards associated with proposals to place human rights duties directly on private actors, particularly businesses. In particular, he points out that, while '[I]nternational law has the *legal* capacity to place direct horizontal duties on all private actors not to violate one another's human rights', '[W]hat it lacks is the *practical* and *political* capacity to *enforce* those duties.'(Knox 2008: 19) Knox argues that very significant practical considerations 'militate toward leaving private violations of human rights to domestic law', (Knox 2008: 20) particularly as 'the obligation of States to *ensure* the right goes beyond merely avoiding direct violation … [I]t requires affirmative action to secure the right, or make it safe from loss or interference', based on a requirement 'that governments use due diligence to ensure that human rights are protected from private interference'. (Knox 2008: 21)

However, he readily concedes that 'powerful reasons argue against leaving private violations *entirely* to domestic law', as 'domestic governments may fail to prevent it [a private actor] from interfering with other's human rights', thus indicating 'the need to use international law where domestic institutions are inadequate' and providing 'international human rights law with a crucial, albeit limited, role in specifying, placing, and enforcing private duties'. (Knox 2008: 20)

International Investment Law

Recent developments in international investment law, the body of rules that functions to provide protection to private sector investors operating in foreign jurisdictions against arbitrary interference with their property or business interests by the sovereign actions of host States, suggest that this area may offer some clarity in respect of the requirements imposed by the human right to water on host States and private actors in cases of water services privatisation. Though inviting foreign investment in formerly public water and sanitation services is frequently intended to improve the quality and coverage of such services, and thus to further the realisation of the human right to water, it is quite clear that this body of law can function to restrict the development and application of human rights norms. There remains considerable scope for foreign investors to rely on international investment rules to challenge human rights-inspired regulatory measures taken by host States before international investment tribunals. Though, in principle,

> "in such cases states should be able to invoke their human rights obligations under international treaties in defence of their regulatory measures since states have an obligation to protect human rights under both customary and conventional international law," (Subedi 2008: 168)

the situation is complicated by 'deficiencies in the clarity of the substantive norms protecting human rights on the one hand and foreign investments on the other hand' as well as by a lack of 'interfaces between the supervision mechanisms of international investment law and international human rights'.(Kriebaum 2006: 3–4)

The rules on the protection of foreign investors are principally set out in a complex and extensive web of bilateral investments treaties (BITs) and regional investment treaties (RITs). Over 2,300 BITs have now been concluded between States, of which over 1,700 are currently in force.[32] Such agreements will normally provide for, *inter alia*, the fair and equitable treatment of investors, the payment of compensation in the event of direct or indirect expropriation of the investment, and mandatory dispute settlement before an investment arbitration tribunal. General international law in respect of indirect, regulatory expropriation remains somewhat unclear. Obviously, States enjoy the freedom, as a function of the fundamental principle of State sovereignty, to regulate commercial enterprises, including regulation for economic and social purposes, and in many instances non-discriminatory and *bona fide* regulation will be regarded as falling within the accepted 'police powers' of States and therefore perfectly lawful and not incurring the liability of the host State.(Kriebaum 2006: 15) On the other hand, international tribunals have found that international rules on expropriations may 'cover non-discriminatory regulation that might be said to fall within an exercise of a state's so-called police powers'.[33] Kriebaum reports that the jurisprudence of international arbitral tribunals in regulatory expropriations is, somewhat confusingly, dominated by two opposed theoretical approaches. Under the "sole-effects" doctrine, in determining whether an indirect expropriation has occurred, the tribunal will only consider the effect of the governmental measure on the investment and will entirely disregard the purpose of the measure.[34] In contrast, under the "police powers" doctrine, the tribunal will also consider a range of other factors including, in particular, the purpose of the measure.[35] Clearly, where the police powers doctrine is employed, particularly in its more extreme form, the regulatory capacity of host States in the area of human rights protection would be very considerable and the protection offered to foreign investors significantly reduced. Though Kriebaum favours the intermediate approach adopted in *Tecmed*, whereby the Tribunal first examined the impact of the measure on the investment and then assessed whether that impact was proportional to the measure's stated public purpose,[36] she confirms that a 'binary system' appears to operate in terms of these theoretical approaches and that 'it is impossible to predict which strand a tribunal will follow and which outcome it will reach' so that, under relevant customary and general international law, 'neither investors nor states are in a position to foresee the consequences of their actions'.[37]

In addition, the principle of 'fair and equitable treatment', whether included under a BIT or invoked under customary international law, might operate to restrict the capacity of host States to legislate for the realisation of human rights. Subedi points out that the scope of this 'cardinal principle of foreign investment law' has recently been expanded 'in a broad manner to catch a wide variety of state activities that have detrimental consequences for foreign investors'.[38] In this regard, he cites the recent ICSID Tribunal case of *Enron v. Argentina*, where the Tribunal concluded that 'a key element of fair and equitable treatment is the requirement of a "stable framework for the investment"' and stated further that 'this interpretation has been considered "an emerging standard of fair and equitable treatment in international law"'.[39] In respect of the legal capacity of host States to take measures for the realisation of human rights, such as the human right to water, Subedi is concerned to ensure that 'investment tribunals do not unduly stretch the protection under the principle to include all sorts of governmental activities, of both omission and commission', in a manner inconsistent with the established tenets of general international law.(Subedi 2008) He notes that this is an absolutely central issue in recent investment arbitration cases.[40]

However, as noted above, the specific rules applying to the protection of foreign investors are usually set out in BITs and RITs, negotiated between the host State(s) and the home State(s) in which the investor is based. In addition, the investor and the host State will normally conclude an investment contract, particularly in the case of infrastructure projects or concessions relating to the provision of public services, such as water supply or sanitation services, which typically 'provide far higher levels of protection to investors than typical commercial contracts by insulating them from future changes in the domestic law of the host country'.(Subedi 2008: 171) Such "stabilisation clauses" provide that 'the law in force in the host state at the time the contract takes effect is the law that will apply to the terms of the contract, regardless of future changes in the legal regime of the host country'.(Subedi 2008) A stabilisation clause has been similarly described as one which 'purports to obligate the State to maintain, for a long-term period, the regulatory framework that applies to the investment at the time the investment is made'(Harten 2007: 381) and might function 'to frustrate the protection of human rights in the country'. (Harten 2007: 382) Van Harten further describes stabilisation clauses as 'a legal mechanism to secure the economic interests of an economic combine against the danger posed by the right of a self-governing people to determine their future'. (Harten 2007: 382) Likewise, Subedi observes that '[S]uch provisions are liable not only to limit the sovereignty of states but also undermine the object and purpose of future international human rights and environmental treaties'(Subedi 2008: 171) and questions the consistency of such provisions with the principles of economic sovereignty and the right of economic self-determination of States, which he describes as 'serious contenders to qualify as principles of *jus cogens*, which override all other rules, whether treaty-based or otherwise'. (Subedi 2008: 170) However, he explains that 'due to the narrow commercial focus of international investment or commercial arbitration tribunals, adequate attention to such competing principles of international law has not been given in resolving investment disputes', nor to 'this actual and potential conflict'. (Subedi 2008: 172; Kriebaum 2006: 22)

At any rate, such stabilisation clauses have been enforced by arbitral tribunals and have formed the principal basis of arbitration awards.[41] Indeed, even in the *Methanex* case, cited by Kriebaum as an example of an investment arbitration tribunal applying a radical police powers doctrine, the NAFTA Tribunal stated categorically that

> "As a matter of general international law, a non-discriminatory regulation for a public purpose … is not deemed expropriatory or compensable *unless specific commitments had been given by the regulating government to the then putative foreign investor contemplating investment that the government would refrain from such regulation.*"[42]

Therefore, international investment arbitration tribunals would appear to be quite prepared to apply stabilisation clauses so as to restrict exercise of the otherwise legitimate police powers of host States, even perhaps where such exercise is intended to further the realisation of human rights in line with the international obligations of host States.

The enforceability against host States in international law of such stabilisation clauses, or of any other contractual clauses entered into with investors, may be greatly enhanced through the inclusion of so-called 'umbrella clauses' in investment treaties. According to Van Harten, '[M]any treaties contain such clauses, which typically provide for the States Parties to "respect" or "observe" or "abide by" all of their obligations and commitments to foreign investors'.[43] Such clauses have been relied upon by arbitral tribunals[44] and may be interpreted broadly as one which 'transforms a breach of contract by the State into an outright violation of the treaty'.(Van Harten 2007: 388) Therefore, international investment law might well be expected to operate to insulate and exempt foreign investors from the effect of regulatory measures introduced by host States, subsequent to the agreement of a water services concession, for the purposes of ensuring realisation of the human right to water.[45]

However, it is interesting to note that the concept of a human right of access to water and sanitation would appear to have received solid, if implicit, support in recent statements of international arbitration tribunals engaged in the settlement of investor–State disputes over water and sanitation service contracts. The International Centre for the Settlement of Investment Disputes (ICSID), the most commonly used arbitration institution in investor–State cases,(Tienhaara 2007: 230) has reversed its previous position to refer pointedly to the human rights implications of such contracts while weighing the element of public interest at stake in the particular dispute in question. In the course of a very high profile and controversial dispute over the privatisation of water services in the Bolivian city of Cochabamba, an ICSID Arbitral Tribunal decided in January 2003[46] that, without the agreement of the parties to the dispute, it did not have jurisdiction to grant a request from interested third parties to, *inter alia*, attend the proceedings and to have access to all submissions made to the Tribunal.[47] Further, the Tribunal did not believe that any useful purpose could be served by permitting the Petitioners to make submissions at that particular point.(Tienhaara 2007: 234) This decision suggests that the Tribunal was not at that time prepared to accord much significance to the public interest aspects, and by inference to the human rights elements, of the dispute. However, in a landmark decision in May 2005, concerning a petition from five NGOs to participate as *amici curiae* in investor–State proceedings concerning a water distribution and sewage services contract in Argentina, the ICSID would appear to have recognised that the provision of water and sanitation services inevitably involves questions of human rights.[48] In assessing the appropriateness of the subject matter of the case for *amicus curiae* submissions,[49] the ICSID reasoned that

> "The factor that gives this case particular public interest is that the investment dispute centres around the water distribution and sewage systems of a large metropolitan area, the city of Buenos Aires and surrounding municipalities. Those systems provide basic public services to millions of people and as a result may raise a variety of complex public and international law questions, *including human rights considerations*."[50]

This position was followed, almost *verbatim*, by an Arbitral Tribunal composed of the same members in a March 2006 decision concerning a petition to submit an *amicus* brief in very similar proceedings from an NGO and three named individuals, variously described as having expertise in law, human rights, and sustainable development.[51] It seems logical to conclude that the Tribunal could only have had in mind the emerging concept of the human right to water. In justifying the possibility of accepting *amicus* submissions from appropriate non-parties, which 'may be able to afford the Tribunal perspectives, arguments, and expertise that will help it arrive at a correct decision',[52] the Tribunal distinguished the matters of public interest arising therein from the 'matters of public interest … present in virtually all cases of investment treaty arbitration under ICSID jurisdiction'. What set this case apart from other 'ostensibly private litigation' before the ICSID was quite clearly the 'particular public interest' flowing from the fact that it involved the provision of water and sanitation services.[53]

Indeed, in the February 2007 decision of the ICSID Tribunal in respect of the petition by five NGOs to make an *amicus curiae* submission in the proceedings of the case relating to the provision of water supply and sewage services in Buenos Aires, resubmitted in light of the Tribunal's May 2005 Order,[54] the Tribunal appreciated the potential impact that its final decision might have for the realisation of elements of the human right to water for the millions of people within the area covered by the contract.[55] In response to the claimants' argument that the withdrawal from the claim of AASA, the former operator of the water and sewage system, changed the nature of the subject matter of the case, so that it no longer involved sufficient matters of public interest to make it appropriate for an *amicus* submission, the Tribunal confirmed its finding of May 2005, stating that

> "Even if its decision is limited to ruling on a monetary claim, to make such a ruling the Tribunal will have to assess the international responsibility of Argentina. In this respect, it will have to consider matters involving the provision of "basic public services to millions of people". To do so, *it may have to resolve "complex public and international law questions, including human rights considerations*" (Order of May 19, 2005, para. 19). It is true that the forthcoming decision will not be binding on the current operator of the water and sewage system of Buenos Aires. *It may nonetheless have an impact on how that system should and will be operated.*"[56]

This statement clearly anticipates the possibility that consideration of human rights requirements might result in a Tribunal decision that stipulates basic standards of water and sanitation service provision that any contracted private operator ought to meet, irrespective of any standards set out in a concession contract. Further, the Tribunal recognised the potential significance of its own pending final decision in this case for the identification of minimum acceptable standards of water and sanitation service provision generally, and thus for the implementation of core elements of the human right to water:

> "More generally, because of the high stakes in this arbitration and the wide publicity of ICSID awards, one cannot rule out that *the forthcoming decision may have some influence on how governments and foreign investor operators of the water industry approach concessions and interact when faced with difficulties*."[57]

At the very least, the Tribunal is acknowledging the potential significance of its decision in this case for ensuring basic standards in the case of water services privatisations involving foreign investors. However, taken at its broadest, this statement could also amount to recognition of the importance of ICSID practice for the application and development in international law of the human right to water.

The ICSID took a similar approach in another decision in February 2007 allowing a petition by five NGOs to make an *amicus* submission in the course of an investor–State dispute concerning a private sector lease contract for water supply in Tanzania.[58] In seeking to establish their credentials for making an *amicus* submission, the petitioners emphasised their interests, *inter alia*, in 'ensuring that the constitutional and environmental rights of the Tanzanian people are secured and realized by all',[59] in 'issues relating to access to water, especially for the poor and women',[60] and in using 'international law, institutions and processes to protect the environment, human health and human rights'.[61] Also, they included among the reasons for their petition the fact that 'this arbitration process goes far beyond merely resolving commercial or private conflicts, but rather has a substantial influence on the population's ability to enjoy basic human rights'[62] and 'raises a number of issues of vital concern to the local community in Tanzania, and a wide range of potential issues of concern to developing countries (and indeed all countries) that have privatised, or are contemplating a possible privatisation of, water or other infrastructure services'.[63] In considering the Petitioners' related request to have access to the key arbitration documents, the Tribunal described the Petitioners' proposed role in the proceedings, stating that 'it is envisaged that the Petitioners will address broad policy issues concerning sustainable development, environment, *human rights and governmental policy*'.[64] In response to the Claimant's argument that sufficient issues of public interest do not arise in the present case as the claimant has exited Tanzania and the operator is defunct, the Tribunal, finding for the Petitioners, found that 'this is not determinative of the issue, since any decision by the Arbitral Tribunal *still has the potential to impact upon the same wider interests*'.[65]

Recognition by ICSID tribunals of the relevance of human rights considerations in international investment disputes concerning water services, though implicit, is significant in terms of the recognition of the human right to water in general international law, especially as the incidence of investor–State disputes has risen significantly in recent years.(Tienhaara 2007: 230) Also, human rights considerations are increasingly likely to be raised before investment dispute tribunals in light to the trend towards allowing non-disputant third party participation where such disputes raise serious questions of public interest.(Tienhaara 2007: 231–238) As examples of this trend, Tienhaara lists the 2001 decision of the UNCITRAL *ad hoc* Arbitral Tribunal to accept *amicus curiae* briefs in the claim made under Chapter 11 of NAFTA[66] in the *Methanex* case[67] and the subsequent 2003 statement of the Free Trade Commission of NAFTA on third party participation in Chapter 11 Disputes;[68] the new format 'model' bilateral investment treaties (BITs) adopted by Canada in 2003[69] and by the US in 2004[70] expressly providing for third party participation in investment dispute proceedings; the adoption of new ICSID Rules of

Procedure for Arbitration Proceedings in 2006;[71] and the 2006 decision of UNCITRAL that its Working Group II on International Arbitration and Conciliation should commence work on revision of the UNCITRAL Arbitration Rules in respect of, *inter alia*, third party intervention and disclosure of awards.[72] One might add to this list the binding interpretative note adopted by the NAFTA Parties in 2001 providing public access to relevant documents submitted to or issued by NAFTA Chapter 11 tribunals and the making public of awards.[73]

Though the ICSID decisions cited above do not make clear on precisely what sources of law the tribunals have based the conclusion that they are entitled or required to consider questions of human rights requirements or entitlements, it is clear that in deliberating on such investment disputes the tribunals may generally have regard to rules contained under

> "the ICSID Convention, multilateral investment treaties, bilateral investment treaties (including separate investment chapters in 'economic integration agreements'), customary international law, general principles of law, specific agreements or decisions, and national legislation." (Fauchald 2008: 2 of 74 online)

In his painstaking empirical study of the sources of law applied by ICSID tribunals, Fauchald explains that

> "The instrument referring the case to ICSID arbitration often determines the rules to be applied. Otherwise, Article 42(1) of the ICSID Convention applies [which provides that] 'the Tribunal shall apply the law of the Contracting State party to the dispute … *and such rules of international law as may be applicable*'." (Fauchald 2008 online)

Further, in examining the approach of ICSID tribunals to the application of customary international law and general principles of law in particular, he observes that

> "ICSID tribunals may be faced with the application of customary international law and general principles of law where there is a reference to such rules in a relevant treaty, where the treaty does not address the issue in question, *i.e.,* where there is a legal *lacuna* in the treaty, and where customary international law replaces a clause in a treaty (*i.e.,* it develops after the treaty has been concluded)." (Fauchald 2008: 6 of 74 online)

In respect of the first situation listed above, he notes that it is 'quite common to include clauses referring to customary international law in relevant treaties',(Fauchald 2008: 7 of 74 online) including, for example, Article 1131(1) of NAFTA. Therefore, the tribunals may be inferring the human rights dimension in the above cited cases from a variety of legal sources, including customary international law and general principles of law as listed in Article 38(a)–(c) of the Statute of the International Court of Justice. Thus it would be open to ICSID tribunals to identify and apply elements of the human right to water as leading-edge principles and concepts of customary or general international law. By appearing to interpret and apply the applicable law in a manner consistent with the key parameters of the human right to water, the ICSID approach is entirely consistent with Paragraph 35 of General Comment 15 which provides that '[A]greements concerning trade liberalization should not curtail or inhibit a country's capacity to ensure the full realization of the right to water'.

Clearly, it is not difficult to imagine a situation where a poor State, desperate to gain access to the capital and expertise of a foreign private sector water services operator, might be tempted to limit its policy options, at least in respect of that particular investor, in exchange for such private investment. Indeed, ample evidence suggests that such stabilisation clauses are very much more prevalent in investor contracts with poorer States.[74] Where a stabilisation clause is included in a water services concession contract, the State's freedom to introduce subsequent measures intended to ensure the realisation of the human right to water would appear to be severely limited, and this situation would be exacerbated where an umbrella clause is included in the relevant investment treaty. This suggests that one potentially effective means of facilitating the adoption of such national measures by host States would involve the inclusion of an express reference to human rights considerations, or at least to the human right to water, in bilateral or regional investments treaties. After all, a number of international investment disputes based on BITs have already arisen concerning water services concessions.[75] Such a reference might recognise or make a limited exceptions for, the requirements of the human right to water.[76] However, in addition to introducing considerable uncertainty into investment treaties, due to the substantive uncertainty of the requirements inherent in the concept, such a treaty provision could only impact on concession contracts concluded subsequent to entry into force of the treaty. It remains, however, an approach that human rights groups might advocate to States that have declared support for the concept, whenever such States are negotiating or reviewing bilateral investment treaties, especially with developing countries. Equally, host States negotiating water services concession contracts with foreign investors might seek inclusion of such a limited exception to any stabilisation clause. At any rate, the inclusion of an express reference to, or exceptions for, considerations relating to the human right to water in bilateral or regional investments treaties would be entirely consistent with the exhortations of the CESCR in Article 35 of General Comment 15, which provides in full:

> "States parties should ensure that the right to water is given due attention in international agreements and, to that end, should consider the development of further legal instruments. With regard to the conclusion and implementation of other international and regional agreements, States parties should take steps to ensure that these instruments do not adversely impact upon the right to water. Agreements concerning trade liberalization should not curtail or inhibit a country's capacity to ensure the full realization of the right to water."

Ultimately, States seeking foreign private investment in water services that are seriously committed to ensuring realisation of the human right to water would be well advised to introduce measures setting out the appropriate regulatory framework prior to conclusion of any concession contract. It is telling that the UN SRSG should recently report that '[A]s a next step, the Special Representative is exploring the feasibility of developing guidance on "responsible contracting" for host Government agreements in relation to human rights.'[77]

Conclusion

A human rights-related dispute might typically arise in the context of water services privatisation where the privatised service is not operated in accordance with human rights requirements, for example where the operator fails to extend the service to all those that require it or where those unable to pay for the service are denied access to a minimum quantity of water. Though the emergence of the human right of access to water is unlikely, at least under international law, to impose specific obligations directly on private entities involved in the provision of water supply and sanitation services, the concept would appear recently to have received implicit support in the statements of international arbitration tribunals engaged in the settlement of investor–State disputes over water and sanitation service contracts. The International Centre for the Settlement of Investment Disputes (ICSID), the most commonly used arbitration institution in investor–State cases, has on a number of occasions pointedly referred to the human rights implications of water and sanitation concession contracts while weighing the element of public interest at stake in the particular dispute in question. ICSID tribunals have appreciated that their decisions may be expected to influence the practice of States in framing legal and contractual arrangements for the participation of the private sector in water services and, thus, their potential impact on the realisation of elements of the human right to water for the millions of people living within areas covered by such contracts. Such recognition by ICSID tribunals, though implicit, is significant on a number of levels in terms of the recognition of the human right to water in general international law. The incidence of investor–State disputes has risen significantly in recent years and human rights considerations are increasingly likely to be raised before investment dispute tribunals in light to the trend towards allowing non-disputant third party participation where such disputes raise serious questions of public interest.

Irrespective of how the acknowledged human rights considerations are to be taken account of in international investment disputes, one potentially effective means of facilitating the adoption of national measures intended to ensure the realisation of the human right to water by host States would involve the inclusion of an express reference to human rights considerations, or at least to the human right to water, in bilateral or regional investments treaties. Human rights groups might advocate the inclusion of certain limited exceptions to States that have declared support for the concept, whenever such States are negotiating or reviewing bilateral investment treaties, especially with developing countries. Equally, host States negotiating water services concession contracts with foreign investors might seek inclusion of such a limited exception to any stabilisation clause employed. The opportunities for effective intervention by advocates of the human right to water during the negotiation and conclusion of international investment treaties are patently obvious. In seeking to build on the advances hinted at by ICSID tribunals, such advocates need only to refer to the examples of a State's violation of the obligation to fulfil the right to water listed under paragraph 44(c)(vii) of General Comment 15, which include the 'failure of a State to take into account its international legal obligations regarding the right to water when entering into agreement with other States or with international organizations.'

References

Backer, L. Catá, 'Multinational Corporations, Transnational Law: The United Nations' Norms on the Responsibilities of Transnational Corporations as a Harbinger of Corporate Social Responsibility in International Law', (2006) 37 *Columbia Human Rights Law Review*.

Fauchald, O. K., 'The Legal Reasoning of ICSID Tribunals – An Empirical Analysis', (2008)19/2 *European Journal of International Law* 301–364, page 2 of 74 of online version. (References omitted).

Knox, J. H., 'Horizontal Human Rights Law', (2008) 102: 1 *American Journal of International Law* 1, at 19.

Kriebaum, U., 'Privatizing Human Rights – The Interface Between International Investment Protection and Human Rights', (2006) 3/5 *Transnational Dispute Management* 1, at 3–4.

Mann, C. C., 'The Rise of Big Water', (May 2007) *Vanity Fair*.

Morgan, B., 'The Regulatory Face of the Human Right to Water' (2004) 15/5 *Water Law*.

—. 'Turning Off the Tap: Urban Water Service Delivery and the Social Construction of Global Administrative Law', (2006) 17/1 *European Journal of International Law* 215–246, at 218.

Morgan, Bronwyn, "Turning Off the Tap: Urban Water Service Delivery and the Social Construction of Global Administrative Law", *European Journal of International Law* 17 (2006)

Murphy, S. D., 'Taking Multinational Corporate Codes to the Next Level', (2005) 43 *Columbia Journal of Transnational Law* 389.

Petrova, V., 'At the Frontiers of the Rush for Blue Gold: Water Privatization and the Human Right to Water' (2006) 31 *Brook. Journal of International Law* 577, at 584.

Prasad, N., ''Privatisation Results: Private Sector Participation in Water Services After 15 Years', (2006) 24 *Development Policy Review* 669; C. C. Mann, 'The Rise of Big Water', (May 2007) *Vanity Fair*, at 122.

Ratner, S. R., 'Corporations and Human Rights: A Theory of Legal Responsibility', (2001) 111 *Yale Law Journal* 443.

Schreiber, W., 'Realizing the Right to Water in International Investment Law: An Interdisciplinary Approach to BIT Obligations', [Spring 2008] 48 *Natural Resources Journal* 431–478, at 434.

Smets, Henri, "Economics of Water Services and the Right to Water" in *Fresh Water and International Economic Law*, ed. Edith Brown Weiss, Laurence Boisson de Chazournes and Nathalie Bernasconi-Osterwalder, (Oxford: Oxford University Press, 2005)

Subedi, S. P., *International Investment Law: Reconciling Policy and Principle* (Hart Publishing, Oxford / Portland, 2008), at 168.

Tienhaara, K., 'Third Party Participation in Investment – Environment Disputes: Recent Developments', (2007) 16/2 *Review of European Community and International Environmental Law* 230, at 230.

Van Harten, G., 'The Public–Private Distinction in the International Arbitration of Individual Claims Against the State' (2007) 56 *International and Comparative Law Quarterly* 371, at 381.

Van Hofwegen, *Report of the Task Force on Financing Water for All* (2006).

Weissbrodt, D. and M. Kruger, 'Norms on the Responsibilities of Transnational Corporations and Other Business Enterprises with Regard to Human Rights', (2003) 97 *American Journal of International Law* 901.

Williams, M., 'Privatization and the Human Right to Water: Challenges for the New Century', (2007) 28 *Michigan Journal of International Law* 469, at 493

Winpenny, J., *Financing Water for All: Report of the World Panel on Financing Water Infrastructure* (2003).

Notes

[1] See http://www.water2001.de/outcome/BonnRecommendations/Bonn_ Recommendations.pdf

[2] See, United Nations Development Programme, *Human Development Report 2006. Beyond Scarcity: Power, poverty and the global water crisis* (UNDP, 2006), at 77; 3rd UN World Water Development Report: *Water in a Changing World* (UNESCO / Earthscan, 2009), at 9, 62–63, 262, 287.

[3] UNDP, 2nd UN World Water Development Report (2006).

[4] See Committee on Economic, Social and Cultural Rights, *General Comment No. 15, The Right to Water (Articles 11 and 12 of the International Covenant on Economic, Social and Cultural Rights)*, UN DOC. E/C.12/2002/11, 26 November 2002, paras. 23 and 24. Hereinafter, General Comment 15.

[5] General Comment 15, *ibid.,* paras. 25–29.

[6] General Comment 15, *ibid.,* para. 33.

[7] RWE AG Corporate Code of Conduct, at 7–8. See, Williams, 2007, at 488, who also points out that RWE is listed as a participant in the UN Global Compact.

[8] *Promotion of All Human Rights, Civil, Political, Economic, Social and Cultural Rights, Including the Right to Development–Business and human rights: Towards operationalizing the "protect, respect and remedy" framework,* A/HRC/11/13 (22 April 2009), at 8, para. 21.

[9] *See* Organisation for Economic Co-operation and Development, OECD Guidelines for Multinational Enterprises, June 21, 1976, available at

http://www.oecd.org/dataoecd/56/36/1922428.pdf (revised 2000).
[10] Draft Declaration, Article 1.
[11] On the background to this ISO initiative, see generally, Morgan, 2004, at 182–183. See also Morgan, 2006, at 224–227.
[12] ECOSOC, Commission on Human Rights, Sub-Commission on the Promotion and Protection of Human Rights, *Norms on the Responsibilities of Transnational Corporations and Other Business Entities with Regard to Human Rights*, UN Doc. E/CN.4/Sub.2/2003/12/Rev.2 (26 August 2003).
[13] ECOSOC, Commission on Human Rights, Sub-Commission on the Promotion and Protection of Human Rights, *Norms on the Responsibilities of Transnational Corporations and Other Business Entities with Regard to Human Rights*, UN Doc. E/CN.4/Sub.2/2003/12/Rev.2 (26 August 2003).
[14] Article H, para. 19.
[15] Article E, para. 12.
[16] On the drafting history and legal status of the UN Draft norms generally, see L. Catá Backer, 2006: 287.
[17] UN Doc. E/CN.4/2006/97 (22 February, 2006), para. 69. See Williams, *ibid.* See further, J. G. Ruggie, 'Business and Human Rights: The Evolving International Agenda', (2007) 101 *American Journal of International Law* 819, at 826–827. See Knox, 2008, at 41–42.
[18] UN Doc. A/HRC/11/13 (22 April, 2009), *supra,* n. 8, at 16, para. 58.
[19] UN Doc. A/HRC/8/5. See UN Doc. A/HRC/11/13, *ibid.,* at 3, para. 1.
[20] UN Doc. A/HRC/11/13, *ibid.,* at 3–4, paras. 3–5.
[21] *Ibid.,* at 6–7, para. 13.
[22] *Ibid.,* at 7, para. 14.
[23] *Ibid.,* at 13, para. 46.
[24] *Ibid.,* at 14, para. 48.
[25] *Ibid.,* at 14, para. 52.
[26] *Ibid.,* at 14, para. 49.
[27] *Ibid.,* at 18, para. 73.
[28] *Ibid.,* at 21, para. 87.
[29] *Ibid.,* at 21, para. 89.
[30] *Ibid.,* at 23, para. 99. Originally proposed under the SRSG's 2008 Report, UN Doc. A/HRC/8/5, para. 92.
[31] A number of commentators highlight the hazards associated with proposals to place human rights duties directly on private actors, particularly businesses.
[32] UNCTAD Research Note, *Recent developments in international investment agreements*, UNCTAD/WEB/ITE/IIT/2005/1, at 1 and 7, cited by Kriebaum, 2006, at 2.
[33] NAFTA, *Pope and Talbot v. Canada* (Award, 26 June 2000), 7 ICSID Reports 69. See also, NAFTA, *SD Myers Inc. v. Govt. of Canada* (Partial Award, 13 November 2000), 40 ILM (2001) 1408. See further Kriebaum, *ibid.,* at 16.
[34] *Ibid.,* citing NAFTA, *Metalclad Corp. v. United Mexican States* (Award, 30 August 2000), 5 ICSID Reports 226.

[35] *Ibid.*, citing, at 18, NAFTA, *Methanex Corporation v. USA*, (Final Award, 3 August 2005) and UNCITRAL, *Saluka Investments BV v. Czech Republic*, (Partial Award, 17 March 2006), which emphasised the fact that measures were for a public purpose commonly accepted as being within the police powers of States in order to find no expropriation, as an 'extreme form of the police powers doctrine', and ICSID, *Tecnicas Medioambientales Tecmed SA v.United Mexican States,* (29 May 2003, 43 ILM (2004) 133, as an 'intermediate approach'.

[36] *Ibid.,* at 18, 20.

[37] *Ibid.*, at 20.

[38] Subedi, 2008, at 173. He points out, at 172, that:

> "traditionally the principle of fair and equitable treatment is supposed to be concerned with the obligation not to deny justice in criminal, civil or administrative adjudicatory proceedings in accordance with the principle of due process embodied in the principal legal systems of the world."

[39] *Enron Corporation and Ponderosa LP v. Argentine Republic,* ICSID Case No. ARB/01/3, Award of 22 May 2007, para. 260. See Subedi, 2008, at 172.

[40] Subedi, 2008, at 175. Indeed, he further points out, at 174, that

> "the expansion of the principle of fair and equitable treatment was one of the reasons given by Bolivia for its withdrawal from ICSID in May 2007, along with the allegation that ICSID tribunals have tailored the principle of fair and equitable treatment in favour of foreign corporations."

[41] Van Harten, 2007, at 384, cites the example of *Texaco Overseas Petroleum Co. and California Asiatic Oil Co. v. Government of the Libyan Arab Republic* (Merits) (19 January 1977), 17 *ILM* 4. See also, *AGIP SpA v. People's Republic of Congo* (Merits) (30 November 1979) 1 ICSID Rep. 306; 21 *ILM* 726, para. 85.

[42] *Supra*, n. 35, Part IV–Chapter D, at 4, para. 4. See Kriebaum, 2006, at 19. (Emphasis added.)

[43] *Ibid.,* at 388, citing A. C. Sinclair, 'The Origins of the Umbrella Clause in the International Law of Investment Protection', (2004) 20 *Arbitration International* 411, at 412–413.

[44] See *SGS Société Général de Surveillance v. Philippines* (Jurisdiction) (29 January 2004), 16 *World Trade and Arbitration Materials* 91, cited in Van Harten, 2007, at 389.

[45] On the essentially regulatory nature of measures required for the practical implementation of key elements of the human right to water, see generally, B. Morgan, 'The Regulatory Face of the Human Right to Water' (2004) 15/5 *Water Law* 179–186.

[46] ICSID Case No. ARB/02/3 *Aguas del Tunari SA v. Republic of Bolivia,* Letter from the President of the Tribunal Responding to the Petition (29 January 2003).

[47] ICSID Case No. ARB/02/3 *Aguas del Tunari SA v. Republic of Bolivia,* Petition of La Coordinadora Para la Defensa del Agua y Vida, La Federation Departmental Cochabamba de Organizaciones Regantes, Sempa Sur, Friends of the Earth Netherlands, Oscar Olivera, Omar Fernandez, Father Luis Sanchez, and Congressman Jorge Alvarado to the Arbitral Tribunal (29 August 2002).

[48] ICSID Case No. ARB/03/19, *Aguas Argentinas S.A., Suez, Sociedad General de Aguas de Barcelona S.A. and Vivendi Universal S.A. v. The Argentine Republic,* Order in Response to a Petition for Transparency and Participation as Amicus Curiae (19 May 2005).
[49] The Tribunal considered three basic criteria in considering whether to exercise its power to accept *amicus* submissions: a) the appropriateness of the subject matter of the case; b) the suitability of a given non-party to act as *amicus curiae* in that case; and c) the procedure by which the *amicus* submission is made and considered. *Ibid.,* para. 17.
[50] *Ibid.*, para. 19. (Emphasis added).
[51] ICSID Case No. ARB/03/17, *Aguas Provinciales de Santa Fe S.A., Suez, Sociedad General de Aguas de Barcelona S.A. and InterAguas Servicios Integrales del Agua S.A. v. The Argentine Republic,* Order in Response to a Petition for Participation as *Amicus Curiae* (17 March 2006).
[52] *Ibid.,* para. 20.
[53] *Ibid.,* para. 18
[54] *Supra,* n. 49.
[55] ICSID Case No. ARB/03/19, *Suez, Sociedad General de Aguas de Barcelona S.A. and Vivendi Universal S.A. v. The Argentine Republic,* Order in Response to a Petition by Five Non-Governmental Organizations for Permission to Make an *Amicus Curiae* Submission (12 February 2007).
[56] *Ibid.,* para. 18. (Emphasis added).
[57] *Ibid.* (Emphasis added).
[58] ICSID Case No. ARB/05/22, *Biwater Gauff (Tanzania) Ltd. V. United Republic of Tanzania*, Procedural Order No. 5 (2 February 2007).
[59] *Ibid.*, para. 11(a), regarding the Lawyers' Environmental Action Team (LEAT).
[60] *Ibid.*, para. 11(c), regarding the Tanzania Gender Networking Programme (TGNP).
[61] *Ibid.*, para. 11(d), regarding the Center for International Environmental Law (CIEL). The Petitioner NGOs are further described, at para. 20, as having

> "specialized interests and expertise in human rights, environmental and good governance issues locally in Tanzania, and *in the multiple critical inter-relationships between international investment law and sustainable development at the international level*."

(Emphasis added.)
[62] *Ibid.*, para. 14.
[63] *Ibid.*, para. 12.
[64] *Supra,* n. 58, at para. 64. (Emphasis added).
[65] *Ibid.,* para. 53. (Emphasis added).
[66] North American Free Trade Agreement (17 December 1992), 32 *ILM* 289.
[67] *Methanex Corp. v. United States of America,* Decision of the Tribunal on Petitions from Third Parties to Intervene as 'Amici Curiae' (15 January 2001). See further, S. Saha, 'Methanex Corporation and the USA: The Final NAFTA Tribunal Ruling', (2006) 15/1 *Review of European Community and International Environmental Law* 110; H. Mann, 'Opening the Doors, At Least a Little:

Comment on the *Amicus* Decision in *Methanex v. United States*', (2001) 10/2 *Review of European Community and International Environmental Law* 241. It is worth noting that the case involved a challenge by a Canadian investor to a Californian measure aimed at protecting ground water from contamination through the phasing out of a petroleum additive and that the non-disputant *amicus* briefs permitted in *Methanex* concerned environmental arguments additional to the public health arguments raised by the parties to the dispute. Clearly, similar public health and environmental grounds could lend support to elements of the human right to water.

68 Statement of the Free Trade Commission on Non-Disputing Party Participation (Montreal, 1 October 2003).

69 The Canadian model Foreign Investment Promotion and Protection Agreement (FIPA) is available at www.dfait-maeci.gc.ca/tna-nac/documents/2004-FIPA-model-en.pdf

70 The US Model BIT is available at www.state.gov/documents/ organization /29031.doc

71 *ICSID Convention, Regulations and Rules* (ICSID, April 2006), available at www.worldbank.org/icsid/basicdoc/basicdoc.htm

72 UNCITRAL, 39th Session (June/July 2006).

73 Notes of Interpretation of certain NAFTA Provisions (31 July 2001). See N. Bernasconi-Osterwalder and E. Brown Weiss, 'International Investment Rules and Water: Learning from the NAFTA Experience', in E. Brown Weiss, L. Boisson de Chazournes and N. Bernasconi-Osterwalder, *Fresh Water and International Economic Law* (2005, Oxford University Press) 263, at 286–287.

74 See, SRSG Report (22 April, 2009), *supra,* n. 8, at 11, para. 32.

75 The *Aguas del Tunari* dispute, ICSID Case No. ARB/02/3, is based on the BIT between the Netherlands and Bolivia, the *Aguas del Aconquija* dispute, ICSID Case No. ARB/97/3, is based on the BIT between France and Argentina, and the *Azurix* dispute, ICSID Case No. ARB/01/12, is based on the 1991 BIT between the USA and Argentina.

76 See, for example, the 2007 Norwegian Model BIT.

77 *Supra*, n. 8, at 12, para. 36.

CHAPTER EIGHTEEN

INTEGRATED AND ADAPTIVE WATER MANAGEMENT AS PART OF THE CLIMATE CHANGE ADAPTATIONS STRATEGIES – THE LEGAL APPROACH

VICKY KARAGEORGOU

Introduction: The Impact of Climate Change on Water Resources

The world is facing changes at a faster rate than ever seen before. Changes, such as population growth, migration, urbanization and climate change will have significant impacts on the way that water resources need to be managed in the future. Climate change is regarded, though, as the major catalyst for re-orienting water management, because it will impact significantly the water cycle, the water resources and the relevant services worldwide. It has been estimated[1] that even relatively small changes in few degrees centigrade can lead to an increase of water availability by 10–30% in some regions, while in others water availability will be decreased by 10–30%. Globally, the negative impacts of future climate change on freshwater systems are expected to outweigh the benefits, as it is projected that the area of land subject to increasing water stress due to climate change will be more than double that with decreasing water. Furthermore, there are sound reasons to expect an increase in the frequency and intensity of extreme weather events[2], such as floods and droughts, as well as long-term shifts on water availability.

Water is a critical core sector, so that changes in water availability and quality are expected to have significant impact on key economic activities, such as agriculture (increased demand for irrigation), energy (reduced hydropower potential and cooling water availability) and tourism (recreational activities). Climate associated changes in the available water resources and their quality are responsible for increased health risks[3] and are also expected to aggravate the impacts of other stresses and pressures, such as changing consumption and productions patterns and land-use changes. It becomes evident that the broader dynamics of the national economies can be influenced and the capacity of societies to pursue sustainable development can be put at risk. The challenges of climate change to water sources have thus to be understood and addressed in an interactive way, if societies are to adapt effectively to climate induced changes. Therefore, just as the gradual change of our energy habits, namely the way that societies produce and use their energy, constitutes the cornerstone of any legal and policy framework aiming at avoiding or even limiting the consequences of climate change (mitigation policies), a series of fundamental changes of the way that societies use and manage their water sources has to become an integral part of any policies aiming at coping with the unavoidable effects of climate change (adaptation policies)[4].

At this stage, a critical question comes to the fore. This question relates to whether the existing legislative and institutional framework, on which water management practices and protection measures are based, is sufficiently adapted to cope with the newly emerged climate-induced challenges on water resources. As a starting point for answering this question it should be mentioned that the majority of the current water practices, not only in developing but also in many developed countries, are – despite the different degrees of their effectiveness – not robust enough to cope with the impacts of climate change on water supply availability, flood risk, agriculture and aquatic ecosystems. This means subsequently that also the legislative framework in place at international, regional and national level should be to some extent reformed, in order to provide the necessary impetus for dealing with these changes.

The scope of the chapter is to analyze the capacity of the existing water-relevant legislative framework at international, regional and local level vis-à-vis its capacity to support adaptation, with the aim of providing certain suggestions about how and to which extent water legislation should be reformed in order to become sufficiently adapted. The structure of the analysis can be described as follows: The first section of the chapter focuses on adaptation as a strategic policy response to climate change as well as its linkages with water issues. In the second section a preliminary analysis of the existing legislative framework for freshwater at international and regional level is conducted, with the aim of searching for the directions given and the approaches adopted in order to make water legislation adaptive. The third section of the chapter pays attention to the paradigm shift that has already started taking place in environmental law in general and in water law in particular and has been accelerated due to climate change. In this context, the enhancement of the flexibility of the water allocation regimes at national and transboundary level is presented as one key area, on which efforts for making water legislation adaptive should focus. Finally, certain remarks are made, which relate to the challenges that the water law of the 21st century is facing.

I. Adaptation as a Policy Response to Climate Change and Linkages to Water Management

A. The international level

Since its inception, the international climate policy has focused predominantly on striving as far as possible to reduce greenhouse gases in order to prevent dangerous climate change. This policy approach can be described under the key word "mitigation". However, the existing legal framework for climate change, namely the United Nations Framework Convention on Climate Change (UNFCCC)[5] and the Kyoto Protocol[6], does not provide an official definition for mitigation. Therefore, the relevant definition given by the Intergovernmental Panel on Climate Change[7] (IPCC), which defines mitigation "an anthropogenic intervention to reduce the *sources* or enhance the *sinks* of greenhouse gases", has become predominant in the international literature.

Increasingly, however, the climate-driven changes are being recognized as irreversible and societies must take necessary measures to adapt to the impacts of these changes. This policy approach is described under the keyword "adaptation". The action to be taken to deal with the unavoidable effects of climate change is, though, not only a matter of necessity. It is also a matter of equity, as the impacts fail disproportionately on those least able to bear them[8]. The international climate change regime recognizes the need for adaptation, by setting several obligations to the countries to take adaptive action. As the legislative framework in place does not provides an official definition for adaptation, the definition given by the Intergovernmental Panel on Climate Change (IPCC) also is chosen in this case for the purpose of the analysis[9]. In particular, IPCC defines adaptation as "adjustment in natural or human systems in response to the actual or expected climate stimuli or their effects, which moderates harm or exploits beneficial opportunities".

From a systematic point of view, mitigation and adaptation should be thus seen as complementary policy approaches, both necessary for an effective climate policy[10]. Mitigation and Adaptation strategies employ institutional, technological and behavioral options, the use of a wide range of regulatory and economic instruments with aim to encourage the use of certain mainly behavioral and technological options as well as research and development initiatives for reducing uncertainty and increasing resilience of the ecosystems and the communities, although they have to some extent different orientations. Furthermore, the comparative analysis of the chosen definitions can be – to some extent – enlightening concerning the specific characteristics of each policy approach that can be described as follows: 1) the mitigation reduces both negative and positive impacts, while adaptation is selective in terms of taking advantage of the positive ones and reducing the negative 2) while mitigation has global benefits, adaptation works mainly on the scale of an impacted system, which is mostly local or even regional 3) the effects of adaptive actions are more to difficult to measure in a metric way in comparison to the results of the mitigation efforts[11].

Besides the guidance provided by the definitions and the relevant analysis, the legislative framework establishing certain adaptation-related obligations for the Contracting Parties constitutes the critical source, where several directions of regulatory nature can be extracted concerning the *modus* and the *focus* of adaptation. The majority of the provisions related to adaptation can be found in Article 4 of the UNFCCC, as described bellow: 1) Article 4 par. 1 b in association with par. 8 and 9 of the same Article establish the obligation for the Parties to develop national plans to facilitate adequate adaptation to climate change (National Adaptation Plans) 2) Article 4 par. 1 e requires the cooperation of Parties in preparing for adaptation and requests to elaborate appropriate and integrated plans for coastal zone management, water resources and agriculture as well as for the protection and rehabilitation of areas affected by floods or drought and desertification 3) Article 4 par. 4 establishes the obligation to the developed country Parties and other developed countries included in the Annex II of the Convention to assist developing country parties that are particularly vulnerable to the adverse effects of climate change. Furthermore, relevant in terms of providing the legal basis for establishing mechanisms to finance adaptation are also the provisions included in Article 11 of the Convention[12] as well as Article 11 par. 2 b of the Kyoto Protocol[13] to the Convention.

It is worth mentioning that there are also two main UNFCCC Work Programmes, namely the Buenos Aires Programme of Work on Adaptation and Response Measures and the Nairobi Work Programme[14], that deal more broadly with adaptation and its interdependencies with several sectors, such as the coastal zones, agriculture and water resources. The main aim of the programmes is to help all countries to improve their understanding of the impacts of climate change and to make informed decisions on the appropriate adaptive measures.

The systematic analysis of the aforementioned provisions of the Convention as well as the methodologies and the tools developed by the Work Programmes can thus, be helpful to draw up to certain conclusions relating to the approach and the directions to be followed, when elaborating and implementing adaptation strategies. The first conclusion relates to the recognition of *vulnerability* as the most crucial factor for assessing the adaptive capacity of the societies in the sense that society' exposure to climate change and subsequently its capacity to adapt are closely related to the level of its development[15]. There is no official definition for vulnerability in the UNFCCC and the definition given by IPPC[16], although enlightening, does not address the factors that determine vulnerability. With the help of the systematic analysis of the relevant provisions of the UNFCCC (article 4 par. 4) and the Kyoto Protocol (Article 12 par. 8), one can reach the conclusion that the vulnerability concept within the broader UNFCCC framework is used to describe those Contracting Parties that are in need of financial, technological and other forms of assistance. In this context, it is to emphasize that prioritizing adaptation to those mostly in need at country and community level is also regarded as a human rights obligation, because climate change has already or will adversely affect many people in poor regions in terms of securing their basic economic and social rights, such as the right to food and the right to water[17]. The second conclusion relates to the importance of adopting a strategic and integrated approach to deal with adaptation, which is a cross-cutting issue. In particular, a strategic response to the increased impacts of climate change should reach the full breadth of economic and development decision-making processes and the adaptation priorities and concerns should be subsequently streamlined into sectoral policies, such as energy, tourism and agriculture[18]. The National Adaptation Plans of Action (NAPA), which have to be elaborated in accordance with UNFCCC and contain the basic directions and the specific adaptation measures, should subsequently reflect the aforementioned strategic and cross-sectoral approach. Finally, the third conclusion relates to the requirements that the adaptation strategies in several sectors as part of the general strategic framework (NAPA) should fulfill as well as their specific measures that should contain. In particular, as vulnerability is the key factor to assessing the adaptive capacity of communities or eco-systems, adaptation strategies and measures should be based on the results of the

Vulnerability Assessments[19], which have to identify vulnerable areas and immediate and future adaptation priorities as well. Furthermore, in view of the scientific recommendations[20] that have been made, an adaptation strategy can be successful only if includes measures in all the steps of adaptation chain: 1) prevention measures 2) measures for improving resilience of the impacted system or sector 3) preparation measures 4) reaction measures and 5) recovery measures[21].

The question that can be raised at that point is whether the UNFCCC framework (legislative provisions and COP decisions) provide the regulatory basis for the integration of water issues into the core of the adaptation policies, taking into consideration that water is not only a source of life but also a key transmitter of climate change effects. The answer to the question cannot be any other than negative.

The lack of integration between water issues and adaptation policies within the existing legal and policy framework at the international level becomes obvious after a systematic analysis of the relevant legal provisions within the UNFCCC and the COP Decisions. In particular, only a reference for the elaboration of integrated plans for water resources as a means for adaptation can be found among the relevant provisions of the Convention, while there is not one single mention of water in any of the outcome documents from COP-13 and COP-14[22]. Furthermore, although the Nairobi Work Programme and other initiatives in this field constitute a remarkable effort to integrate water issues in the core of the adaptation policies, their contribution is not enough for placing water in the center of the adaptation agenda.

Therefore, the COP-15 negotiations are of crucial importance in terms of raising the profile of water issues within the adaptation policy framework and ensuring the inclusion of a clear reference in the new climate agreement that is hoped to be agreed next December in Copenhagen. In this direction, the Nairobi Statement on Land and Water Management for Adaptation to Climate Change[23] can provide valuable guidance for the inclusion and the context of relevant provisions that clearly recognize water and land management as integral parts of any adaptation strategy.

B. The regional (European) level

As was the case at the international level, the focus of the European climate policy at its initial phases was on mitigation. The gradual recognition of the need for strengthening the adaptive capacity of the European countries to the adverse impacts of climate change has led to the publication of the Green Paper "Adapting to Climate Change: Possibilities for EU action"[24]. This Green Paper was a first attempt to examine the impacts of climate change effects in several European regions and sectors and to identify possible adaptation actions at the EU level. Adaptation was recognized as a cross-cutting issue, so that it was acknowledged that the expected outcomes cannot be delivered by a single policy. Instead, concerted action in different policy areas deemed required. With regard to water policy, the Green Paper emphasized the importance of applying economic instruments and the user pays principle across all the water-related sectors, with the aim of providing incentives for reducing water consumption and increasing the efficiency of water use. Furthermore, flood protection as well as measures based on the sustainable land-use and spatial planning were regarded necessary for increasing the resilience of the water eco-systems.

Following the Green Paper, the European Commission has published the White Paper on Adaptation to Climate change[25], which aims to increase EU' s resilience to the impacts of climate change by establishing a cross-cutting policy framework. In this context, a series of actions, focusing, among others, on human capital (e.g awareness raising, capacity building) and green infrastructure[26] are proposed. The White Paper recognizes water, including linkages to coastal and marine waters, as a key sector for applying adaptation measures and stresses the need for the proper implementation of the existing legal instruments (EU Water Framework Directive and EU Floods Directive) for increasing the resilience of water eco-systems[27]. It is also worth mentioning that the Commission Communication on Water Scarcity and Droughts[28] is another EU policy document, which underlies the relationship between climate change adaptation and water management, as it provides certain recommendations to the competent authorities of the EU Member States on how to develop drought plans with mitigation and prevention measures in order to minimize the negative impacts caused by droughts.

At this point it has become evident that the linkage between climate change adaptation and sustainable water management is clearly recognized in the majority of the EU adaptation policy documents in comparison to what is the case at the international level.

II. Freshwater Regulation at International and Regional Level: Directions and Trends

Water Law creates the backbone of an effective water management at all different levels, from the international (mainly transboundary) to the local level[29]. A systematic analysis of the many regional and national water regimes leads to the conclusion that water law has at least to some extent incorporated the paradigm shift that has already started taking place in the water sector due to economic, societal and environmental changes (population growth and increasing water demand as well as increased water pollution). This gradual adoption of a more integrated regulatory approach in the protection and management of water resources has to be supplemented by new mechanisms and tools or by the enhancement of the existing ones, so that the legal and institutional regimes in place can provide the framework for dealing effectively with the new risks and uncertainties caused by climate-induced changes. At this point, a preliminary research in some important International and Regional Agreements as well as the relevant EU legislation is conducted in order to assess the impetus given to the national legislators to reform water law, where necessary, in order to support adaptation.

A. The global level

The management and protection of shared water is regulated by an emerging regime of international law on freshwater, which is based on International Treaties, soft-law documents[30], such as declarations, customary law and general principles of international law. In terms of geographical coverage, there is only one *global* agreement with regulative focus on the management of international watercourses, their conservation and use for purposes other than navigation. It is the United Nations Agreement on the Law of the Non-Navigational Use of International Water Courses ("UN Watercourses Convention") [31], which has not yet entered into force. Although it is not legally binding[32], the Convention provides guidelines for establishing bilateral or multilateral transboundary water regimes, as it has codified a series of principles of customary law for management and protection of shared water, with the principle of equitable and reasonable utilization and participation (article 5) and the no harm rule (article 7) being the most important ones[33]. From a systematic point of view it was correctly argued[34], the drafters of the Convention limited their intensions and their efforts mainly in codifying existing customary law and did not focus on developing it progressively.

In this context, there is no any explicit reference for establishing an obligation to the riparian states to support adaptation within transboundary water management. This is quite understandable, especially when man takes into consideration the long period for the drafting of the Convention at which the climate change impacts on water sources had not been – even to a limited extent – recognized. The concept of optimal and sustainable utilization of a shared watercourse set in Article 5 of the Convention ("In particular, an international watercourse shall be used and developed by watercourse States with a view to attaining optimal and sustainable utilization thereof") is, though, open-ended, so that it can constitute the legal basis for any adaptation strategies and measures elaborated and implemented by the riparian states.

B. The regional level

An important example of regulation of transboundary water courses at regional level is the 1992 Convention on the Protection and Use of Transboundary Watercourses and International Lakes ("Helsinki Convention"). In particular, the "Helsinki Convention" provides a sound legal framework for cooperation on shared water resources within the UNECE Region[35], which is more specific and sophisticated in comparison to the rather abstract and general rules of the UN Watercourse Convention[36]. The Convention has, thus, influenced the drafting of a number of bilateral and multilateral Conventions on transboundary waters in the UNECE region[37] as well as the regulative concept of the EU Water Framework Directive.

The "Helsinki Convention[38]" does not provide any explicit reference relating to the impacts of climate change on water resources and subsequently the obligation for the Contracting Parties to take action. Despite the lack of a clear reference, the set of the guiding principles established in the Convention as well as the specific obligations foreseen for the Contracting Parties can constitute the basis for taking measures to support adaptation. In particular, the basic concept of the Convention relates to the establishment of a specific framework of cooperation with the aim to prevent, control and reduce the transboundary impacts relating to international watercourses and lakes (Article 2 par. 1 and 2 and Articles 9–16 of the Convention). To the transboundary impacts defined as significant adverse effects on the environment resulting for a change in the condition of the transboundary waters caused by a *human activity* or physical origin (Article 1 nr. 2) are also included those impacts related to adaptation to or mitigation of climate change[39], although this is not mentioned explicitly. This is due to the fact that there is a far-reaching consensus in the scientific community that climate change is at least to certain extent caused by human activities. In this context, adaptation measures, such as construction of dams and reservoirs, can contribute substantially to avoiding negative transboundary impacts and to generate the best possible effects for the shared basin. Furthermore, the complex normative setting of the two most important principles of International Water Law, namely the principle of reasonable and equitable utilization of the shared water sources and the no harm principle into one normative setting in Article 2 of the Convention[40] can also provide a basis of any decision on adaptation measures within a transboundary river basin.

Moreover, a strong regulatory basis for taking measures in the direction of the so-called proactive adaptation[41] provides the recognition of the precautionary principle (Article 2 par.5 lit a) as one of the guiding principles for the transboundary water management. This principle, which requires that action should be taken even before adverse effects are fully scientifically proven, is of immense importance in climate change adaptation, as high uncertainties exist regarding the exact impacts on water availability and quality. Relevant in terms of supporting a proactive adaptation approach in the field of transboundary water management is also the explicit recognition of the sustainability principle both in its broader context (Article 2 par. 5 c) with a distinctive future dimension[42] and in its narrow context as ecological sustainability (Article 2 par. 2 lit.b). Finally, a series of specific obligations relating among others to setting joint water quality objectives (Article 9 par.2 lit. e) and establishing joint institutional bodies, to exchanging information about the current and expected conditions of transboundary waters (Article 13), to establishing joint programmes for monitoring and assessment (Article 11), set the framework for the riparian States to cooperate in elaborating adaptation strategies in transboundary context.[43].

Another characteristic legal instrument for the regulation of freshwaters with regional scope of application, is the EU Water Framework Directive[44]. Although the WFD sets a consistent framework for Integrated and Sustainable Water Management[45] it does not, however, include any obligation for adaptation to climate change. The Directive provides, though, the framework to address climate change impacts through the river basin planning process for a variety of reasons. First of all, it obliges Member States to assess environmental pressures in each river basin district (article 5 and Annex II), including subsequently pressures, which constitute the direct or indirect impact of climate change. Furthermore, the Programmes of Measures which constitute an integral part of the River Basin Management Plans and are designed to achieve the environmental quality objectives set for each river basin district, can include specific measures for climate change adaptation and also give emphasis to "no regret measures. Finally, it provides that for the calculation of the level of the cost-recovery of water services (Article 9) long term forecasts of supply and demand for water in each river basin district have to be taken into account.

III. Making Water Legislation Adaptive

A. The paradigm shift in environmental law and its general influence

One of the major challenges that the environmental law currently faces, relates to the *modus* to be adopted and the instruments to be introduced, so that the law can develop its regulatory impacts ("Steurungwirkung[46]") for pursuing legitimized objectives, such as environmental protection and public health under changing circumstances and a high level of uncertainty. In particular, law came up with a change in the "paradigm", on which environmental regulations at the initial phase were grounded and was based on the condition that natural ecosystems maintained a "relatively stable dynamic equilibrium"[47]. When the so-called "equilibrium paradigm" became invalid, the then laws fostering environmental protection were confronted with the challenge to set a framework that takes into account the inherent uncertainties associated with the complex and non-linear functions of the ecosystems. Environmental regulations, which are grounded on the "non-equilibrium paradigm", should be subsequently flexible enough to accommodate the ongoing environmental challenges and to adapt to unforeseen changes in the future. Moreover, modern environmental law has to regulate the risks[48] that constitute the "byproduct" of the scientific and technological development, a fact that has also influenced substantially its regulative concept.

Climate change and its wide-ranging effects related to precipitation, climate variability and extreme weather events have substantially intensified the challenges posed to the legal systems to cope with risk and uncertainty[49]. Furthermore, being viewed as an environmental crisis substantially different from any other relevant crisis that humanity has faced until recently, climate change requires major legal reforms in several fields, where its effects, if not managed properly, can have catastrophic consequences for the well-being of the societies. This becomes more than evident in the case of water law due to the already mentioned effect on water resources.

One of the major challenges that water law currently faces is to facilitate equitable and consistent adaptation to climate-induced changes in the water sector by incorporating new elements or strengthening existing ones, where necessary. The regulative context of water laws should also reflect in a clear way the adoption of a new "paradigm" for regulating issues regarding the protection and management of natural resources, as it will be described in the next section of the analysis.

B. Conceptual elements of water legislation within the framework of adaptive management

From a systematic point of view, it should be at first underlined that any efforts undertaken to reform national water laws have firstly to take into consideration the directions contained in the relevant legislative provisions as well as the methodologies and approaches developed in the scientific community about the concept of the adaptation strategies and the types of measures to be used (preventive or recovery measures or measures to improve resilience). Furthermore, the coherent, comprehensive and progressive freshwater regimes at regional level, with the EU Water Framework Directive and the UNECE Convention on Transboundary Waters and International Lakes being viewed as leading examples, can play a role model function in terms of providing a framework concept, which leaves room for embedding adaptive elements into water legislation. The efforts to seek synergies between the directions given under the broader context of the international adaptation regime and the regulatory trends for water management and protection at the international, regional and local level can lead to certain conclusions relating to the approach to be adopted and the strategic priorities to be achieved, when reforming national water laws with the aim to become sufficiently adapted.

The first conclusion relates to the importance of adopting a strategic approach instead of undertaking piecemeal adaptation efforts. This should be reflected not only in the coordination of the designed measures but also in the streamlining of the regulative concept of water laws with the relevant laws regulating land-use or other natural resources. The second conclusion relates to the strategic priority goals[50] that have to be achieved, when reforming national water laws. These priority goals, which also indicate the basic directions on which adaptive water legislation should focus, are the following: 1) The reduction of the vulnerability of the societies and particularly of its more marginalized parts to shifts in hydro-meteorological trends and extreme weather event[51]. 2) The protection and restoration of ecosystems, which are critical for water resources and services[52]. 3) The matching of water demand with water supply mainly by increasing water efficiency[53].

Moreover, as already mentioned, the described concept of water law should reflect the new "paradigm" on which the laws regulating the protection and management of natural resources should be based. This new paradigm, except for the consideration of the non-linear function of the eco-system , should incorporate especially in the field of water law, the philosophy and the methodology of the so-called "adaptive management" as an approach of dealing not only with environmental but also with quantifiable uncertainties. The concept of "adaptive management", which is applied in the field of eco-system management, is based on the fundamental assumption that the human ability to predict the future key drivers of any given eco-system and the system behavior and responses is inherently limited[54]. This assumption becomes even more relevant due to the added uncertainty caused by climate change. As a response, adaptive management requires a structured process of active learning by all stakeholders and the continuous improvement of management practices and measures by incorporating the insights of the experience gained from the outcomes of the implemented policies[55].

Reasoning from the logic of adaptive management in general in conjunction with the concept of Integrated Water Management[56], as stipulated in several documents, one can reach the conclusion that key features of adaptive and integrated water legislation can be the following 1) the setting of certain quality objectives that have to be in line also with the carrying capacity of the aquatic ecosystems and the limited water supplies 2) the establishment of a management regime preferably at river basin level to achieve the quality objectives, which should be characterized by an experimental character 3) the setting of data collection and monitoring mechanisms for the water quality and availability 4) the establishment of "intelligent" institutions that can go beyond day to day management and can also have an open dialogue with the scientific community and society 5) the introduction of policies and instruments flexible enough to incorporate and respond to the continuous flow of scientific information. In particular, among the variety of instruments used, water allocation regimes should be assessed and, where necessary reformed, in order to be more responsive to change[57]. It becomes obvious that evident that the legislation and the policies introduced have an "experimental" and "provisional" nature in the sense that they are substantially based on a "learning by doing" process[58]. After the description of the directions and the conceptual elements that adaptive water legislation should include, emphasis is placed on searching for directions with the aim to enhance the flexibility of water allocation regimes at national and transboundary level. This is due to the fact that it is being viewed as one of the most difficult tasks within the broader framework of adaptive water management.

C. Flexible Water Allocation Regimes as a Cornerstone of Adaptive Water Legislation

1. The national level

As the most profound impacts of climate change on water resources relate to the changes in precipitation and subsequently to the reduction of the water available, flexible and equitable allocation of this scarce natural source becomes one of the hardest tasks. In this context, existing allocation regimes established in the majority of national water laws worldwide should be assessed in terms of their capacity to adapt to annual or seasonal changes in water availability. A systematic overview of the different allocation regimes leads to the conclusion that there are two principal approaches to water rights. The first approach is that of the private water rights that confer on the holder the full set of property rights and are the closest approximation to the so-called “ownership” of the water. One constellation of this system of water rights is applied in the south-western USA and is governed by the “prior appropriation” doctrine, which gives priority to the older rights (“first in time, first in right”). This means in particular that the first rights holder acquires the superior legal right, while the junior water rights holders are the first to lose access in times of water shortages[59]. Although at this point no comprehensive analysis for the advantages and disadvantages of this concept from an economic, social or environmental perspective can take place, it should be underlined that serious doubts can be raised about its compatibility with the concept of adaptive and sustainable water management, mainly because it cannot enable the consideration of ethical, environmental and legal limitations relating to water use and consumption. In particular, one of its most important disadvantages is that it cannot provide the framework for setting limits to over-abstraction and subsequently cannot contribute to avoiding the quality degradation of the scarce resource[60]. Moreover, it does not seem flexible enough to respond to changes in water availability and to the environmental needs of the aquatic eco-systems[61] (i.e environmental flows). It can be thus assumed that due to the incapability of the private water rights regimes to support sustainable and adaptive water management, new or reformed water laws should keep water rights under the public domain[62].

The second approach is that of the use or usufruct water rights, which is based on the concept of the "public ownership" of water. The right of water abstraction is granted through an allocation licensing system[63] and is thus relative and conditional, because the licenses given determine clearly how much water can be taken and for which purposes. Licenses are also of limited duration. It becomes thus obvious that the concept of the usufruct rights with the licensing system as its cornerstone sets certain limits and conditions to the water use and can unquestionably contribute to the conservation and protection of waters by preventing over-allocation and pollution. Furthermore, abstraction licensing system provides the appropriate framework for an enhanced implementation of any decided prioritization of water uses, which is a crucial component of water policies as a response to the unforeseen climate-induced changes relating to water availability. Moreover, it can make licenses for water abstraction more flexible by introducing options for relief, in cases where water levels are low and by incorporating low flow values, below which abstraction would be not possible[64]. It can even provide for the prohibition of abstraction for a certain time period until water levels recover, in cases where the aquifers are under stress. Finally, it allows for the possibility to consider the necessity to conserve of the aquatic ecosystems by defining acceptable environmental flows[65]. Taking into consideration the strengths of the concept of the usufruct water rights, it can be assumed that this regulatory regime has a high degree of compatibility with the model of sustainable and adaptive water management. This does not imply that all the existing licensing allocation schemes are already flexible enough but only that the philosophy and the methodology of the concept will allow build in flexible options without further constraints.

In this context, one of the major tasks of the national legislators when introducing or reforming water laws is to establish a transparent water licensing system, which not only enables the orderly allocation of a scarce resource but also leaves room for modifications of the sharing water rules as response to changing circumstances. The ultimate goal of any effort to introduce or mainly reform water allocation schemes is thus to establish effective and equitable risk regulations, being characterized to some extent by a provisional and experimental character. Towards this direction, central elements of a flexible and simultaneously fair allocation regime could be the following: 1) the setting of clear-cut criteria on which the prioritization of water uses should be based, including also the consideration of social and environmental needs 2) the detail setting of the standards conditions for granting licenses 3) the introduction of the periodic review of water allocation as the rule governing allocation regimes in association with the possibility to include special provisions in water licenses addressing both quantity and quality variability 4) the establishment of registries of the granted licenses and of strong mechanisms to regularly monitor the compliance of the users' withdrawals with the terms set in the licenses as well as the condition of the scarce resource 5) the introduction of penalties mainly of administrative nature provided for the cases of breaching the licenses' terms[66]. Finally, it should be emphasized that the legislative framework in place also has to include certain provisions so that a fine balance is struck between the flexibility in water allocation and the relative security of the water-related investments[67].

2. The transboundary level

Similar issues relating to the necessity to build in flexibility are also raised within the concept of transboundary water agreements, which not only establish the legal and institutional framework for the management and the protection of the shared resource but also often set the amount of water the upstream riparian delivers to the downstream[68]. Typically, such treaties are used to allocate water on the grounds of the assumptions of water availability drawn from historical precipitation patterns. These patterns are, though, no longer valid. Since either climate-induced changes or other socio-economic factors can hinder the (upstream) riparian's ability to supply the volume of the water provided in the transboundary water agreement, it is of crucial importance to establish the appropriate mechanisms within agreements in order to accommodate the changes in water availability[69]. The inclusion of mechanisms into the treaties for enhancing the flexibility to cope with water variability is, though, a task even harder than the similar task at the national level. This is due to the fact that the difficulties caused by uncertainty and climate-induced changes are added to those that are the consequence of the historical, cultural and legal differences already existing in the shared basins, which are regulated by the relevant transboundary water agreements. Therefore, the absence of such mechanisms in the majority of the transboundary agreements can be interpreted mainly as a lack of political will and not as a lack of awareness[70]. Yet as water crisis (raising numbers of extreme weather events and changes in water availability), have started to emerge in the shared basins, political and legal approaches that do not take sufficiently into account the new circumstances, would not only be ineffective, but they could even lead to major political crises.

Although at this point no extensive analysis about the variety of the relevant mechanisms as established in the Treaties can take place, certain remarks are made, which focus on the *modus* to be adopted and the options available towards enhancing the flexibility of water allocation regimes. The first remark relates to the significance of considering not only the optimistic water availability scenarios but also the extreme hydrological values, when negotiating or reviewing existing transboundary water agreements[71]. Furthermore, options for making more flexible the water allocations to be delivered from the upstream to downstream riparian countries should be examined during the negotiation process and then consensus for one of them should be reached. For instance, an option could be that the agreement specifies that the upstream riparian can deliver less water (but at least a defined minimum amount) for a certain period and with acceptable justification, such as a long drought period[72]. With the view to avoiding future conflicts mainly by setting clear rules of the game, an explicit escape clause can also be introduced, so that the water quantity to be delivered by the upstream riparian in cases of extreme weather events is specified. Moreover, as it is difficult for the downstream riparian to accept such an escape clause, its introduction should be accompanied by compensation mechanisms, which for example compel the upstream riparian to return the water, when the extreme event (i.e drought) ends[73]. Another option, which reflects the approach of the water allocation rules regarded as risks regulations and subsequently emphasizes the "periodical" and "experimental" character of the norms, is the one that provides for the periodic review of the allocation rules by setting a specific provision in the Treaty[74]. Finally, it should be underlined that any effort to pursue a more flexible and equitable water allocation at transboundary level can be successful only if it is undertaken and implemented by joint institutions (i.e river commissions) in the shared basin. These institutions should be granted a strong mandate and a wide scope and jurisdiction, which goes far beyond their basic competencies (i.e control and monitoring of water uses or the regulation of water infrastructure development) and can even include a conflict-resolution mechanism, in order to have the capacity to handle effectively delicate issues with important political implications[75]. The adoption of a participatory management approach by the joint bodies and the promotion of a continuous and active dialogue with all relevant stakeholders is also important in terms of increasing the legitimacy and the efficiency of the decisions taken.

IV. Epilogue

Climate change and its wide-ranging effects have accelerated the paradigm shift that has already started taking place in environmental law. This is more than evident in the field of water law, where the assumptions on which regulations were based are no longer valid due to the intensification of extreme weather events and the unforeseen changes in water availability. Although climate change, which is regarded as the most important pressure on water resources, as well as other factors (i.e increasing population and increasing water demand), pose significant difficulties on the water management institutions, these difficulties should not be seen as threats but rather as catalysts for better water management.

Water law is the most important means for pursuing the goals of efficient, equitable and sustainable water management under the newly emerging circumstances. Towards this direction, national water laws should adopt a more integrated and "learning by doing" approach. The relevant regulations have thus to be reformed substantially in order to improve their responsiveness to climate change impacts and to provide a flexible and effective framework for robust management practices, while taking strongly into account social and environmental needs. At the transboundary level, the water crises that are expected to be intensified as a consequence of the increasing water scarcity, make it more than necessary to promote a culture of trust among the riparians. This should eventually result in the adoption of clear-cut and simultaneously flexible (allocation) rules within the transboundary agreements and in the establishment of joint bodies that not only have strong mandates but also promote new forms of collaborative governance.

Notes

[1] IPCC Report 2007, *Climate Change 2007: Climate Change Impacts, Adaptation and Vulnerability*. Contribution of Working Group II to the Fourth Assessment Report of the IPCC.

[2] United Nation Development Programm (UNDP), Human Development Report 2006: *Beyond Scarcity – Power, Poverty and the Global Water Crisis*, Macmillan Palgrave 2006, p. 8.

[3] World Health Organization, Climate Change and Human Health, available at: http://www.who.int/global change/climate/en (last accessed on 3 November 2009).

[4] Global Water Partnership, Technical Advisory Committee (TAC), *Climate Change Adaptation and Integrated Water Resource Management-An Initial Overview*, Policy Brief No 5, 2008, p. 6.

[5] 1992 United Nations Convention Framework Convention on Climate Change available at: http://unfccc.int /2860htp (last accessed on 3 November 2009)

[6] 1997 Kyoto Protocol to United Nations Convention Framework Convention on Climate Change, available at http://unfccc.int/kyoto-protocol/items/2830.php (last accessed on 3 November 2009).

[7] IPCC, *Climate Change 2001 – Impacts, Adaptation and Vulnerability*, Cambridge University Press, p. 992

[8] See W.N Adger/J. Paavola/S.Huq/M.J Mace (Eds), 2006, *Fairness in adaptation to climate change*, MIT Press, p. 335pp.

[9] IPCC, *Climate Change 2001 – Impacts, Adaptation and Vulnerability*, Cambridge University Press, p. 992.

[10] For the complementary relationship between mitigation and adaptation see R.J. Klein/S. Huq/F. Denton/T.E Dowing/ R.G Richels/J.B Robinson/F.L. Toth, 2007, *Inter-relationships between adaptation and mitigation,* Climate Change 2007 – Impacts, Adaptation and Vulnerability, Contribution of Working Group II to the Fourth Assessment Report of the Intergovernmental Panel on climate change, Cambridge University Press, p. 750.

[11] See I.M Goklany, *A climate policy for short and medium term: stabilization or adaptation*? Energy and Environment, Vol.16, 2002, p. 670.

[12] Article 11 of the Convention provides for the necessity of establishing a mechanism for the provision of financial resources on a grant or concessional basis, including the transfer of technology. It is stated that this mechanism will function under the guidance of, and is responsible to the Conference of the Parties.

[13] Article 11 par. 2 of the Kyoto Protocol establishes the obligation to the developed country Parties to provide financial resources, including the transfer of technology, to meet the agreed full incremental costs incurred by developing countries in advancing the implementation of the commitments of Article 4 of the Convention. The afore-mentioned costs mainly include the adaptation cost to the adverse effects of climate change. In this context, it is worth mentioning that there are three funding mechanisms operating under the broader framework of the UNFCCC and the Kyoto Protocol to finance adaptation actions especially in the developing countries. These are the following: 1) The Special Climate Change Fund, which operates under the framework of the UNFCCC and finances long-term adaptation measures that increase resilience of national development sectors, such as water resources management, agriculture and health to the impacts of climate change 2) The Least Developed Countries Fund, which also operates under the UNFCCC umbrella, and finances the world's least developed countries to elaborate and implement the National Adaptation Programmes of Action 3) The Adaptation Fund, which is a quite innovative funding mechanism, firstly because its revenue is generated to certain extent through a 2% levy on the emission reduction projects under the Kyoto Protocol' s Clean Development Mechanism and secondly because there is a strong representation of the developing countries in its executive body, namely the Adaptation Board. For an assessment of the effectiveness of the Funds operating under the UNFCCC umbrella see M.J. Mace, *Funding for Adaptation to Climate Change: UNFCCC and GEF Developments*

since COP-7, RECIEL, vol. 14, no. 3, 2005, p. 236-240. For a first assessment of the Adaptation Fund see B. Müller, *The Nairobi Climate Change Conference: A breakthrough for adaptation funding*, Oxford Energy and Environment Comment, January 2007

[14] UNFCCC 2007, Background Broschure on the Nairobi Work Programme.

[15] See Christoplos/S. Anderson/M. Arnold/V. Gataz/M. Hedger/R.Klein/K.de. Goulven, *The Human Dimensions of Climate Adaptation: The importance of local and institutional issues,* Stockholm Environment Institute, April 2009, p. 14ff.

[16] The IPPC defines vulnerability as "the degree to which a system is susceptible to, or unable to cope with, adverse effects of climate change, including climate variability and extremes. IPCC, *Climate Change 2001 – Impacts, Adaptation and Vulnerability*, Cambridge University Press.

[17] For the human rights argument to support adaptation see German Watch/Brot für die Welt, *Making the Adaptation Fund work for the most vulnerable people*, December 2008, p. 6-7. For the definition of the right to food see Narula, *The Right to Food: holding global actors accountable under International Law*, Columbia Journal of Transnational Law, vol. 44, 2006, p. 691-796; Food and Agricultural Organization of the United Nations (FAO), *The Right to Food and Access to Natural Resources*, Right to Food Series, Rome 2008. For the definition of the right to water and implicit or explicit references for its recognition sees J. Scanlon/A.Cassar/N.Nemes, *Water as a Human Right* ?, IUCN Environmental Policy and Law Paper No. 51, IUCN 2004; M.A Salman/S.McInerney-Lanford, *Human Right to Water-Legal and Policy Dimensions*, World Bank- Law, Justice and Development Series, 2004. It is also mention that although the human rights approach concerning food and water has gained some recognition at the international level, there are still uncertainties about their meaning and their practical implications.

[18] See I.Burton/E. Diringer/J. Smith, *Adaptation to Climate Change: International Policy Options*, Pew Center on Global Climate Change, November 2006, p. 12, available at: www.Pewcenter.org/docUploads/PEW _Adaptation.pdf, (last accessed on 3 November 2009).

[19] See Draft Guidance on Water and Climate Adaptation elaborated by the Task Force on Water and Climate of the Convention of the Protection and Use of the Transboundary Watercourses and International Lakes, Geneva 24.4.2009, p. 61, where as vulnerability assessments defined the systematic efforts to delineate the places, human groups and eco-systems that are at highest risk, the sources of their vulnerability and how the risk can be diminished or eliminated. The draft Guidance is available at: www.unece.org/env/water/meetings/Water.and.Climate/second.meet/Guidance_on _Water_and_Climate_Adaptation_draft_for_Task_Force_03_04.doc (last accessed on 5 November 2009).

[20] See Draft Guidance on Water and Climate Adaptation, *supra* note 19, p. 67; I.Burton/E.Malone/H. Saleemul, *Adaptation Policy Frameworks for Climate Change: Developing Strategies, Policies and Measures* in: B.Lim/E.Spranger-

Siegfried (Eds), United Nations Development Programme, Cambridge University Press 2004, (Technical Paper 8 on Formulation an Adaptation Strategy).

[21] Measures for prevention and for improving resilience are related both to the gradual impacts of climate change and the extreme weather events, while preparation (for example drought management plans in the water sector), reactive (safe sanitation provision and emergency medical care) and recovery (for example clean up activities) measures are applied mainly to the cases of extreme climate events.

[22] See Global Public Policy Network on Water Management on Water Management, *Water and Climate Change Adaptation-Key Messages for the 15th Conference of the Contracting Parties of the UNFCCC,* March 2009, available at: http://gppn.stakeholderforum.org (last accessed on 3 November 2009).

[23] The Nairobi Statement on Land and Water Management for Adaptation to Climate Change, Nairobi 17.04.2009. It is a soft law document and constitutes the final product of a dialogue Initiative of the Danish Ministry of Foreign Affairs in collaboration with other International partners. The statement is available at:http://www.landwaterdialogue.um.dk (last accessed on 3 November 2009).

[24] Green Paper from the Commission to the Council, the European Parliament, the European Economic and Social Committee and the Committee of Regions, *Adapting to Climate Change-Possibilities for EU Action,* COM 2007, 354 final, Brussels 29.06.2007.

[25] European Commission, White Paper, *Adapting to Climate Change-Towards a European Framework for Action,* COM (2009), 147 final, Brussels 1.04.2009.

[26] As green infrastructure is defined in the White Paper the interconnected network of natural areas, including some agricultural land, such as greenways, wetlands, parks, forest preserves and native plant communities, and marine areas that naturally regulate storm flows, temperatures, flooding risk, and water, air and ecosystem quality. See White Paper, *supra* note 25, p. 5.

[27] Furthermore, options for further regulating the standards for water using equipment and water performance in certain sectors as well as for boosting ecosystem storage capacity through the introduction of the appropriate measures are left open. See White Paper, *supra* note 25, p. 11

[28] Communication from the Commission to the European Parliament and to the Council, Addressing the challenge of water scarcity and droughts in the European Union, COM (2007) 414 final, Brussels 18.07.2007.

[29] See A. Iza/R. Stein (Eds), RULE-Reforming Water Governance, IUCN, Gland-Switcher land 2009, p. 49 ff.

[30] Since the early 1990s international discourse has started focusing on the management and protection of water resources at transboundary and national level in quite a different way by adopting a more integrated approach and by incorporating sustainability into water management. This new approach is captured in a series of soft law documents, such as the Dublin Statement on Water and Sustainable Development, the Chapter 18 of the Agenda 21 and the Berlin Rules on Water Resources. Among the different soft-law documents, the Berlin Rules on Water Resources adopted by the International Law Association to "replace" the

Helsinki Rules is – in my opinion – the most characteristic soft-law document in terms of incorporating the paradigm shift for a more integrated and participatory management that deals with all water (international and national). See J. Dellapenna, *The Berlin Rules on Water Resources: The new Paradigm for International Water Law,* available at:http://www.ualg.pt (last accessed on 4 November 2009).

[31] The Convention was adopted by the United Nations General Assembly on 21 May 1997. The concept of the Convention was based – to a significant extent – on the Helsinki Rules on the Uses of the Water of International Rivers developed by the International Law Association. For the background of the evolution of the Convention see G. Eckstein, *Development of International Water Law and the UN Watercourse Convention* in: A. Turton/R. Henwood (Eds), Hydropolitics in the developing world-A Southern African Perspective, African Water Issues Research Unit 2002, p. 81-96.

[32] The International Court of Justice cited the Convention as evidence of International Customary Law and underscored its importance when it referred to a number of its provisions within the Gabcikovo-Nagymaros case, a dispute between Hungary and Slovakia over Danube. See Case concerning the Gabcikovo-Nagymaros (hung/Slov) 1997, I.C.L 92 (Sept. 25), 37 I. L.M 162 (1998), available at: http://www.icj-cij.org/idocket/his/ihsjudgment/ihsdocument/html7 (last accessed on 4 November 2009).

[33] The principle of equitable and reasonable utilization, which constitutes the cornerstone of international water law and was codified in the UN Convention, is based on the recognition of the right of all riparian States to use water from a shared source only to the extent that these uses do not come in contradiction with the uses in another riparian State. Article 6 of the UN Convention sets a list of factors that have to be taken into account when balancing reasonable uses of a shared water resource. The principle of no harm, which is set in a quite weak formulation in Article 7, requires that riparian States have to take the appropriate measures to prevent the causing of any significant harm to the neighbouring states. At this point it is not possible to engage into the question of whether one of the two principles is hierarchically higher than the other, an issue on which there is extensive literature. In my opinion though, the concrete formulation of the two principles within the concept of the Convention can probably lead to the conclusion that the no harm principle is subordinated to the principle of equitable and reasonable utilization. In this direction see J. Dellapenna, *The Customary International Law of Transboundary Fresh waters*, International Journal of Global Environmental Issues, Vol. 1, Nos 3/4, 2001, p. 264 (284ff).

[34] See J. Dellapenna, *supra* note 33, p. 288; J. Dellapenna, supra note 30.

[35] The Convention was concluded under the auspices of the United Nations Economic Commission of Europe ("UNECE"). UNECE brings together 56 countries located in the European Union, non-EU Western and Eastern Europe, South-East Europe and Commonwealth of Independent States (CIS) and North America (USA and Canada). The Convention has entered into force on October 1996 and has 36 Contracting Parties so far. The text of the Convention is available

at: http://www.unece.org/env/water/pdf/watercon.pdf (last accessed on 5 November 2009).

[36] See P. Patronos, *International Law, the EC Directive 2000/60 and the Protection and Use of Transboundary Watercourses*, paper available at: www.nomosphysis.org.gr/articles.php?artid=1182&lang=2&catpid=12 (last accessed on 5 November 2009). In this context, it should be mentioned the Helsinki Convention adopted the eco-system approach as its normative aim (Article 3 par. 1 lit.(i), introduced eco-standards consisting of best available technologies and best environmental practices (Article 3 par. 1 lit (g) and (f) respectively as means for preventing any transboundary impact and emphasized the importance of institutional cooperation to be realized by the establishment of joint bodies (for example river commissions) based on an international agreement or another arrangement (Article 9) .

[37] A characteristic example of a multilateral Convention influenced substantially by the context of the Helsinki Convention is the new Convention on the Protection of the Rhine, which was signed in Bern on 12.4.1999 and entered into force on 1st January 2003. The Convention aims to preserve and improve the eco-system of Rhine through strengthening the cooperation of the riparian states among themselves as well as with the European Commission. The text of the Convention is available at www.iksr.org/fileadmin/user_upload/Dokumente_en/convention_on_tthe_protection_of__the_rhine.pdf (last accessed on 5 November 2009).

[38] The Helsinki Convention is supplemented by the Protocol on Water and Health, which was adopted in 1999. The primary target of the Protocol to be achieved by the Contracting Parties is access to drinking water and the provision of sanitation to everyone, having as ultimate goal to protect human health and well-being through preventing, controlling and reducing water-related diseases. The Protocol is an important tool to address climate change impacts on water resources and particularly on water services, such as water supply and sanitation. The text of the Protocol is available at: www.unece.org/env/documents/2000/wat/mp.wat.2000.1.e.pdf (last accessed on 5 November 2009)

[39] See J. Timmerman/F. Bernadini, *Adapting to Climate Change in transboundary water management*, Perspective Paper developed to be presented in the 5th World Water Forum, p. 5, available at www.waterandclimate.org/index.php?id=5thWorldWaterForumpublications810 (last accessed on 5 November 2009).

[40] According to Article 2 par. 1: "The Parties shall take all appropriate measures to prevent, control and reduce any transboundary impact " (The no harm principle"). According to Article par.2 lit.c: "Parties shall, in particular, take all the appropriate measures to ensure that transboundary waters are used in a reasonable and equitable way, taking into account their transboundary character in the case of activities which cause or are to cause transboundary impact" (principle of reasonable and equitable utilization).

[41] The *stimulus* is the crucial criterion for defining adaptive action as proactive or reactive. In particular, when actions are in response to observed climate impacts, we speak about reactive adaptation, while in cases that action is taken in anticipation of future climate change impacts, we speak about proactive adaptation. For the distinction see I.Burton/E. Diringer/J. Smith, *supra* note 18, p.10.
[42] Article 2 par.5 c foresees the following: "Water resources shall be managed so that the needs of the present generation are met without compromising the ability of future generations to meet their own needs".
[43] It is worth-mentioning that a Task Force on Water and Climate was established in 2006 under the Convention, which is entrusted to elaborate a Guidance on Water and Climate Adaptation. The final draft of the Guidance on Water and Climate Adaptation is to be adopted by the Contracting Parties in November 2009. See Draft Guidance on Water and Climate Adaptation, *supra* note 19.
[44] Directive 2000/60/EC of the European Parliament and of the Council of 23 October 2000 establishing a framework for Community Action in the field of water policy, O J L 327, 22.10.2000, P.1-73.
[45] The basic regulatory choices of the Directive can be described as follows: 1) the adoption an integrated approach of all water-related aspects in terms of regulation, planning and governance 2) the protection of all kinds of water (surface and groundwater) 3) the choice of the river basin district as the organizing unit for river basin management 4) the setting of ambitious environmental quality objectives to be achieved within a defined timeframe (2015) 5) the introduction of an integrated planning process at each river basin district, which will result in the adoption of a management plan (river basin management plan-RBMP) 6) the participation of the public and the stakeholders in the planning and management process and 7) the recognition of water as an economic good (Article 9- the cost recovery principle). As regards transboundary river basins, the Directive provides (Article 17) that join environmental objectives should be set and that the elaboration and the implementation of Programmes of measures should be coordinated for the basin as a whole.
[46] The approach of the regulative impact (Steurungswirkung) of the legal norms was developed in the German legal theory. The basic idea on which this concept is grounded, focuses on the impacts which the implementation of the legal norms "produces", which in some cases are not those that the legislator had intended to, when the laws were adopted. This comes true, when the legal norms set do not meet the society' s expectations related to context of the regulation or when the implementation comes in contradiction with special economic or social interests that try to set obstacles. In accordance with the philosophy of this approach, one of the major challenges that the law has to come up with, relates to the formation of the legal norms in such a way, so that the "desired" effects are realized. See G.F Schuppert, *Grenzen und Alternativen von Steurung im Recht* in: Grimm (Hrsg), Wachsende Staatsaufgaben-Sinkende Steurungsfähigkeit des Rechts, 1990, p. 217ff; E. Schmidt-Aßman, *Das Allgemeine Verwaltungsrecht als Ordnungsidee*, Baden-Baden 1998, p. 133; W.Hoffmann-Riem, *Vorüberlegungen zur rechtswissenschaftlichen Innovationsforschung* in: W.Hoffmann-Riem/J.P. Schneider

(Hrsg), Rechtswissenschaftliche Innovationsforschung, Baden-Baden 1998, p. 387 (405ff).

[47] For the shift from the "equilibrium" to the "nonequilibrium" paradigm and its consequences on the regulatory context of the environmental regulations see Dan. A. Tarlok, *The "Non Equilibrium Paradigm" in Ecology and the Partial Unraveling of Environmental Law*, Loyala of Los Angeles Law Review, Vol. 27, 1994, p. 1121 ff; J. B Ruhl, *Thinking of Environmental Law as a Complex Adaptive System: How to Clean up the Environment by Making Mess the Environmental Law,* Houston Law Review, Vol. 34, 1997, p. 933 ff.

[48] For the definition of risk see U. Beck, *Risikogesellschaft-Auf dem Weg in eine andere Moderne*, Frankfurt a. Main, 1986, p. 25 ff; European Environmental Agency (Eds), Late lessons from early warnings, 2001, p. 168f. For the role of law to the management of risks see C. Sunstein, *Risk and Reason, Safety, Law and Environment*, Cambridge University Press 2002; A. Scherzberg, *Risikosteurung durch Verwaltungsrecht: Ermöglichung oder Begrenzung von Innovationen*?, VVDStRL 63 (2003), p. 214ff.

[49] See B.H Thompson, Jr., *Tragically difficult: The obstacles to governing the commons*, Environmental Law, Vol. 30, 2002, p. 241 (253); M. Zinn, *Adapting to Climate Change: Environmental Law in a warmer world*, Ecology Law Quarterly, Vol. 34, 2007, p. 82.

[50] See G. Bergkamp/B. Orlando/I. Burton, *Change Adaptation of Water Management to Climate Change*, IUCN, Gland, Switzerland and Cambridge, 2003, p. 21-26.

[51] The rationale for taking such measures recognizes than extreme weather events of both types (droughts and floods) tend to become chronic conditions rather than emergencies, so that preparedness strategies and proactive measures have to be taken to avoid irreversible damage to people and the ecosystems. Towards this direction, better coordination between land and water planning in association with measures, such as risk zoning for floods or droughts, the use of early warning systems and the flood and drought preparedness programmes seems to be necessary. See G. Bergkamp/B. Orlando/I. Burton, *supra* note 50, p. 22.

[52] The rationale for taking such measures lies not only in their function as natural buffers against extreme events as well as in their importance for the balance of the ecosystems in the river basin area from a holistic approach but also because they are critical for the provisions of goods and services. To achieve this goal measures to restore downstream wetlands, such as flood storage areas and restoring or preserving mountain forests to reduce erosions and peak flows should be introduced. See G. Bergkamp/B. Orlando/I. Burton, *supra* note 50, p. 23.

[53] The rationale for taking such measures is based on the assumption that given the fact that conventional water supply is no longer feasible, demand management with emphasis on increasing water efficiency becomes more than important. Towards this direction, the prioritization of water uses, the reform of water allocation regimes, where necessary and other demand-side measures, such as the re-use of return flows and desalination are deemed necessary. See G. Bergkamp/B. Orlando/I. Burton, *supra* note 50, p. 25..

[54] See C. S Holling, *Adaptive Environmental Assessment and Management*, New York 1978; C.J Walters, *Adaptive Management of Renewable Resources*, New York 1986; W.T Coleman, *Legal Barriers to the Restoration of Aquatic Systems and Adaptive Management,* Vermont Law Review, Vol. 23, 1998, p. 177ff.

[55] See C. Paul-Holst, *Requirements for adaptive water management* in: C. Paul-Holst/ P. Kabat/ J. Möltgen (Eds), Adaptive and Integrated Water Management: Coping with complexity and uncertainty, Springer, Berlin 2007, p. 1-22; G.T Raadgever/E. Mostert/N. Kranz/E. Interwies/J. Timmerman, *Assessing Management Regimes in Transboundary Rivers: Do they support adaptive management* ?, Ecology and Society, Vol. 13 No.1, Art. 14, available at: http: //www.ecologyandsociety.org/vol13/iss1/art14 (last accessed on 5 November 2009).

[56] Integrated Water Resource Management is defined as "a process that promotes the coordinated development and management of water, land and related resources to maximize the economic and social welfare in an equitable manner without compromising the sustainability of vital ecosystems". See Global Water Partnership, Technical Advisory Committee (TAC), *Integrated Water Resource Management*, Background Paper No 4, 2000.

[57] See J. C. Neuman, *Adaptive Management: How Water Law needs to change,* Environmental Law Reporter, Vol. 31, 2001, p. 11432 (11434).

[58] See J. C. Neuman, *supra* note 57, p. 11434; C. Bruch, *Adaptive Water Management: Strengthening Laws to cope with Uncertainty* in: A.K. Biswas/C. Tortajada/R. Izquiero (Eds), Water Management Beyond 2020, Springer Verlag 2009, p. 89 ff.

[59] See St. J. Shupe, *Waste in Western Water Law*: *A Blueprint for Change*, Oregon Law Review, Vol. 61, 1982, p. 483ff; D. Tarlok, *How well can International Water Allocation Regimes Adapt to Global Climate Change,* Journal of Land Use and Environmental Law, Vol. 15, 2000, p. 423 (429).

[60] See J.C Neuman, *Beneficial Use, Waste and Forfeiture: The inefficient search for Efficiency in Western Water Use,* Environmental Law, Vol. 28, 1998, p. 919; N.Hall/B. Stunz/R.H Abrams, *Climate Change and Freshwater Resources,* Natural Resources and the Environment, Vol. 22, No3, 2008, p. 30 (34).

[61] See T. Le Quense/G. Pegram/C. von der Heyden, Allocating Scarce Water, A primer on water allocation, water rights and water markets, WWF UK, 2007, Godalming UK, available at: http://assets.wwf.org.uk/downloads/scarce_water.pdf (last accessed on 5 November 2009).

[62] . See A. Iza/R. Stein (Eds), *supra* note 29, p. 54.

[63] The European Water Framework Directive requires every Member State to have an abstraction licensing system in place (Article 11 par. 3 lit (e)).

[64] See E.Levina/H. Adams, *OECD, Domestic Policy Frameworks for Adaptation to Climate Change in Water Sector,Part I:Annex I Countries*, (COM/ENV?EPOC/IEA/STL (2006) 2, May 2006. p. 32.

[65] As "Environmental Flow" can be defined the water regime provided within a river, wetland or a coastal zone to maintain eco-systems and their benefits, where there are competing water uses and where flows are regulated. See M.

Dyson/G.Bergkamp/J. Scanlon (Eds), *The Flow-The essentials of environmental flows*, 2nd edition Gland, Switcher land , IUCN, 2008, p. 17.

[66] See A. Iza/R. Stein (Eds), *supra* note 29, p. 55 ff; E.Levina/H.Adams, OECD, supra note 64, p. 32-33.

[67] See T. Le Quesne/G. Pegram/C. von der Heyden, WWF UK, *supra* note 61.

[68] See E. Benvenisti, *Collective action in the utilization of shared freshwater: the challenges of international water resources law*, The American Journal of International Law, Vol. 90, 1996, p. 384-415; I. Fischhendler, *Legal and institutional adaptation to climate uncertainty: a study of international rivers*, Water Policy, Vol. 6, 2004, p. 281 (282-284).

[69] See P. Gleick, *Water and conflict, freshwater resources and international security*, International Security, Vol. 18, p. 79ff; I. Fischhendler, *supra* note 68, p. 282.

[70] See I. Fischhendler, *supra* note 68, p. 298.

[71] See Draft Guidance on Water and Climate Adaptation, *supra* note 19, p. 33.

[72] See I. Fischhendler, *supra* note 68, p. 283; A. Drieschova/M.Giordano/I. Fischhendler, *Governance Mechanisms to address flow variability in water treaties*, Global Environmental Change, Vol. 18, Issue 2, 2008, p. 285 (287ff).

[73] See I. Fischhendler, *supra* note 68, p. 283.

[74] See G.T. Raadgever/E. Mostert/N. Kranz/E. Intewies/J.G. Timmerman, *supra* note 55

[75] See E. Feitelson/M. Haddad, *Identification in Joint Management Structures for Shared Aquifers, A comparative Palestinian-Israeli Effort*, World Bank Technical Paper No 415, The World Bank, Washington D C 1998, p. 10ff. Swedish Ministry for Foreign Affairs *Transboundary Water Management as an International Public Good,* Stockholm, 2001, available at: http://docs.watsan.net/Downloaded_Files/PDF/ODI-2001-Transboundary.pdf (last accessed 9 November 2009); I. I. Fischhendler, *supra* note 68, p. 283-284.

CHAPTER NINETEEN

ETHICAL ISSUES RAISED BY CARBON CAP-AND-TRADE REGIMES

DONALD A. BROWN

Introduction

This chapter examines ethical issues raised by the cap-and-trade regimes that have emerged to solve the climate change crisis in the last decade. These regimes have emerged:

1. at the international level under the Kyoto Protocol;
2. at the regional international level, including in the European Union and between US states and the Canadian provinces; and
3. at the subnational level including among Northeastern U.S states.

There is also a large voluntary carbon trading market that has emerged around the world that is not the focus of this chapter although these regimes raise many ethical issues considered here.

At the international level, under the Kyoto Protocol to the United Nations Framework Convention on Climate Change there are three different trading regimes. They are:

(a) *Emissions Trading (ET):* A mechanism that allows a nation with a Kyoto target to buy d allowances from a country with a Kyoto target that does need all of its allowances.
(b) *Joint Implementation (JI):* A mechanism that allows project financing by nation with a Kyoto target in another country with a Kyoto target.

(c) *Clean Development Mechanism (CDM):* A mechanism that allows nations with Kyoto targets to finance projects in developing countries that reduce greenhouse gas emissions.

The goal of this chapter is to spot the major ethical issues raised by carbon cap-and-trade regimes that have emerged around world, not necessarily to resolve these issues.

Spotting ethical issues raised by cap-and-trade regimes will not necessarily lead to a consensus about what should be done about these issues because there are competing ethical theories that might reach different conclusions about these trading regimes including utilitarian, rights, virtue, relationship, and ecological based theories among others. However, for some issues there may be an overlapping consensus among ethical theories about what ethics requires (Brown et al., 2006). For other issues there may be agreement among ethical theories that some positions on cap-and-trade issues are ethically problematic no matter what ethical theory is applied to analyze the issue under consideration. Therefore, spotting ethical issues raised by carbon cap-and-trade regimes may be practically valuable despite the inability for some issues to state unambiguously what ethics demands if spotting the ethical questions leads to eliminating from consideration some positions on these issues that fail to pass minimum ethical scrutiny (Brown et al., 2006).

The purpose of this chapter is not to resolve all ethical issues entailed by cap-and-trade regimes but to encourage further ethical reflection about these issues. Cap-and-trade regimes have in a very short time reached wide support around the world with only very limited ethical reflection on the issues discussed in this chapter. Because uncritical acceptance of these existing regimes may lead to significant injustices and given that there may be opportunity to change and restructure existing regimes in the future, this chapter is meant to encourage ethical reflection on existing as well as future cap-and-trade regimes.

Although existing cap-and-trade regimes differ in many of their details, they all have the following common steps. First a government establishes an emission limitation for total emissions from the government's jurisdiction and then permits or allowances are either given away or auctioned off and in this way create a society-wide "cap." The cap is expected to be "tightened," that is, reduced over the years, thus increasing the costs over time of future allowances. The permits allow holders to emit greenhouse gases (GHGs), usually in tons of carbon for a specific period. To enable trading, rules are established that allow those entities with caps to meet their obligations either by purchasing unneeded allowances from others that have caps, funding projects that reduce emissions at places under the control of others, or purchasing offsets created by carbon reduction projects somewhere in the world.

Cap-and-trade regimes are usually justified on several different grounds including:

- Trading provides a mechanism for making carbon emissions reductions at lowest possible cost. Because carbon is well mixed in the atmosphere, it doesn't make any difference where reductions are made in the world to lower future atmospheric concentrations of GHGs. Therefore emitters of carbon can finance inexpensive projects to reduce carbon emissions and apply reductions achieved by these projects against their reduction obligations and in so doing reduce GHG emissions that cause climate change. In this way, cap-and-trade regimes maximize the efficiency of carbon reductions. And so cap-and-trade regimes are usually supported on the grounds that they provide the flexibility to achieve the greatest reductions at lowest possible cost.
- Proponents of cap-and-trade regimes often point to a successful program still in place in the US that has been declared to reduce 40% sulfur emissions (SO_x) by coal-burning power plants in the period 1990–2004 (EDF, 2009).
- Cap-and-trade regimes provide economic benefits to developing countries through CDM credits and other "offsets" thus helping developing countries economically.
- Cap-and-trade regimes keep high-cost emitters in the political game because they can reduce their emissions at low cost and thereby help minimize political opposition for climate change legislation.

This chapter next looks at the following ethical issues entailed by cap-and-trade regimes:

a) Justice of the Cap
b) Allocating Global Commons Resources as Property Rights.
c) Environmental Effectiveness
d) Distributive Justice
e) Procedural Justice

Ethical Issue One: Justice of the Cap

A cap-and-trade regime is just only if the cap is just both in the amount of the cap and how the cap is distributed among emitters. If the total society-wide cap, before it is allocated among emitters within the jurisdiction of the government allocating the cap, is less than the government's fair share of safe global emissions, then the cap is not environmentally just particularly to those who are vulnerable to climate change (Brown, 2009). Because each nation or government has a duty to reduce its emissions to its fair share of safe global emissions (Brown et al, 2006), justice requires that caps be consistent with the government's GHG reduction obligations. Almost all nations have agreed that they should reduce their emissions at levels to prevent dangerous anthropogenic interference with the climate change system based upon equity under the United Nations Framework Convention on Climate Change (UN, 1992). Yet many cap-and-trade regimes do not derive the quantity of the cap from these international obligations (Brown, 2009).

A strong case can be made that the world is not currently on an emissions reduction pathway needed to prevent dangerous climate change (Bear and Athanasiou, 2005; Athanasiou and Bear, 2002). In fact some scientists are argue that is already too late to prevent catastrophic climate change (Hamilton, 2009). Therefore an argument can be made that most existing and proposed cap-and-trade regimes are not sufficient to prevent dangerous climate change and therefore are unjust.

If the cap upon which the trading regime is based does not constitute a level of carbon reduction consistent with the nation's duty to reduce emissions, the entire cap-and-trade regime is unjust. In such cases, the regime is particularly unjust to those who are most vulnerable to climate change because the regimes gives rights to pollute at levels above what justice requires. Since many parts of the world including sub-Saharan Africa, Bangladesh, small island developing states in the Pacific and Indian Ocean, large parts of South-East Asia among others are particularly vulnerable to climate change impacts, a cap that is not sufficiently environmentally protective is unjust to these vulnerable countries and peoples (Argwal and Narain, 1991).

In addition to national governments, each individual subnational government, business, organization, and entity also has a duty to reduce its emissions to its fair share of safe global emissions. If the allowances that entities are allocated by the government are not sufficient to assure that the entity's emissions are no greater than its fair share of safe global emissions, an argument can be made that their individual allocation is unjust.

Because some emitters are often not included within the scope of a cap-and-trade regime, a strong case can also be made that those excluded from the cap are emitting at unjust levels if they are not emitting at levels less than the emitter's fair share of safe global emissions. For this reason, the creation of a cap-and-trade regime that excludes classes or categories of GHG emitters should not necessarily be understood to satisfy what justice requires if some emitters within the government's jurisdiction are allowed to emit at unjust levels because they are excluded from regime. However, a counter argument can be made that a regime is just if total emissions from the area within the jurisdiction of the government are below the government's fair share of safe global emission regardless of whether some emitters are not covered by the government's GHG allocation because governments have the right to make decisions distributing the burdens and benefits of government policies within their jurisdiction.

Ethical Issue Two: Allocating Global Commons Resources for Private Consumption

Cap-and-trade regimes can be understood to give emitters of GHGs a property right to emit at a level consistent with their allocation. Yet the atmosphere is usually understood to be a global commons. One feature of global commons resources is that they are owned by all people and as a consequence no person or entity can acquire a private property right that excludes others from the use of that resource (Ott and Sachs, 2000). It is also generally understood that governments have a duty to protect commons resources for the benefit of all peoples and that such responsibilities include obligations to prevent entities from using the commons for private uses. And so there is a potential conflict with assigning property rights to pollute in a carbon cap-and-trade regime and the duty of governments to protect commons resources for the benefit of all.

One way to reconcile the conflict between honoring the prohibition against allocating a commons resource for private interests and the need to give a GHG emitter some authorization to pollute for a specific amount of time is to make sure that the polluters right to emit at any time under the cap may be rescinded if it is latter determined that an existing allocation is not sufficiently protective of human health or the environment. That is, trading regimes should make it clear that rights to emit are temporary and can be adjusted for cause. Under such an approach, an allowance is understood to be only a temporary property right like a rent subject to specific conditions and reservations, not a permanent right to emit at the level prescribed by the allowance (Ott and Sachs, 200). If this approach were taken the common practice of allowing those who have allowances to bank unused portions of the allowance indefinitely would need to be abandoned. Such an approach would also limit the ability of entities to purchase pollution rights that could be used at any future time as a credit against any future emissions obligations.

Ethical Issue Three: Environmental Effectiveness

As we have seen, a cap-and-trade regime can be unjust if it is does not sufficiently protect those who are vulnerable to climate change. Many design features of cap-and-trade regimes can lead to inadequacies of the environmental protection effectiveness of the regime even if the cap is initially sufficiently protective. Some of these design issues follow:

- *Conflicts between transaction costs and environmental effectives in monitoring and verification of carbon projects.* Because there is a tension between keeping the transaction costs low and the environmental effectiveness of the cap-and-trade regime, rules often fail to assure that carbon will be reduced at levels described in the trading proposal. This is so because, for example, there are often huge scientific questions about how much carbon storage will be achieved by forests which are comprised of mixed species given that you it is virtually impossible to measure the carbon that will be stored in every tree or stored in forest soil in light of the fact that there are practical limits on how much measurement and compliance data can be obtained. To have high levels of confidence of the carbon storage capacity of forests and other biological systems that are relied upon to sequester carbon is often scientifically challenging because of large amounts of sampling, monitoring, and verification data that are needed to reduce uncertainties. Yet monitoring and verification costs can be prohibitively expensive, and as a result rules have sometimes been created that err on the side of keeping transaction costs low by limiting the amount of data that will be required. This threatens the environmental effectiveness of the project.
- *Permanence.* Many proposed projects for carbon trading raise serious questions about whether the carbon reduced by a

project will stay out of the atmosphere forever. Yet permanent storage of carbon is needed to assure equivalence between emissions reductions avoided if no credits were issued and atmospheric carbon reductions attributable to a project which creates carbon credits. This is so because emissions reductions guarantee that some quantity of GHGs will not wind up in the atmosphere, yet some projects which are used to substitute for emissions reductions have difficulty in demonstrating that the quantities of carbon reductions projected will actually be achieved. For instance, carbon stored in forests, soils, or geological carbon sequestration projects could be released to the atmosphere under the certain conditions. For example, rapid temperature change could kill trees thus releasing back into the atmosphere carbon stored in the trees. This problem is usually referred to as the problem of "permanence" of carbon reduction projects. For this reason, only projects that assure permanent reduction of carbon in the atmosphere can be categorized as environmentally effective projects.

- *Leakage.* Many proposed projects for carbon trading raise serious questions about whether carbon reduced by a project at one location will result in actual reductions in emissions because the activity which is the subject of the trade is resumed at another location. For example, paying people to plant trees in location A is not environmentally effective if these same people that receive the money chop down trees at place B. This is the problem usually referred to as "leakage." Forestry and other kinds of bio-sequestration projects that sequester carbon, in particular, often create leakage challenges. Industrial projects can also create leakage problems if the industry gets credit for reducing carbon at one industrial plant while moving the carbon producing activities to another place. If leakage occurs, then the trade is not environmentally effective.
- *Additionality.* A project subject of a trade will also not be environmentally effective if the project would have happened anyway for other reasons. This is so because a trading regime assumes that a GHG emitter should get credit because of their willingness to invest in projects that reduce carbon emissions that would not happen without the incentive to get credit for carbon reductions. If the project would happen without the investment of the emitter, than the investment in the project is not "additional"

to business as usual. This is the problem usually referred to as the "additionality" problem.

- *Enforcement of trading regime***.** A trading regime is environmentally ineffective if its conditions can not be enforced. Although enforcement of trading regimes is sometimes practical when the project on which the trade is based is within the jurisdiction of the government issuing the allowances, enforcement is particularly challenging when the project is located outside of allowance issuing government. In such cases, enforcements must be "out-sourced" to other institutions. In addition, while many hundreds of millions of dollars are being invested in setting up emissions trading schemes all over the world, virtually no resources are being channeled into their enforcement of verification. Although most cap-and-trade regimes have built-in carbon reduction verification steps, verification remains extremely difficult for many types of carbon reduction projects for which credits are being issued because of the lack of enforcement or long-term verification potential. This enforcement challenge is exacerbated when projects for which credits are issued are in poor countries without the technical capability to enforce or verify. Because of this, a strong case can be made that those who desire to rely on projects that have dubious enforcement and verification potential should have the burden of demonstrating enforcement and verification potential before they may obtain credits generated from these projects..
- *Allowing Delay In Investing In New Technology.* The ability to buy credits in a trade can create incentives to delay in making the transition away from fossil fuels that is likely necessary to avoid dangerous climate change. As emissions reductions pathways that stabilize atmospheric GHG concentrations at safe levels require that global emissions peak in the next few years followed by large reductions in emissions yearly, there is an immediate need to invest in sustainable energy. Yet the ability to trade for cheap emissions reductions credits reduces the pressure on high-emitting entities to change behavior now. In addition cap-and-trade regimes do not invest in sociological, political, and historical analysis of how to make historical change required to achieve the civilization challenging GHG reductions needed

Ethical Issue Four: Distributive Justice

Cap-and-trade regimes can create many distributive justice issues. They include the following:

- *Diminishing cheap projects in developing countries.* Under the CDM international trading system established by the Kyoto Protocol, developed nations can fund carbon reduction projects in developing countries without targets. If developing countries in some future international climate regime are asked to meet targets, they run the risk of having sold their cheapest carbon reduction projects to developed country interests by participating in the CDM, thus making achievement of any national target more expensive. Over 4000 CDM projects have been funded in developing countries and these projects have often been selected by investors because they produce the greatest reductions at lowest cost. Some of these projects, such as hydroelectric dams, are now not options for developing countries to meet future domestic GHG emissions reductions obligations.
- *Diminishing ODA. S*ome countries have announced that they will move official development assistance monies (ODA) into CDM projects thus moving funds that would benefit the quality of life in the developing country to energy projects. This could have significant impact on monies that would be otherwise available to help poor nations with such dire emergencies such as health care or nutrition. As a result, a problem created largely by developed countries may lead to worsening human health problems in countries that are minor causes of climate change when ODA flows are applied to GHG emissions projects which largely benefit emitters in developed countries.
- *Distributive justice and internal allocation of government-wide cap.* How a cap is allocated among entities within a government creates many potential distributive justice problems. Governments sometimes distribute a cap they have by giving away allowances, auctioning allowances, and other ad hoc considerations that often take into account political feasibility. Each of these methods of distributing a cap raises distributive justice issues that are often ignored for political reasons. For instance, both auctioning allowances and giving away allowances could be significantly regressive, making higher-income households better off while making lower-income households

worse off. Auctioning could also be regressive if the most wealthy get the most permits forcing those without the financial resources into non-polluting options. Sometimes governments choose to allocate the cap by placing caps on "upstream" carbon users such as coal and petroleum companies and ignoring "down tream" carbon emitters such as coal fired industrial users. A decision to place a cap upstream makes the climate change regime easier to administer but could have regressive effects on those least able to afford increased fuel costs. An upstream cap also can create little incentives for those who can afford to waste energy to change behavior. In contrast, downstream caps puts responsibility on energy users. There is no ethically neutral way to decide these design quetions.

- *Distributive justice and revenue from allowances.* When allowances are auctioned or otherwise purchased, governments must make decisions about how to use allowance revenues. These decisions raise a host of distributive justice issues that are often ignored for political reasons. Some governments have chosen, for instance to use allowance revenues to fund climate change technology research, to meet international obligations to fund climate change adaptation projects in developing countries, to fund programs to reduce deforestation projects in developing countries, to buy off politically powerful opponents to climate change legislation, to help those least able to cope with rising energy costs, or to subsidize nuclear power, geologic carbon sequestration, or renewable energy. Decisions about how to allocate revenues from distributing allowances raise distributive justice issues.

Ethical Issue Five: Procedural Justice

Cap-and-trade regimes raise a host of procedural justice issues including the following:

- *The right to participate in cap decisions*. Because carbon trading design has distributive justice implications, those whose interests may be affected by design considerations have procedural justice rights to participate in decisions on trading design. In representative democracies, an argument can be made that individual procedural rights are satisfied by duly elected representatives who participate in government decisions on cap-

and-trade. However, some representative governments do not adequately represent some populations such as indigenous peoples. Even where governments are controlled by representatives of the people, procedural rights to comment on cap-and-trade rules as they are developed are often inadequate. Some developing nations have been accused of ignoring rights of local people when they permit carbon reduction projects which benefit only a small group at the expense of interests of citizens.

- *The technical ability to participate in cap-and-trade regimes*. The Kyoto Protocol has created a staggeringly complex system for those who seek to sell or purchase carbon credits. Only those who can hire consultants and experts can effectively participate in the creation of trading projects and the development of the rules of the trading regime. For this reason, design of trading regimes that have been created often reflect the economic interests of those with the financial ability to participate in the design of much more than the interests of those who cant afford to hire experts and lobbyists. Because of the complexity of cap-and-trade regimes, strong cases can be made that governments should fund capacity building of groups that don't have the financial resources to hire experts.

References

Agarwal, Anil, and Sunita Narain. (1991). *Global Warming in an Unequal World: A Case of Environmental Colonialism*, Centre for Science and Environment, New Delhi.

Athanasiou, Tom, and Paul Baer. (2002). *Dead Heat: Global Justice and Global Warming*. New York: Seven Stories Press

Baer, Paul and Tom. Athansiou. (2005). *Honesty About Dangerous Climate Change*, Ecoequity, www.ecoequity.org/2005/11/honesty/) (last visited, Nov 8, 2009)

Brown, Donald, Nancy Tuana. Marilyn Averill, Paul Bear, Rubens Born, Carlos Eduardo Lessa Brandão, Marco Túlio S. Cabral, Robert Frodeman, Christiaan Hogenhuis, Thomas Heyd, John Lemons, Robert McKinstry, Mark Lutes, Benito Meulller, José Domingos Gonzalez Miguez, Mohan Munasinghe, Maria Silvia Muylaert de Araujo, Carlos Nobre, Konrad Ott, Jouni Paavola, Christiano Pires de Campos, Luiz Pinguelli Rosa, Jon Rosales, Adam Rose, Edward Wells, Laura Westra. (2006) *White Paper on the Ethical Dimensions of Climate Change,* The Collaborative Program on the Ethical Dimensions of

Climate Change, Rock Ethics Institute, Penn State University, http://rockethics.psu.edu/climate/whitepaper/whitepaper-intro.shtml, (last visited, Nov 5, 2009)

Brown, D., (2009), *Minimum Ethical Criteria For All Post-Kyoto Regime Proposals: What Does Ethics Require of A Copenhagen Outcome,* ClimateEthics.org, http://climateethics.org/?p=50 (last visited, Nov 5, 2009)

Environmental Defense Fund (EDF). (2009). *The Cap an Trade Success Story,* www.edf.org/page.cfm?tagID=1085 (last visited Nov.08, 2009)

Hamilton, Clive.(2009) *Is It Too Late To Prevent Dangerous Climate Change.* www.clivehamilton.net.au/cms/media/documents/articles/rsa_lecture.pdf(last visited Nov.08, 2009).

Ott, Herman, Wolfgang Sachs, *(2000) Ethical Aspects of Emissions Trading,* Wuppertal Institute, www.wupperinst.org/uploads/tx_wibeitrag/WP110.pdf. (last visited Nov.08, 2009)

United Nations (UN), (1992), United Nations Framework Convention on Climate Change Nations, UN Document, A:AC237/18.

PART V –

DEMOCRACY AND CIVIL SOCIETY: NEOCOLONIALISM V. NATURE PROTECTION

INTRODUCTION

LAURA WESTRA

This volume includes an almost dizzying array of issues, problems, disciplines and approaches. Starting from Part I with a number of chapters discussing the ideal and the reality of democracy, and the goal of ecological integrity against a background of contrasting approaches, Part II then proposed some alternative forms of governance, based on the beliefs and traditional values of Indigenous Peoples from various regions.

Part III approached the difficult question of ecological health and human responsibility in both ethics and law in relation to the limits of property rights, while Part IV returned to the details of both rights and responsibilities, through a consideration of some specific issues, including air (climate change), water, and food; hence, through the components of basic human rights.

Part V, therefore, should draw some conclusions linking the issues to our main focus: the right to, and the responsibility for, ecological integrity in a globalized world. Agnès Michelot perhaps comes closest to our main topic by discussing the UNESCO Man and the Biosphere Programme on protected areas. From the beginning of our joint work, the Global Ecological Integrity Group (GEIG) has discussed the question of the size and details of protected areas, and both "core areas" and "buffer zones" have been the main focus of our research, given their importance for the protection and support of both integrity and biodiversity.

Thus two chapters address the international legal aspects of the respect and preservation not only of natural diversity, but also of the cultural diversity of traditions and languages that Tullio Scovazzi considers as he writes of the Convention for the Safeguarding of the Intangible Cultural Heritage (UNESCO, 17 October 2003). How does the Convention relate to the central focus of our work – that is, the protection of integrity? As many of us have argued (see, for instance, the chapters by Bradford Morse and Ramy Bulan in this volume), the protection and preservation of their territories' integrity is an integral part of Indigenous Peoples' cultural and traditional rights. Hence, although the Convention's intent is not, directly, the protection of nature, nature is "indirectly protected" by its mandates.

From the consideration of these legal instruments, we need to turn to an assessment of the global situation and governance, and the moral, economic, psychological and historical background that has led humanity to this impasse. Andrew Brennan and Norva Lo address our general propensity toward unrestrained consumerism, which they term "affluenza", given the many traits it shares with both disease and addiction. The two main resulting crises – climate change and the global financial "meltdown" – finally came to a head almost at the same time, but they evoked vastly different responses from the international community in general, and the most affected states in particular.

While the first was denied and minimized for years, and even now it does not receive the response it requires, large sums were promptly available for the redress of the economic crisis. Trillions of dollars were provided for bail-outs, a hundred times more than the amounts available to ameliorate the effects of climate change and the environmental problems we face. The ongoing "affluenza" increases the force of the "private realm" (that is, the individuals' concern with their own interests), in clear conflict with true democratic citizens' concern for the good of their community.

From this ethical and psychological analysis, we turn to Sheila Collins's historical overview of the "bad turn" from the origins of natural law to its use by Locke, for instance, and others who abandoned its true foundation in nature, in order to insist on the primacy of the "right to property", as Locke understood it. Through Locke, the Western conception of natural law ignored the approach of Indigenous Peoples, which is based instead on a more holistic vision and on a "deep understanding of the earth systems". These visions were repressed through colonialism, but they survived in the thought of St Francis, for instance, in the experience of Islamic mystics, and in the writings of many American environmentalists.

Finally, we return to the problem of "globalization and extended eco-footprints" in the final chapter by William Rees. Globalization is based on a number of assumptions, such as the notion that "consumption always increases personal well-being", or that "growth will reduce or eliminate poverty", or that "the economy has no important connection to the environment", and that there are "no biophysical impediments to continuous economic growth"; all of which, Rees shows, are nothing but myths. These well-entrenched myths contribute substantially to the fact that "neoliberal economics and trade theory are essentially divorced from the social and ecological contexts". As a market-based society, we tolerate without qualms both ecological and social dumping, fostered by the "increasing entanglement of nations", which hides the reality from states who continue to exceed their limits.

These states contribute significantly to climate change disasters, to droughts and glacial melts, but also to forced mass migrations and to what Rees terms "geopolitical chaos", through new forms of colonization. Rees's final words acknowledge that "humanity's long-term collective interests converge with the individual and even national interest, for the first time in history". That should provide humanity with a unique incentive to work together and cooperate as never before, because the alternative is unthinkable. What remains to be considered is whether novel forms of traditional governance and principles of "thick" democracy throughout the globalized world will be sufficient to support and enforce human cooperation for survival.

CHAPTER TWENTY

LES RÉSERVES DE BIOSPHÈRE DU PROGRAMME SUR L'HOMME ET LA BIOSPHÈRE DE L'UNESCO: AU DELÀ DES AIRES PROTÉGÉES, UN MODÈLE DE SOCIÉTÉ DURABLE?

AGNÈS MICHELOT

Abstract

Biosphere reserves are sites recognized under UNESCO's Man and the Biosphere Programme (MAB) which innovates and demonstrates approach to conservation and sustainable management. Launched in 1970s the MAB offers an interdisciplinary research agenda and capacity to improve the relationship of people with their environment. It raises biosphere reserves to be the principal internationally designated areas dedicated to sustainable development in the 21st century. The MAB has developed a world network of biosphere reserves to promote knowledge-sharing, research and monitoring, education and training, and participatory decision-making. Based on an adaptative concept implying a zonation (core area, buffer zone, transition area), biosphere reserves promote solutions to reconcile the conservation of biodiversity with its sustainable use. These highly innovative sites have three interrelated functions: conservation of landscapes, ecosystems, species and genetic variation; economic and culturally adapted human development; and logistic support through research, monitoring, environmental education and training.

They rely on a ethical approach recognizing that human is part of nature and that nature is changing in contact with human. More generally, the biosphere reserve concept appears as a project of "sustainable society" based on social equity (benefits sharing, redistribution of ecological services, building capacity of adaptation, sharing knowledge on conservation) and on the ecosystem approach. The world network of biosphere reserves gives local working examples to accommodate adaptations to society's call for sustainable development. It helps to provide tools to deal with the complex and dynamic nature of ecosystem, and to face climate change and human development challenges such as urbanization.

Biosphere reserves represent a means not only to promote conservation and sustainable use in a equitable way, but also a pact between communities and society as a whole to open the way to a fair and sustainable future.

Le Programme sur l'homme et la biosphère (MAB) est destiné à promouvoir la mission essentielle de l'UNESCO consistant à favoriser la paix et la sécurité à travers des activités de recherche et de renforcement de capacités menées pour susciter des réponses locales à des problèmes de dimension mondiale[1]. Ce programme lancé par l'UNESCO en 1971 est une réponse à la prise de conscience des menaces qui pèsent sur l'environnement. Centré sur l'interaction entre l'homme et son environnement et sur l'utilisation durable des ressources existantes, il a été conçu initialement comme un programme de recherche interdisciplinaire appliqué en écologie, en sciences sociales et naturelles et pour la formation et la démonstration. Il s'appuie donc sur une approche dialectique des relations entre l'homme et la nature en tenant compte des liens d'interdépendance qui les unissent, et sur une approche intersectorielle permettant d'intégrer de nouvelles connaissances parfois issues de disciplines émergentes comme l'économie écologique ou la socio-écologie. La réalisation du Programme MAB va rapidement passer par la mise en pratique d'un concept tout à fait novateur développé dans les années 1970, le concept de réserve de biosphère. L'idée est de créer des réserves destinées à expérimenter les moyens de concilier deux objectifs a priori inconciliables: la conservation des ressources naturelles et la promotion du développement économique et social pour les populations locales.

Les premières réserves de biosphère vont être désignées en lien avec des parcs nationaux ou des sites de recherche intensive mais peu à peu le concept va évoluer dans ses applications concrètes. Son caractère souple et multidimensionnel applicable à des contextes variés et des perceptions culturelles différentes en matière de protection de l'environnement comme de développement, lui permet de s'adapter dans toutes les régions du monde. A présent, le réseau des réserves de biosphère comprend 553 sites répartis dans 107 Etats et contribue à la Décennie des Nations unies pour l'éducation en vue du développement durable (2005–2014) dont l'UNESCO assure le pilotage. Plus largement, le programme MAB représente une plate-forme importante pour le développement de plusieurs partenariats notamment avec les structures en charge de la mise en œuvre des plus importantes conventions multilatérales relatives à la conservation de la diversité biologique et sert à atteindre les Objectifs du Millénaire pour le développement. Dans cette perspective, le réseau mondial de réserves de biosphère représente la contribution phare du MAB à la recherche, à l'apprentissage et au renforcement de capacités pour parvenir au développement durable.

L'évolution du concept de réserve de biosphère apparaît comme la manifestation de nouvelles orientations permettant de dégager un projet de société humaine durable articulé d'une part sur la responsabilité de l'être humain quant à la préservation de la nature, et d'autre part sur l'amélioration du bien-être des populations humaines[2]. Le concept permet d'envisager des principes et des modèles de gouvernance susceptibles de maintenir un équilibre entre la conservation de la nature, la protection locale et globale de l'environnement et les progrès sociaux, économiques, techniques et scientifiques accélérés par le processus de mondialisation.

I. Un Concept à la Recherche de Modèles de Développement Durable

Plusieurs facteurs et constats ont conduit l'UNESCO a lancé le programme MAB destiné à assurer la continuité de l'existence de l'homme dans son environnement tout en poursuivant la conservation et le renouvellement des ressources naturelles. Les connaissances des ressources étaient inégalement diffusées dans et selon les différentes zones géographiques, ainsi les pays développés disposaient-ils de données plus importantes sur leurs zones climatiques tempérées. De manière générale, un manque aigu de scientifiques spécialistes de l'identification et de la gestion des ressources naturelles était constaté dans les pays en développement. Les planificateurs, les décideurs et les utilisateurs de ces ressources ne disposaient pas de la formation ni des compétences requises pour s'occuper des problèmes liés à la pénurie de ressources naturelles. Enfin, l'UNESCO cherchait à encourager la coopération internationale dans le domaine des sciences de la vie.

Or, le Programme MAB, en initiant le concept des réserves de biosphère, va chercher à mettre en réseau des expériences et des structures permettant de disposer de résultats scientifiques, de personnels compétents et de mécanismes d'échanges d'information utiles. Par ailleurs, le concept va conduire à dépasser l'analyse des données environnementales liées à la survie de l'homme. La réserve de biosphère représente un instrument susceptible de démontrer les relations globales existant entre l'homme et son environnement et de mettre en évidence les conséquences des actions de l'homme sur l'avenir.

A. Histoire d'un concept

Le concept de réserve de biosphère a été élaboré en 1974 par un groupe de travail du MAB et, par la suite, en 1976, l'UNESCO lance le réseau des réserves de biosphère pour créer des sites où l'on expérimente des mesures et plans de gestion visant à obtenir un équilibre entre les nécessités de la conservation, la promotion du développement économique et la sauvegarde de valeurs culturelles. En 1983, se tient le premier Congrès international sur les réserves de biosphère organisé, à Minsk, par l'UNESCO et le Programme des Nations unies pour l'environnement (PNUE) en coopération avec l'Organisation des Nations unies pour l'agriculture (FAO) et l'Union mondiale pour la nature (UICN). Cet événement permet d'apprécier les progrès réalisés sur les questions de conservation du patrimoine naturel et conduit à adopter un plan d'action pour les réserves de biosphère unanimement approuvé par la Conférence générale de l'UNESCO et le Conseil d'administration du PNUE. Afin d'en évaluer l'efficacité et conformément à la résolution 27 C/2.3 de sa Conférence générale, l'UNESCO organise la Conférence sur les réserves de biosphère en 1995 à Séville qui va aboutir à l'adoption de deux textes essentiels pour la reconnaissance internationale et harmonisée du statut des réserves de biosphère: la Stratégie de Séville et le cadre statutaire.

Si la Stratégie définit les objectifs et recommande les actions à entreprendre aux plans international, national et local, le Cadre statutaire prévoit quant à lui les règles liées aux fonctions et au fonctionnement des réserves. Il fixe la procédure de désignation, d'appui et de promotion des réserves en tenant compte de la diversité des situations nationales[3].

Dès lors, la réserve de biosphère, qui n'était qu'un projet au sein du Programme MAB, prend une autre dimension. En effet, même si les deux textes n'ont pas valeur juridique contraignante, ils ont fait l'objet d'un large consensus et d'une adoption par la Conférence générale de l'UNESCO qui atteste de l'engagement des Etats membres de l'organisation de respecter les critères, objectifs et modes de désignation des réserves de biosphère[4].

B. Des principes structurant un concept souple

Selon le cadre statutaire du Réseau mondial des réserves de biosphère, les réserves de biosphère sont des *« aires portant sur des écosystèmes ou une combinaison d'écosystèmes terrestres et côtiers/marins reconnues au niveau international dans le cadre du Programme de l'UNESCO sur l'Homme et la biosphère (MAB) »*[5]. La désignation officielle d'une réserve de biosphère par le Conseil du MAB après examen de la proposition faîte par l'Etat, implique à la fois reconnaissance de la mise en œuvre d'un concept tel que défini par l'UNESCO et l'intégration du site dans le réseau mondial.

Le concept repose sur plusieurs principes acceptés par les Etats et qui marquent l'originalité et la souplesse de la démarche entreprise au sein du Programme MAB.

1) Le respect de la souveraineté territoriale pour la création et le suivi des réserves

Dans le cadre du programme MAB les réserves sont établies à la demande des Etats et, après leur intégration dans le réseau mondial des réserves de biosphère, continuent de relever de la souveraineté de l'Etat sur lequel elles se trouvent[6]. Le cadre statutaire adopté à Séville a pour objet de diffuser les bonnes expériences de gestion et chaque Etat est encouragé à mettre en place des réserves qui tiennent compte de son contexte particulier. Il peut ainsi arriver que des réserves de biosphère soient établies en application de législations nationales sans pour autant que les Etats en demandent l'intégration dans le réseau mondial. Dans ce cas de figure, elles ont le statut de réserves de biosphère nationales et n'ont pas à remplir la procédure ni les critères liés à ce « label international » accordé par l'UNESCO[7]. Les Etats qui disposent de réserves de biosphère reconnues par l'UNESCO doivent quant à eux prendre les mesures juridiques nécessaires et appropriées pour assurer le respect du Cadre statutaire et se soumettre à toutes les obligations relatives au statut international de réserve de biosphère. L'une d'elle consiste à accepter un examen périodique tous les dix ans sur la base d'un rapport adressé par l'Etat au secrétariat[8]. Enfin, un Etat reste libre de retirer du Réseau une réserve de biosphère placée sous sa juridiction[9].

2) Les trois fonctions ou l'approche intégrée

Dans la mesure où les réserves de biosphère s'efforcent de constituer des sites modèles d'étude et de démonstration susceptibles de promouvoir une relation équilibrée entre activités humaines et protection de l'environnement, elles doivent remplir trois importantes fonctions complémentaires qui interagissent à niveau égal.

La fonction de conservation s'attache à maintenir les écosystèmes, à protéger les paysages, les espèces et la diversité génétique. La perspective est de protéger les terres contre toutes formes de dégradation et d'assurer la protection de la diversité biologique. A cette fin, les réserves de la biosphère établissent un certain équilibre au sein de l'écosystème en évitant des modifications de populations ou des perturbations dans le renouvellement des ressources génétiques.

Cette fonction est étroitement liée à la deuxième fonction des réserves de biosphère, la fonction de développement destinée à soutenir les actions de développement économique et humain durable des points de vue socioculturel et écologique[10]. Ainsi les expériences de lutte contre la désertification menées en Afrique avec des programmes de réhabilitation des terres dégradées dans les réserves de biosphère illustrent cette relation étroite entre conservation et développement tout en renvoyant à la fonction logistique.

Cette troisième fonction est celle qui fournit les moyens pour des activités d'éducation environnementale et de formation, de recherche et de surveillance continue sur des questions de conservation et de développement durable. Cette fonction a connu de sérieux approfondissements. Elle touche à des enjeux fondamentaux en particulier pour les pays en développement demandeurs de formations adaptées à la gestion des réserves de biosphère. Les lacunes au niveau des capacités locales pour l'acquisition des connaissances nécessaires ont été relevées comme l'un des problèmes majeurs pour la mise en œuvre des objectifs et fonctions des réserves de biosphère[11]. Par ailleurs, l'activité scientifique permet d'améliorer les connaissances sur les interactions entre l'homme et la biosphère et donc de mieux faire face aux obstacles pesant sur le développement durable[12]. L'objectif III de la Stratégie de Séville souligne ainsi l'intérêt des programmes internationaux de recherche relatifs à la diversité biologique[13], la désertification, le cycle de l'eau, l'ethnobiologie et les changements globaux. Les exemples sont nombreux du soutien de la recherche à la conservation et aux capacités de développement. Le Programme MAB a notamment soutenu des programmes de recherche sur la régénération des écosystèmes pour faire face à différents besoins des populations locales et permettre le maintien de processus reproductifs[14]. Enfin, l'éducation environnementale est une action phare des réserves de biosphère, elle peut prendre la forme d'ateliers de formation à destination des communautés locales et, le plus souvent, implique une collaboration étroite avec les établissements scolaires pour organiser des visites commentées de la zone centrale de la réserve de biosphère. La Stratégie de Séville prévoit également d'inclure dans les programmes scolaires et les manuels d'enseignement ainsi que dans les médias des informations sur la conservation et l'utilisation durable des ressources telles que pratiquées dans la réserve de biosphère[15]. La connaissance de leur patrimoine naturel par les populations situées dans les zones tampons et de transition est souvent un atout non seulement pour le maintien des processus écologiques et le développement d'activités écologiquement durables mais également pour la surveillance des sites.

Plus largement, l'amélioration de la formation des spécialistes et des gestionnaires ainsi que la sensibilisation du public sont susceptibles d'être améliorées et dynamisées par des échanges que peuvent soutenir des réseaux et programmes internationaux.

3) Le zonage

La réserve de biosphère remplit ses fonctions à partir d'un zonage basé sur l'identification de trois zones avec des statuts et des objectifs différents. L'aire centrale est consacrée à la protection à long terme et doit par conséquent bénéficier d'un statut juridique particulier attribué par l'Etat. Elle permet d'assurer la fonction de conservation de la réserve et de mener des recherches et d'autres activités n'ayant pas d'impact perturbateur pour l'écosystème. La zone tampon est également une zone clairement identifiée et délimitée qui entoure ou jouxte l'aire centrale avec pour particularité d'accueillir des activités et des pratiques écologiquement viables. Elle peut notamment être utilisée pour des activités de coopération et d'éducation environnementale, les loisirs, le tourisme et la recherche. Enfin la zone de transition est une aire définit plus souplement. Elle peut accueillir des activités humaines d'exploitation comme des activités agricoles et des établissements humains. Sur ce territoire, délimité souplement, de nombreux partenaires, organisations publiques ou privées, agences scientifiques, groupes culturels et communautés locales participent à la gestion et à l'élaboration de projets de développement durable.

Initialement ces trois zones ont été conçues comme une série de cercles concentriques dont le centre est occupé par la zone centrale. Cependant, chacune d'elle peut bénéficier d'un régime juridique différent en fonction des législation et plans nationaux d'aménagement des territoires. Une même zone peut couvrir des situations différentes[16] et un grand nombre de réserves de biosphère ont une partie de leur territoire reconnue au niveau international notamment comme patrimoine mondial ou comme sites de la Convention de Ramsar sur les zones humides[17]. Par ailleurs, les zones peuvent relever du domaine public ou de la propriété privée. La réserve de biosphère est en fait un ensemble hétérogène d'unités socio-territoriales regroupées dans un projet de développement durable cohérent.

On notera que le zonage ne correspond pas précisément aux trois fonctions de la réserve de biosphère dans la mesure où les trois zones peuvent les remplir à des niveaux variés. Les zones tampons sont liées à leur contribution aux objectifs de conservation et l'aire de transition est plus directement consacrée au développement et à des modes d'exploitation durable.

Enfin, de nouvelles formes de réserves de biosphère se sont développées comme par exemple les « réserves en grappe » et les réserves transfrontières[18] Selon les objectifs poursuivis, la configuration des zones et leur articulation varient. Cette flexibilité a donné au concept de réserve de biosphère toute sa capacité d'adaptation face à des besoins locaux et des contextes socio-économiques variés. Quoiqu'il en soit les réserves de biosphère ne peuvent être considérées comme de simples aires protégées car elles couvrent une zone multifonctionnelle avec des objectifs diversifiés d'utilisation des ressources et d'aménagement du territoire[19]. Pour cette raison la réserve de biosphère ne peut pas facilement apparaître dans une catégorisation nationale d'aires protégées ni même être intégrée dans une classification internationale telle que celle de l'UICN.

4) La gestion participative: une méthodologie omniprésente

La gestion participative est évoquée dans le cadre statutaire comme un critère de désignation de réserve de biosphère. Les pouvoirs publics, les communautés locales et les intérêts privés doivent être associés à la conception et à la mise en œuvre des fonctions de la réserve de biosphère[20]. La Stratégie de Séville détaille quant à elle les problèmes liés au soutien et à la participation des populations locales en énonçant: la résolution des conflits, l'octroi des bénéfices localement, et la participation des partenaires à la prise de décision et à la responsabilité de la gestion[21].

Chaque aspect renvoie à des problématiques et, potentiellement, à des acteurs différents. La participation dans le processus de création d'une réserve de biosphère va reposer sur la mobilisation de l'ensemble des collectivités locales, des partenaires institutionnels et des représentants de la société civile. Les acteurs doivent se saisir des enjeux de développement durable et pour cela la construction d'une perspective commune doit intervenir en amont du processus de création de la réserve. L'implication de tous apporte une légitimation locale au projet de réserve de biosphère ce qui va favoriser un réel engagement dans une gestion durable de l'espace concerné. Par la suite, la réserve de biosphère devra susciter l'émergence d'espaces de dialogues à différentes échelles:[22] au niveau de la structure de gestion de la réserve par une table de concertation ou au niveau d'un secteur ou d'une ressource (tourisme, industrie forestière...). Cette approche doit aussi permettre de prévenir les conflits et notamment les conflits d'intérêts liés à des usages concurrents.

La gestion participative est entendue dans ses différentes implications. Elle correspond à une méthodologie de fonctionnement de la réserve basée sur le dialogue et la concertation. Cette méthode permet à l'ensemble des acteurs impliqués de clarifier par eux-mêmes les potentialités et les enjeux du territoire. La gestion participative doit également conduire à un partage équitable des bénéfices provenant de l'exploitation des ressources naturelles. Au delà des procédures légalisées de participation, c'est une véritable culture de la participation que le concept de réserve de biosphère encourage[23].

5) Les structures, réseaux et mécanismes institutionnels

Le programme MAB dispose de structures internationales et s'appuie sur des comités nationaux dont vont bénéficier les réserves de biosphère.

Les activités du MAB sont coordonnées par un conseil international de coordination(CIC) qui établit les grandes orientations du programme et représente le principal organe de décision[24]. Un bureau est élu lors de chaque session par le CIC pour assurer un suivi entre deux sessions et un secrétariat assure la coordination globale à l'échelle internationale. A l'échelon national, les comités MAB nationaux soutiennent les activités du programme et le développement des réserves de biosphère. Ces comités sont reliés par des réseaux sous-régionaux à caractère biogéographique ou thématique et sont intégrés à des réseaux intra-continentaux plus larges (AFRIMAB, ASIAMAB, EUROMAB...). De manière générale, on estime qu'il existe quatre niveaux de formation de réseau pour le programme MAB: un niveau de projets opérationnels entre réserves de biosphère, un niveau sous-régional entre des réseaux sous-régionaux à caractère biogéographique, un niveau régional entre les régions telles qu'elles sont identifiées et un niveau international entre les régions à l'intérieur ou en dehors du même continent[25].

Enfin, la Stratégie de Séville recommande de mettre en place des mécanismes institutionnels permettant de gérer, coordonner et intégrer les programmes et les activités de la réserve[26]. Le cadre statutaire quant à lui prévoit qu'une autorité ou un mécanisme soit établi pour mettre en œuvre le plan ou la politique de gestion de l'ensemble de la réserve de biosphère.

Le concept de biosphère permet d'expérimenter des modèles de développement durable qui tiennent compte de tous les besoins sociaux, culturels, spirituels et économiques de la société en s'appuyant sur des connaissances scientifiques solides, complètes et diversifiées. Dans cette perspective, il participe à la construction d'un projet de société plus vaste.

II. La Voie Vers une «Société Durable»

Suite au Congrès de Minsk de 1983, une réflexion importante portant sur l'ensemble des aires protégées ainsi que sur les réserves de biosphère s'est poursuivie notamment dans la perspective d'améliorer la liaison entre la conservation de la biodiversité et les besoins des communautés locales. Ainsi, lors du quatrième congrès mondial sur les parcs nationaux et les aires protégées tenu à Caracas en 1992, les responsables et gestionnaires des aires protégées ont intégré plusieurs des orientations et principes d'action testés dans les réserves de biosphère. Par la suite, la conférence internationale sur les Réserves de biosphère qui se tient à Séville en mars 1995 va plus loin en envisageant leur rôle dans le contexte du XXIe siècle.

« *(…) les réserves de biosphère sont sur le point d'assumer un nouveau rôle. Elles seront non seulement un moyen de permettre aux populations qui y vivent, ou qui vivent dans leur périphérie, de s'épanouir en équilibre avec le milieu naturel, elles vont également contribuer aux besoins de la société en général, en montrant la voie d'un avenir plus durable.* »[27]

Le concept de réserve permet d'accompagner la prise de conscience environnementale et d'envisager, au regard d'expériences pratiques, un modèle de « société durable » construit à partir d'une démarche éthique et dans le respect des dimensions humaines et écologiques du développement durable.

A. Une démarche au service du développement durable

Les réserves de biosphère représentent des sites de démonstration du développement durable. A ce titre elles doivent proposer de nouvelles perspectives pour promouvoir simultanément la conservation et le développement nécessitant l'utilisation des ressources naturelles. Pour y parvenir le concept de réserve de biosphère s'inscrit dans une démarche éthique destinée à guider nos actions, nos politiques et stratégies de développement tout en s'appuyant sur une approche écosystémique indispensable pour s'adapter aux réalités complexes et dynamiques des processus naturels.

1) Une éthique du développement à travers une approche dialectique

Le concept de réserve de biosphère est basé sur la reconnaissance des liens que l'homme entretient avec la nature matériellement, culturellement voire spirituellement. D'une certaine manière, il tente de répondre à la question fondamentale « comment devons-nous vivre ? »[28] dans une perspective pratique. Les réserves de biosphère doivent nous permettre d'utiliser au mieux nos connaissances et nos expériences pour assurer un avenir viable pour les générations présentes et futures[29]. Plus fondamentalement, le programme MAB part du constat que l'homme fait partie de la nature et que la nature évolue au contact de l'homme. D'une certaine manière aucune solution ne pourra être apportée aux défis du développement durable sans tenir compte de cette relation dialectique. Le projet proposé par le MAB est à la fois « écologiquement humaniste et « humainement écologique » et dépasse largement une relation dans laquelle l'homme serait le sujet et la nature l'objet. Comme le note Serge Gutwirth: « *L'homme est dedans. Il fait intégralement partie du monde dans lequel il vit. Marqué par le monde, il le marque. Déterminé partiellement par son environnement, il le détermine* (...) »[30].

L'approche particulièrement pragmatique des réserves de biosphère, dans la mesure où elle repose sur l'expérimentation, conduit inévitablement à se confronter à la complexité des réalités humaines et naturelles et donc à développer une vision dialectique des problématiques du développement durable. En effet, seule cette approche permet de prendre en compte toutes les implications d'une relation complexe. Très concrètement cela se manifeste dans l'émergence d'une culture interdisciplinaire prônée et soutenue par le programme MAB qui se décline en plusieurs niveaux d'interdisciplinarité: 1) entre biologistes (systématiciens, généticiens, écologues, physiologistes, éthologistes), 2)au sein de l'ensemble des sciences de la nature (paléontologie, géodynamique, climatologie...), 3)entre sciences de la nature et sciences de l'homme et de la société.

Le concept de réserve de biosphère implique la nécessité d'une interdisciplinarité opérationnelle laquelle est le reflet d'une exigence: celle d'appréhender le plus concrètement et le plus efficacement (et donc de manière dialectique) les enjeux de conservation et de développement.

2) L'approche écosystémique

Les objectifs des réserves de biosphère présentés par la Stratégie de Séville montrent les priorités accordées à la conservation de la diversité naturelle et culturelle (Grand objectif I) et la promotion de modèles et de lieux d'expérimentation du développement durable (Grand objectif II) notamment dans l'intérêt des populations locales. De manière générale, il apparaît que le réseau des réserves de biosphère va répondre aux préoccupations de la Conférence des Nations unies sur l'Environnement et le Développement (CNUED) et plus particulièrement à celles de la Convention de Rio sur la diversité biologique de 1992. De fait, l'approche écosystémique définie dans la convention ainsi que les principes et actions qui y sont associées, sont en parfaite convergence avec le concept de réserve de biosphère. En effet, « *l'approche écosystémique est une stratégie de gestion intégrée des terres, des eaux et des ressources vivantes, qui favorise la conservation et l'utilisation durable d'une manière équitable. (…) L'approche écosystémique repose sur l'application de méthodes scientifiques appropriées aux divers niveaux d'organisation biologique qui incluent les processus, les fonctions et les interactions essentiels entre les organismes et leur environnement. Elle reconnaît que les êtres humains, avec leur diversité culturelle, font partie intégrante des écosystèmes* »[31].

La douzaine de principes qui animent l'approche écosystémique comprennent notamment la gestion décentralisée (c'est-à-dire ramenée au plus près de la base), la prise en compte du contexte économique et des avantages potentiels tirés de la gestion, et l'implication de tous les secteurs sociaux ainsi que de toutes les disciplines scientifiques concernées. On retrouve ici les principes propres au fonctionnement des réserves de biosphère et l'expression de la même cohérence quant à l'importance d'une approche concrète et adaptée aux réalités complexes du développement durable. L'approche écosystémique, pleinement appliquée par le Réseau mondial des réserves de biosphère, symbolise également l'attention accordée à la cohésion des territoires et au respect de l'intégrité des écosystèmes.

3) La dimension humaine du concept de réserve de biosphère: entre équité sociale et solidarité internationale

La conservation permet d'améliorer le bien-être des populations y compris sur le plan économique par le biais d'activités et de technologies appropriées. Elle permet notamment de préserver des espèces animales et végétales et de maintenir des ressources génétiques pour le développement local et national. D'une certaine façon, les réserves de biosphère jouent donc le rôle de réservoirs de ressources pour satisfaire les besoins des populations locales. Cependant elles représentent bien d'autres atouts pour la construction d'une société en harmonie avec ses valeurs de solidarité, de dialogue, d'équité sociale et de respect des valeurs culturelles.

Tout d'abord le concept conduit à forger un pacte entre la communauté locale et la société dans son ensemble au travers d'un projet de gestion et d'aménagement du territoire. Le statut de réserve de biosphère scelle une démarche conjointe des autorités nationales et des communautés locales pour la réalisation d'un projet de gestion de territoires orienté sur l'intérêt des populations « résidentes ». Ensuite, la création d'une réserve liée à un réseau mondial ouvre les possibilités d'échanges et de capitalisation d'expériences à une dimension internationale. Une expérience menée localement peut ainsi bénéficier à d'autres communautés. De la même façon, des techniques traditionnelles comme des technologies modernes peuvent être diffusées au sein du Réseau mondial pour résoudre des problèmes liés à la dégradation de l'environnement. Dans une approche globale, les réserves de biosphère apparaissent comme un instrument permettant à une société de faire face à des défis environnementaux d'envergure en coordonnant réflexion et action[32].

Ainsi, la réserve de biosphère apparaît comme la concrétisation d'un projet d'équité sociale qui se manifeste de différentes façons et notamment par: 1) le partage juste et équitable des bénéfices économiques tirés d'une utilisation rationnelle commune, 2) la redistribution des services écologiques obtenus (qualité de l'eau, protection des sols…), 3) le renforcement de capacités d'adaptation aux changements, perturbations ou pressions extérieures (pressions économiques, sociales et politiques) et 4) le partage de connaissances.

Il a ainsi été mis en évidence que les réserves de biosphère participent à la transmission des méthodes traditionnelles de protection de la diversité biologique comme par exemple des techniques d'utilisation du sol favorisant leur stabilisation et contribuant ainsi à lutter contre la désertification.

Par ailleurs, la réserve de biosphère représente un instrument de dialogue à plusieurs niveaux: tout d'abord entre les acteurs de la réserve et « l'extérieur » – notamment au sein du Réseau mondial, mais aussi entre les gestionnaires, décideurs, scientifiques et partenaires de la réserve et donc entre les acteurs de « l'intérieur ». Les acteurs des réserves de biosphère initient, perpétuent et transmettent une culture du dialogue dans des contextes et à partir de pratiques variés. Ce faisant, le dialogue favorise la diffusion des savoirs et la construction sociale de normes, de principes et de réseaux de communication.

Enfin, la démarche initiée par le MAB via le Réseau des réserves de biosphère s'appuie sur le respect de la diversité naturelle et culturelle. Les valeurs culturelles sont au cœur du concept. Par exemple des écosystèmes africains particulièrement riches sont, de manière indirecte et souvent inconsciente, protégés pas des croyances spirituelles et des pratiques traditionnelles qui alimentent la connaissance[33]. La reconnaissance de bois sacrés a permis de préserver des écosystèmes de toutes formes d'exploitation. Le concept de réserve de biosphère est en conformité avec de tels mécanismes traditionnels de protection et peut encourager la défense de préoccupations culturelles en connexion avec la préservation de certains espaces. Enfin, les grands objectifs poursuivis par les réserves de biosphère favorisent une approche holiste de la gestion des paysages impliquant la conservation de la diversité culturelle et naturelle.

B. Des réponses locales à des problèmes mondiaux ?

Depuis le lancement du programme MAB, le Réseau des réserves de biosphère s'est considérablement étendu et ses activités fonctionnelles se sont élargies.

Dans la perspective de la Conférence de Stockholm sur l'environnement humain de 1972 qui a posé le premier jalon d'un engagement politique international en lien avec la prise de conscience globale des risques pesant sur l'environnement, le MAB avait été conçu comme un programme interdisciplinaire de formation et de démonstration pour aborder l'analyse des relations mutuelles entre l'homme et son environnement. La Conférence de Stockholm a mis en avant la défense du bien-être de l'homme en lien avec la protection de l'environnement et la nécessité d'évaluer l'impact du développement sur l'environnement (et vice-versa). Cette réflexion avait alors dynamisé le programme de l'UNESCO qui a initié le concept de réserve de biosphère. Vingt ans plus tard, la Conférence des Nations unies sur l'environnement et le développement conduit à l'adoption d'accords multilatéraux aux objectifs analogues à ceux du MAB. Les réserves de biosphère vont ainsi participer à la mise en œuvre de conventions globales notamment de la convention sur la diversité biologique et de la convention de lutte contre la désertification. En février 2008, le 3e Congrès mondial des réserves de biosphère adopte le Plan d'action de Madrid qui entend faire des réserves de biosphère des sites de références pour le développement durable au XXIe siècle. Le MAB cherche donc à répondre efficacement à des défis de niveau mondial et notamment à l'accélération du changement climatique, à la perte de biodiversité et de diversité culturelle, et au processus d'urbanisation.

1) Faire face aux changements climatiques

Les changements climatiques représentent une terrible menace pour les équilibres naturels et pour les populations humaines et non humaines subissant les désordres météorologiques, l'augmentation du niveau des mers, l'accélération du processus de désertification… La plupart des services écologiques vont être en situation de stress intense dans les vingt prochaines années et les populations humaines devront trouver des moyens de s'adapter et de faire face à la situation.

Les réponses envisagées par la communauté internationale sont axées sur l'adaptation et l'atténuation. Tout d'abord l'adaptation a tendance à se traduire par une modification de l'occupation et de la gestion des sols. Quant à l'atténuation des effets des changements climatiques, elle suppose l'adoption et la mise en œuvre de mesures en matière de séquestration de carbone, de réduction de gaz à effet de serre, d'efficacité énergétique et plus généralement de modification des modes de vie. Dans tous les cas, les réponses sociétales aux changements climatiques impliquent une surveillance de leur impact et le développement de stratégies en matière de gestion et d'aménagement des territoires. Les réserves de biosphère semblent pouvoir jouer un rôle essentiel pour tester des instruments d'adaptation dans la mesure où elles offrent des sites dédiés à des démonstrations et des apprentissages. Par ailleurs, les zones tampon et les aires de transition des réserves de biosphère peuvent permettre de tester des mesures et des stratégies d'atténuation. C'est le cas notamment des réserves abritant des systèmes forestiers susceptibles de piéger le carbone. Enfin, les expériences relatives à d'autres modes de vie plus écologiques, tentées dans les réserves de biosphère, pourraient inspirer de nouveaux dispositifs d'adaptation et/ou d'atténuation.

2) Maintenir les services des écosystèmes

A l'échelle planétaire, la préoccupation est grande de préserver les services rendus par les écosystèmes à nos sociétés. La première catégorie de services comprend le soutien au cycle de nutrition, et donc aux sols et à tous les mécanismes qui assurent les productions primaires. La seconde vise la subsistance elle-même et donc tout ce qui participe à notre activité vitale (eau douce, nourriture, bois, fibres, énergie). La troisième porte sur la régulation et englobe les services écologiques nous protégeant des catastrophes climatiques, des virus, des inondations ou assurant le filtrage de l'eau. Et enfin, la quatrième inclut tous les services culturels ou sociaux comme l'accès à des services moraux, éducatifs ou de loisirs ou même esthétiques.

Le concept de réserve de biosphère est, par son système de zonage et les critères établis pour son identification, destiné à offrir l'ensemble des services écologiques. Par ailleurs, le Réseau mondial a vocation à préserver des sites particulièrement intéressants sous l'angle écologique[34] tout en organisant une coopération internationale susceptible d'optimiser les échanges d'information, de favoriser les résultats scientifiques et les expériences réussies de gestion rationnelle d'écosystèmes.

3) Gérer l'urbanisation

L'une des idées assez récente du programme MAB consiste à explorer les stratégies d'urbanisation en lien avec le concept de réserve de biosphère. L'objectif est de garantir l'harmonie entre les réserves et les systèmes urbains et périurbains qui gravitent autour en tenant compte des populations qui y vivent. Le facteur démographique étant un facteur souvent déterminant pour la conservation et le développement, les réserves de biosphère doivent l'intégrer dans leur plan de gestion.

Par ailleurs, l'urbanisation est, à une large échelle, un paramètre fondamental des équilibres entre nature et sociétés humaines car il modifie la pression anthropique, les régimes d'utilisation des sols et des ressources ainsi que les pratiques culturelles d'exploitation. Ce phénomène en pleine expansion, (considérant qu'aujourd'hui plus de la moitié de la population mondiale vit en zone urbaine), fait peser de lourdes menaces sur l'équilibre entre protection de l'environnement et objectifs de développement durable recherchés par le programme de l'UNESCO. Parallèlement, c'est un défi lancé à la communauté internationale que le concept de réserve de biosphère peut permettre de relever, au moins en partie. En effet, les paysages urbains sont la manifestation tangible de la mainmise de l'homme sur les processus naturels et offrent à cet égard un terrain privilégié pour expérimenter de nouvelles techniques de gestion[35]. Ils poussent à développer une conception dynamique de l'écologie intégrant les perturbations comme facteurs de structuration des communautés biotiques, et à rechercher des modes nouveaux de cohabitation entre l'homme et la nature[36].

Ainsi, un certain nombre de villes ou de collectivités urbanisées ont déjà appliqué le concept de réserve de biosphère à des zones urbaines significatives de leur territoire et d'autres envisagent de s'appuyer sur le concept pour assurer la planification et la gestion durable de leur développement urbain. D'une certaine manière cette perspective est renforcée par les recommandations apportées par le Plan d'action de Madrid pour les réserves de biosphère (2008–2013) qui souligne l'importance d'un engagement dans des actions socio-écologiques et politiques innovantes associant les trois fonctions des réserves de biosphère.

Le concept de réserve de biosphère a montré l'étendue de ses potentialités notamment grâce à sa flexibilité suffisante pour s'adapter à des contextes extrêmement variés. Tourné vers les expériences innovantes et ancré dans les réalités complexes du développement durable, il présente de nombreux atouts pour préparer les sociétés humaines à s'adapter à des désordres écologiques majeurs. Cependant, ses échecs et ses lacunes sont aussi nombreux. Le renforcement des capacités humaines et institutionnelles par le biais de la formation, du transfert de technologie et les mécanismes de coopération en réseaux thématiques ou géographiques reste souvent largement insuffisant dans les régions les plus vulnérables aux bouleversements environnementaux. Même si le programme MAB couvre des domaines potentiellement très stratégiques pour faire face à un contexte planétaire tendu, il ne saurait remplir ses missions sans des financements adaptés et un réel engagement politique.

Notes

[1] Cf. Préambule, Le Programme MAB su service du développement durable, adopté à la 21e session du Conseil international de coordination du MAB, Jeju (République de Corée), 25–29 mai 2009.

[2] Cf. Déclaration de Madrid sur le Programme MAB et le Réseau mondial de réserves de biosphère (WNBR), Madrid, 8 février, 2008.

[3] Cf. Préambule, Cadre statutaire du réseau mondial de réserves de biosphère adopté par la Résolution 28 C/2.4 de la Conférence générale de l'UNESCO (novembre 1995).

[4] Le cadre statutaire a fait l'objet de négociations difficiles et est le fruit d'un large processus de consultation notamment par correspondance. Il a été soumis au Conseil international de coordination du MAB. Cf. Mireille Jardin, Le cas particulier des réserves de biosphère, in Garnier Lisa (ed.), *Entre l'homme et la nature, une démarche pour des relations durables*, Réserves de biosphère Notes techniques 3–200 , Paris, UNESCO, pp.36–38.
[5] Article1 du cadre statutaire du réseau mondial de réserves de biosphère.
[6] Article 2.3 du cadre statutaire du réseau mondial de réserves de biosphère.
[7] Certains pays comme l'Inde, la Chine et le Mexique ont adopté le concept et le décline dans le cadre de leurs instruments de conservation et d'aménagement du territoire.
[8] Article 9.1 du cadre statutaire du réseau mondial de réserves de biosphère.
[9] Article 9.8 du cadre statutaire du réseau mondial de réserves de biosphère. Dans ce cas, l'Etat doit notifier le retrait au secrétariat et l'aire cesse alors d'appartenir au réseau mondial de réserves de biosphère.
[10] Cf. article 3 (ii) du cadre statutaire du réseau mondial de réserves de biosphère.
[11] Cf. MAB en Afrique, Rétrospective et perspectives pour le 21e siècle, UNESCO Dakar, Unité des Sciences naturelles et de la Technologie/NST, octobre 1997, p. 38.
[12] A cet égard l'UNESCO soutient les initiatives de chercheurs juniors par les Bourses du MAB pour jeunes scientifiques.
[13] L'objectif est d'enrichir la base de connaissances sur les questions de durabilité de l'environnement notamment dans le Projet de survie des grands singes (GRASP) et le Programme de surveillance continue intégrée de réserves de biosphère (BRIM).
[14] Exemple de la Réserve de biosphère d'Omo dans le Sud Ouest du Nigéria qui a eu recours à deux méthodes de régénération artificielles appliquées à la forêt. Autre exemple, celui de la réserve de biosphère de M'passa au Gabon avec l'utilisation de méthodes de régénération naturelles dans les différents écosystèmes de la réserve pour maintenir la mosaïque forestière.
[15] Objectif III.3 de la Stratégie de Séville pour les réserves de biosphère.
[16] Par exemple la zone centrale peut pour partie avoir le statut de réserve naturelle et pour une autre partie le statut de parc national selon la catégorisation d'aires protégées applicable par l'Etat sous la juridiction duquel se trouve la réserve de biosphère.
[17] Un certain nombre de réserves de biosphère comprennent des territoires classés patrimoine mondial de l'UNESCO et sites Ramsar comme la réserve de Maya au Guatemala qui comprend le Tikal National Park classé patrimoine mondial et le Laguna del Tigre classé site Ramsar.
[18] En 2009, le MAB reconnaît 8 réserves de biosphère transfrontières dont 6 en Europe et 2 en Afrique (le Delta du fleuve Sénégal devenu RB en 2005 à cheval entre la Mauritanie et le Sénégal, et le complexe du W à cheval sur le Bénin, le Burkina Faso et le Niger). La plus récente réserve de biosphère transfrontière est la réserve Geres-Xures partagée entre le Portugal et l'Espagne et reconnue par le

MAB en 2009. Cf. Biosphere Reserves Technical notes, *Five transboundary Biosphere Reserves in Europe*, UNESCO, Paris, 2003.

[19] UICN, UNESCO-MAB Programme, *Biosphere reserves – Myth or reality ?* Proceedings for the Workshop on Biosphere Reserves, World Congress, Montreal, 1996, IUCN, MAB, 1998.

[20] Article 4.6 du cadre statutaire du réseau de réserves de biosphère.

[21] Objectif II.1 de la Stratégie de Séville.

[22] Bouamrane M. (ed), *Le dialogue dans les réserves de biosphère. Repères pratiques et expériences,* Réserves de biosphère, Notes techniques, 2, UNESCO, Paris, 2007.

[23] Bouamrane M. (ed.), *Biodiversité et acteurs: des itinéraires de concertation*, Réserves de biosphère Notes techniques 1, UNESCO, Paris, 2006.

[24] Le CIC est composé de représentants scientifiques des pays élus pas la Conférence générale de l'UNESCO et d'observateurs issus d'organisations internationales partenaires.

[25] L'article 7 du cadre statutaire des réserves de biosphère prévoit les conditions de participation au réseau aux niveaux mondial, régional et sous-régional. L'article 8 du même cadre statutaire énonce que le Etats encouragent la coopération au sein de sous-réseaux régionaux et/ou thématiques de réserves de biosphère.

[26] Objectif II.2 Mieux assurer l'ajustement harmonieux des différentes zones de la réserve de biosphère et leur interaction, Stratégie de Séville.

[27] Cf. La vision de Séville pour le XXIe siècle, in MAB, *Réserves de biosphère, Programme sur l'Homme et la Biosphère*, UNESCO, Paris, 1996, p. 5.

[28] Question posée par Stephen R. Sterling, Towards an ecological world view, in Ronald Engel J. and Gibb Engel (ed.), *Ethics of Environment and Development. Global challenge, International response*, The University of Arizona Press, Tucson and London, 1990, p. 77.

[29] Cf. La vision de Séville pour le XXIe siècle in UNESCO, *Réserves de biosphère: La Stratégie de Séville et le cadre statutaire du Réseau mondial*, UNESCO, Paris, 1996, p.5.

[30] Gutwirth Serge, Autour du contrat naturel, in Gérard Philippe, Ost François, Van de Kerchove Michel, *Images et usages de la nature,* Publications universitaires de Saint Louis, Bruxelles, 1993, p. 85.

[31] UNESCO, La solution du puzzle: l'approche écosystémique et les réserves de biosphère, UNESCO, Paris, 2000, p. 3.

[32] Sur cette question on se réfèrera avec intérêt à Karr James R., Attaining a Sustainable Society, in Westra L., Bosselmann, Westra R. (ed by), *Reconciling human existence with ecological integrity*, Earthscan, London, Sterling VA, 2008, pp. 21–36.

[33] Ainsi, à titre d'illustration, dans la culture traditionnelle ghanéenne, la terre considérée comme sacrée doit se reposer un jour par semaine des perturbations générées par l'homme.

[34] Selon l'article 4 du Cadre statutaire, l'aire devrait englober une mosaïque de systèmes écologiques représentatifs de grandes régions biogéographiques.

[35] Prévot Julliard Anne-Caroline, Servais Véronique, La ville, un dénominateur commun, in Garnier Lisa (ed.), *Entre l'homme et la nature, une démarche pour des relations durables*, Réserves de biosphère Notes techniques 3–2008, Paris, UNESCO, pp.86–87.

[36] Pickett S.T.A., Cadenasso M.L., Grove J.M., Nilon C.H., Pouyat R.V., Zipperer W.C., Costanza R., Urban ecological systems: linking terrestrial ecological, physical and socioeconomic components of metropolitan areas, *Annual Review of ecology and Systematics*, 32, 2002, pp. 127–157.

CHAPTER TWENTY-ONE

LA CONVENTION POUR LA SAUVEGARDE DU PATRIMOINE CULTUREL IMMATERIEL

TULLIO SCOVAZZI

Abstract

The Convention for the Safeguarding of the Intangible Cultural Heritage was adopted under the sponsorship of UNESCO on 17 October 2003 and entered into force on 20 April 2006. 116 States are today parties to it. The Convention fills a gap within the UNESCO legal instruments for the protection of culture.

The Convention aims at safeguarding what it defines as:

> "practices, representations, expressions, knowledge, skills – as well as the instruments, objects, artefacts and cultural spaces associated therewith – that communities, groups and, in some cases, individuals recognize as part of their cultural heritage. This intangible cultural heritage, transmitted from generation to generation, is constantly recreated by communities and groups in response to their environment, their interaction with nature and their history, and provides them with a sense of identity and continuity, thus promoting respect for cultural diversity and human creativity".

The heritage is manifested, *inter alia*, in the following domains: oral traditions and expressions, including language as a vehicle of the intangible cultural heritage; performing arts; social practices, rituals and festive events; and knowledge and practices concerning nature and the universe. For the purposes of the Convention, consideration is given solely to such intangible cultural heritage as is compatible with existing international human rights instruments, as well as with the requirements of mutual respect among communities, groups and individuals, and of sustainable development.

The negotiations for the Convention have been relatively easy, because the Convention does not specifically deal with the most crucial issues in the field of the intangible cultural heritage, such as the rights of indigenous populations and the protection of intellectual property rights.

The Convention establishes an Intergovernmental Committee for the Safeguarding of the Intangible Cultural Heritage, composed of 24 States parties.The Committee is in charge, *inter alia*, of inscriptions of elements in two lists, namely the Representative List of the Intangible Cultural Heritage of Humanity and the List of Intangible Cultural Heritage in Need of Urgent Safeguarding. A number of inscriptions on the lists have so far been decided by the Committee.

The Convention does not intend to protect nature in itself. But, in certain cases, nature is indirectly protected. This happens where the intangible cultural heritage is represented by the ways in which nature is understood, invented, utilised or recreated by man. As it appears from the inscriptions already made, the main risks affecting the intangible cultural heritage are given by both natural phenomena, such as deforestation, and social phenomena, such as urbanization, tourism (leading to exploitation of heritage for commercial purposes), acculturation (i.e. the shift to a dominant culture which entails the loss of other cultures). Some relevant examples are provided in this chapter.

Un Nouveau Traité dans le Domaine de la Culture

La Convention pour la sauvegarde du patrimoine culturel immatériel[1] a été adoptée le 17 octobre 2003 à Paris au siège de l'Organisation des Nations Unies pour l'Education, la Science et la Culture (UNESCO)[2] et est entrée en vigueur le 20 avril 2006. Le texte de ce nouvel instrument international a été rédigé à la suite de trois réunions intergouvernementales d'experts tenues en 2002 et 2003 sous la présidence de M. Mohammed Bedjaoui (Algérie)[3].

La Conv. Imm. comble un vide dans le domaine du droit international de la culture[4]. Une trentaine d'années après la Convention pour la protection du patrimoine mondial, culturel et naturel (Paris, 16 novembre 1972)[5], les Etats se sont mis d'accord pour mettre en place un deuxième régime de protection axé sur une conception plus large de patrimoine culturel[6]. Loin d'être limité aux monuments et aux sites naturels, ce patrimoine est censé inclure aussi des éléments immatériels, qui ne peuvent pas être "touchés"[7] (traditions orales, pratiques populaires, savoir-faire artisanal, connaissances locales, etc.), ainsi que les contextes humains et territoriaux dans lesquels ils sont créés et transmis[8].

Dans le domaine de l'immatériel, le précédent le plus direct, c'est-à-dire la Recommandation sur la sauvegarde de la culture traditionnelle et populaire, adoptée par la Conférence Générale de l'UNESCO le 15 novembre 1989, n'a pas de valeur contraignante. Des principes indirects et fragmentaires de protection peuvent résulter de certaines dispositions de traités applicables au niveau mondial, tels que le Pacte international relatif aux droits civils et politiques(New York, 1966)[9] ou la Convention sur la diversité biologique (Rio de Janiero, 1982)[10].

Pour certains Etats, notamment pour les pays en voie de développement, la culture orale et traditionnelle représente la forme principale du patrimoine culturel et constitue une contribution importante au progrès économique et social. Il s'agit toutefois d'un patrimoine vulnérable. L'actuelle tendance à la mondialisation menace la poursuite de pratiques culturelles traditionnelles en détournant les individus, en particulier les jeunes, vers une culture mondiale unifiée sur des modèles surtout de type anglo-américain. Plusieurs Etats se sont déclarés préoccupés par la perte progressive de leur patrimoine culturel immatériel et des valeurs sociales correspondantes, à cause aussi de phénomènes d'intolérance et de marginalisation sociale, et ont insisté sur l'exigence de concilier la mondialisation avec la préservation du patrimoine en question. En effet, on pourrait aussi prendre avantage de ce même phénomène de mondialisation pour diffuser les cultures traditionnelles auprès du grand public et dans des pays éloignés. Comme souligne le préambule de la Conv. Imm.,

> «les processus de mondialisation et de transformation sociale, à côté des conditions qu'ils créent pour un dialogue renouvelé entre les communautés, font, tout comme les phénomènes d'intolérance, également peser de graves menaces de dégradation, de disparition et de destruction sur le patrimoine culturel immatériel, en particulier du fait du manque de moyens de sauvegarde de celui-ci».

Mais comment assurer concrètement une protection adéquate à des capacités extraordinaires mais très hétérogènes, telles que celles des artisans de la soie dans les pays orientaux, des cultivateurs de terrasses de riz aux Philippines, des artistes du théâtre sicilien des marionnettes, des danseurs de tango, des parleurs des langues autochtones, des joueurs de *djembe*, des adeptes des rites woodoo, des chamans et guérisseurs indigènes, etc.? C'est le défi de la Conv. Imm., dont on n'examinera ci-après que certains des aspects les plus significatifs.

La Définition de Patrimoine Culturel Immatériel

Dans le Digeste de l'empereur romain Justinien il est noté que toute définition est dangereuse dans un texte juridique[11]. Dans le cas du patrimoine culturel immatériel, la tâche qui consiste à rédiger une définition s'avère particulièrement complexe, surtout si l'on tient compte de l'exigence d'insérer dans les instruments juridiques des définitions suffisamment synthétiques et facilement applicables sur le plan opérationnel. L'art. 2 de la Conv. Imm. En donne la définition suivante:

> «1. On entend par "patrimoine culturel immatériel" les pratiques, représentations, expressions, connaissances et savoir-faire – ainsi que les instruments, objets, artefacts et espaces culturels qui leur sont associés – que les communautés, les groupes et, le cas échéant, les individus reconnaissent comme faisant partie de leur patrimoine culturel. Ce patrimoine culturel immatériel, transmis de génération en génération, est recréé en permanence par les communautés et groupes en fonction de leur milieu, de leur interaction avec la nature et de leur histoire, et leur procure un sentiment d'identité et de continuité, contribuant ainsi à promouvoir le respect de la diversité culturelle et la créativité humaine. (…)
> 2. Le "patrimoine culturel immatériel", tel qu'il est défini au paragraphe 1 ci-dessus, se manifeste notamment dans les domaines suivants:
> (a) les traditions et expressions orales, y compris la langue comme vecteur du patrimoine culturel immatériel;
> (b) les arts du spectacle;
> (c) les pratiques sociales, rituels et événements festifs;
> (d) les connaissances et les pratiques concernant la nature et l'univers;
> (e) les savoir-faire liés à l'artisanat traditionnel».

Puisqu'il serait trop compliqué d'entrer dans les détails d'un libellé ressemblant plus à une description qu'à une définition, il suffira d'y rattacher quelques considérations.

La définition retenue met en évidence le lien entre les aspects créatifs et culturels du patrimoine immatériel et les aspects localisés, tels que communautés, groupes et individus déterminés, ainsi qu'un milieu déterminé. Un patrimoine qui pourrait être considéré immatériel, mais qui est de fait banalisé à cause de sa diffusion au niveau mondial, comme celui des vêtements ou aliments de grande consommation (les *blue-jeans*, les *hamburgers*, etc.), ne semble pas couvert par cette définition.

La définition ne vise pas uniquement les créations ("les pratiques, représentations et expressions"), mais aussi "les connaissances et savoir-faire", comme, par exemple, les connaissances médicinales, agricoles ou écologiques liées aux ressources naturelles et le savoir-faire requis pour les appliquer[12]. Dans le domaine des connaissances, le patrimoine immatériel, loin de se borner à des contenus artistiques ou esthétiques, fait partie du mode de vie quotidien d'une collectivité donnée. De plus, les connaissances en question constituent un phénomène dynamique et en constante évolution[13]. C'est le mépris des connaissances autres que celles qui sont dominantes (c'est-à-dire celles établies par les universités, les instituts de recherche et les entreprises privées) qui pousse souvent à la marginalisation de certaines collectivités, en particulier celles autochtones.

Les langues, dont on a beaucoup discuté pendant la négociation de la Conv. Imm., ne figurent pas dans la définition en tant que telles, mais en tant que vecteurs du patrimoine culturel immatériel[14]. Les religions sont exclues de la définition, car il ne serait pas approprié de placer au même niveau les doctrines théologiques et morales propres aux différentes confessions et les manifestations extérieures de sentiments religieux. Mais les pratiques et les rituels découlant d'une religion (processions, danses sacrées, etc.) peuvent se retrouver dans le champ d'application de la Conv. Imm. (cf. l'art. 2, par. 2, c).

Comme la définition l'exprime de manière implicite, le patrimoine immatériel ne doit pas être nécessairement situé dans le territoire d'un seul Etat. La Conv. Imm. protège aussi le patrimoine ayant une nature transfrontalière ou bien transcontinentale (comme les danses ou les rites afro-américains, la langue euro-argentine du *lunfardo*, etc.).

A la notion de patrimoine culturel immatériel s'attache sa "sauvegarde", définie à l'art. 2, par. 3, de façon suivante:

> «On entend par "sauvegarde" les mesures visant à assurer la viabilité du patrimoine culturel immatériel, y compris l'identification, la documentation, la recherche, la préservation, la protection, la promotion, la mise en valeur, la transmission, essentiellement par l'éducation formelle et non formelle, ainsi que la revitalisation des différents aspects de ce patrimoine».

Il est suffisamment clair que ce ne sont pas seulement les savants chercheurs qui devraient bénéficier des mesures à adopter sur la base de cette définition, mais aussi les praticiens du patrimoine culturel immatériel[15]. Ces derniers constituent la composante sociale de ce patrimoine, c'est-à-dire le facteur qui fait sa différence par rapport au patrimoine monumental ou naturel protégé par la Conv. Patr. Mond.

Durant la négociation, les anthropologues présents dans les délégations ont longuement discuté sur le sens à attribuer au terme de "revitalisation"[16] parmi les mesures à adopter par les Etats parties. La question qui se pose est si les Etats devraient être astreints à maintenir artificiellement en vie des pratiques en train de disparition à cause d'un manque d'intérêt des communautés locales. L'acquisition d'une documentation complète et scientifiquement correcte, avant la disparition de telles pratiques, ne serait-elle pas une mesure suffisante? Ou est-ce que, les Etats doivent accepter que des manifestations disparues depuis longtemps soient ressuscitées tout en comportant le risque de résultats peu crédibles (par exemple, la récupération d'un tournoi datant du Moyen-Age réinterprété par l'accompagnement d'un défilé de majorettes …)? Ces discussions intéressantes ne sont qu'une conséquence ultérieure du caractère social du patrimoine culturel immatériel.

La Mise en Oeuvre d'un Intérêt Général de l'Humanité

L'obligation fondamentale des Etats parties est énoncée à l'art. 19, par. 2:

> «Sans préjudice des dispositions de leur législation nationale et de leurs droits et pratiques coutumiers, les Etats parties reconnaissent que la sauvegarde du patrimoine culturel immatériel est dans l'intérêt général de l'humanité et s'engagent, à cette fin, à coopérer aux niveaux bilatéral, sous-régional, régional et international».

La Conv. Imm. reprend, *mutatis mutandis*, l'approche générale de la Conv. Patr. Mond., dont le préambule évoque l'importance que présente, pour tous les peuples du monde, la sauvegarde de biens uniques et irremplaçables, et ceci à quelque peuple ils appartiennent. A son tour, le préambule de la Conv. Imm. souligne "la volonté universelle" et "la préoccupation partagée de sauvegarder le patrimoine culturel immatériel de l'humanité". Attendu qu'il s'agit ici de la mise en œuvre de l'intérêt général de l'humanité, aux obligations des Etats parties d'assurer la sauvegarde du patrimoine présent sur leur territoire s'accompagne la possibilité qu'ils ont de bénéficier, à certaines conditions, de formes de coopération et d'assistance à l'échelle internationale.

Les obligations des Etats parties concernent la tenue et la mise à jour d'inventaires du patrimoine culturel immatériel (art. 12)[17], l'adoption de mesures de sauvegarde, le développement et la mise en valeur de ce patrimoine (art. 13), la promotion de programmes d'éducation, de sensibilisation et de renforcement des capacités (art. 14).

Les mesures de sauvegarde du patrimoine culturel immatériel à l'échelle internationale sont costituées par l'échange d'informations et d'expériences et les initiatives communes, ainsi que la mise en place d'un mécanisme d'assistance aux Etats parties dans leurs efforts pour sauvegarder le patrimoine. A cet égard, la Conv. Imm. a crée un "Fonds pour la sauvegarde du patrimoine culturel immatériel"[18], géré par le "Comité intergouvernemental de sauvegarde du patrimoine culturel immatériel"[19] sur la base des orientations de l'Assemblée générale des Etats parties. Les formes de l'assistance internationale comprennent la préparation d'études, la mise à disposition d'experts et de praticiens, la formation du personnel, l'élaboration de mesures normatives ou autres, la création et l'exploitation d'infrastructures, la fourniture d'équipement et de savoir-faire, l'octroi de prêts à faible intérêt et de dons (art. 21). Il est prévu qu'"en règle générale, l'Etat partie bénéficiaire doit participer, dans la mesure de ses moyens, au coût des mesures de sauvegarde pour lesquelles une assistance internationale est fournie" (art. 24, par. 2).

Les Deux Listes

Comme la Conv. Patr. Mond., la Conv. Imm. prévoit l'établissement, la mise à jour et la publication par les soins du Comité de deux listes à l'échelle internationale, à savoir la "Liste représentative du patrimoine culturel immatériel de l'humanité" et la "Liste du patrimoine culturel immatériel nécessitant une sauvegarde urgente".

Les négociateurs de la Conv. Imm. ont préféré éviter la notion de "valeur universelle exceptionnelle", constituant la condition pour l'inscription des biens sur la "Liste du patrimoine mondial " et sur la "Liste du patrimoine mondial en péril" établies par la Conv. Patr. Mond. Dans le cas du patrimoine culturel immatériel, la notion de représentativité prime sur celle d'exceptionnalité. La création d'une sorte de hiérarchie entre les différents éléments du patrimoine immatériel a paru incompatible avec la nature de ce patrimoine, le nombre presqu'infini de ses éléments et la composante sociale de ses manifestations. Aux termes de l'art. 16, par. 1, une liste représentative est en mesure d'"assurer une meilleure visibilité du patrimoine culturel immatériel, faire prendre davantage conscience de son importance et favoriser le dialogue dans le respect de la diversité culturelle". Les critères permettant de figurer sur la Liste représentative ont été élaborés par le Comité et sont soumis à l'approbation de l'Assemblée générale des parties (art. 16, par. 2)[20].

L'inscription sur la Liste du patrimoine nécessitant une sauvegarde urgente est faite à la demande de l'Etat partie concerné[21], sur la base de critères élaborés par le Comité et soumis à l'approbation de l'Assemblée générale des parties (art. 17, par. 2) [22]. D'après l'art. 17, par. 3, "dans des cas d'extrême urgence", le Comité peut inscrire un élément du patrimoine sur cette Liste "en consultation avec l'Etat concerné" (et, donc, même sans nécessité de présentation d'une demande par cet Etat)[23].

Une disposition transitoire (art. 31) assure que les éléments déjà proclamés "Chefs-d'œuvre du patrimoine oral et immatériel de l'humanité" sur la base de la distinction spécifique créée par l'UNESCO en 1998[24] soient intégrés dans la Liste représentative établie par la Conv. Imm. Mais aucune proclamation de chefs-d'œuvre ne pourra être faite après l'entrée en vigueur de la Conv. Imm.

Patrimoine Culturel Immatériel et Droits Humains

Il est évident que la sauvegarde du patrimoine culturel immatériel s'insère dans l'objectif général de la protection des droits humains, entendus dans la plupart des cas au sens de droits des collectivités. C'est donc à bien forte raison que la Conv. Imm. rappelle, dans son préambule, la Déclaration universelle des droits de l'homme et les deux Pactes internationaux (New York, 1966) relatifs, respectivement, aux droits économiques, sociaux et culturels et aux droits civils et politiques. C'est notamment l'art. 27 de ce dernier instrument qui revêt une importance particulière dans le domaine du patrimoine immatériel:

> «Dans les Etats où il existe des minorités ethniques, religieuses ou linguistiques, les personnes appartenant à ces minorités ne peuvent être privées du droit d'avoir, en commun avec les autres membres de leur groupe, leur propre vie culturelle, de professer et de pratiquer leur propre religion, ou d'employer leur propre langue».

On peut toutefois se poser la question sur certaines pratiques traditionnelles (comme la mutilation sexuelle des filles) qui sont en conflit avec les droits humains fondamentaux. La Conv. Imm., par une phrase figurant dans l'art. 2, par. 1[25], exclut que de telles pratiques puissent entrer dans son champ d'application:

> «Aux fins de la présente Convention, seul sera pris en considération le patrimoine culturel immatériel conforme aux instruments internationaux existants relatifs aux droits de l'homme, ainsi qu'à l'exigence du respect mutuel entre communautés, groupes et individus, et d'un développement durable».

En revanche, le seul fait que certaines pratiques aient un caractère secret, étant confidentielles par leur nature même et réservées aux membres d'une communauté ou d'un groupe déterminé, n'exclut pas l'applicabilité de la Conv. Imm.:

> «En vue d'assurer la sauvegarde, le développement et la mise en valeur du patrimoine culturel immatériel présent sur son territoire, chaque Etat partie s'efforce: (…)
>
> (d) d'adopter les mesures juridiques, techniques, administratives et financières visant à: (…)
>
> (ii) garantir l'accès au patrimoine culturel immatériel tout en respectant les pratiques coutumières qui régissent l'accès à des aspects spécifiques de ce patrimoine».

La Question des Droits des Communautés Autochtones

Il faut noter que la négociation de la Conv. Imm. a été menée de manière relativement aisée, par rapport à d'autres traités élaborés dans le cadre de l'UNESCO[26], et ceci grâce au fait que le texte final ne prend pas de position sur les deux questions les plus épineuses concernant le patrimoine culturel immatériel. On peut à cet égard critiquer l'absence de solutions et la considérer comme un facteur d'inachèvement de la Conv. Imm.

La première des questions épineuses concerne l'attribution de droits aux communautés autochtones (ou indigènes ou tribales[27]). Il est bien connu que l'octroi de droits collectifs à certaines peuples ou minorités – qu'il s'agisse de droits culturels ou autres – représente un problème politique très délicat dans certains Etats, qui craignent (à tort ou à raison) le développement de formes d'auto-détermination.

La Conv. Imm. ne fait mention des "communautés autochtones" que dans le préambule:

> «Reconnaissant que les communautés, en particulier les communautés autochtones, les groupes et, le cas échéant, les individus, jouent un rôle important dans la production, la sauvegarde, l'entretien et la recréation du patrimoine culturel immatériel, contribuant ainsi à l'enrichissement de la diversité culturelle et de la créativité humaine».

Dans d'autres cas, le rôle des communautés autochtones (qui, après le préambule, ne figurent plus dans le libellé de la Conv. Imm.) peut être considéré comme étant implicite dans la référence faite aux "collectivités" en général. On peut citer l'art. 11, b, selon lequel "il appartient à chaque Etat partie (…) d'identifier les différents éléments du patrimoine culturel immatériel présents sur son territoire, avec la participation des communautés, des groupes et des organisations non gouvernementales pertinentes", ainsi que l'art. 15, dont le texte est le suivant:

> «Dans le cadre de ses activités de sauvegarde du patrimoine culturel immatériel, chaque Etat partie s'efforce d'assurer la plus large participation possible des communautés, des groupes et, le cas échéant, des individus qui créent, entretiennent et transmettent ce patrimoine, et de les impliquer activement dans sa gestion».

La Question des Droits de la Propriété Intellectuelle

La seconde des questions épineuses écartées d'emblée par la Conv. Imm. concerne les droits de propriété intellectuelle sur les éléments du patrimoine culturel immatériel[28]. La manière dont les principaux droits de propriété intellectuelle ont été conçus et formulés est en conflit avec la plupart des exigences propres au patrimoine culturel immatériel et avec les besoins des communautés qui ont créé et qui perpétuent ce patrimoine, en particulier les communautés autochtones et locales[29].

Par exemple, la condition d'originalité s'avère inappropriée pour une grande partie du patrimoine immatériel, qui par sa nature est constitué par la transmission de pratiques et de connaissances de génération en génération. La condition selon laquelle l'auteur ou l'inventeur est une personne (physique ou juridique) individuelle apparaît tout aussi inappropriée, car la culture des collectivités, et en particulier celle des peuples autochtones, ne peut s'exprimer que sous forme collective, le savoir étant détenu par la communauté entière ou par des groupes y appartenant[30]. En outre, la limite de la durée de la protection d'une période pour un certain nombre d'années après la mort de l'auteur ou après le dépôt d'un brevet n'est pas adaptée au cas d'un patrimoine d'origines anciennes. Celui ci a souvent une signification religieuse et sociale pour une certaine communauté et il ne devrait pas tomber dans le domaine public. Enfin, la condition selon laquelle une expression culturelle est subordonnée, pour sa protection, à sa fixation sur un support matériel (écrit ou autre) ne peut pas toujours s'appliquer à un patrimoine transmis oralement d'une génération à l'autre.

En ce qui concerne les communautés autochtones et locales, la Convention sur la diversité biologique prévoit certaines obligations pour les Etats parties, même si ces obligations sont exprimées par des formules assez vagues. Par exemple, d'après l'art. 8,

> «Chaque Partie contractante, dans la mesure du possible et selon qu'il conviendra: (...)
> j) Sous réserve des dispositions de sa législation nationale, respecte, préserve et maintient les connaissances, innovations et pratiques des communautés autochtones et locales qui incarnent des modes de vie traditionnelle présentant un intérêt pour la conservation durable de la diversité biologique et en favorise l'application sur une plus grande échelle avec l'accord et la participation des dépositaires de ces connaissances, innovations et pratiques et encourage le partage équitable des avantages découlant de l'utilisation de ces connaissances, innovations et pratiques; (...)»[31].

Jusqu'à présent, une division du travail a en pratique pris place entre l'UNESCO et l'Organisation Mondiale de la Propriété Intellectuelle (OMPI), selon laquelle l'UNESCO traite de la question générale de la sauvegarde du patrimoine culturel immatériel, tandis que l'OMPI traite des aspects concernant spécifiquement la propriété intellectuelle. L'art. 3 de la Conv. Imm. ne fait que perpétuer cette division du travail:

> «Rien dans la présente Convention ne peut être interprété comme: (...)
> (b) affectant les droits et obligations des Etats parties découlant de tout instrument international relatif aux droits de la propriété intellectuelle ou à l'usage des ressources biologiques et écologiques auquel ils sont parties».

Bien que l'OMPI travaille depuis un certains nombre d'années aux questions liées à la propriété intellectuelle dans le domaine de l'immatériel (en particulier par le Comité intergouvernemental sur la propriété intellectuelle, les ressources génétiques, la connaissance traditionnelle et le folklore), elle est encore loin de l'objectif de l'élaboration d'un instrument spécifique de protection sur le plan du droit international. Ceci n'est pas dû uniquement à des questions techniques et juridiques, mais surtout – on peut facilement l'imaginer – au peu d'importance que la culture traditionnelle représente pour certains Etats développés et à la résistance tacite d'un certain nombre d'Etats à s'écarter de conceptions établies (et très profitables pour certains milieux) du régime de la propriété intellectuelle, notamment dans les domaines de la biotechnologie et de l'usage médical des plantes.

Malgré l'adoption de la Conv. Imm., les praticiens de l'immatériel restent pourtant exposés au risque de l'exploitation commerciale de leur pratiques et connaissances de la part de tiers (individus ou sociétés commerciales), mais aussi, de façon plus générale, au danger de déformation, de dénaturation et de falsification de leur patrimoine.

Patrimoine Culturel Immatériel et Protection de la Nature

Un des aspects les plus intéressants dans le domaine du patrimoine immatériel est constitué par le rapport existant entre ce patrimoine et la protection de l'environnement naturel.

La Conv. Imm. ne vise pas directement à la protection de la nature et ne pourrait pas être classée parmi le traités relatifs à l'environnement. Le but de la convention est de protéger la créativité humaine et la diversité culturelle et, par conséquent, les différentes manières par lesquelles l'homme est capable de comprendre, interpréter, inventer, créer et recréer un certain contexte. Toutefois, le contexte en question est, dans plusieurs cas (même si non pas dans la totalité des cas), composé par des éléments naturels qui sont par conséquent indirectement impliqués dans la Conv. Imm. A cet égard, on peut envisager deux situations différentes.

A) Comme on l'a déjà vu[32], la Conv. Imm. ne prend en considération que le patrimoine culturel immatériel conforme à l'exigence d'un développement durable (art. 2, par. 1). Il s'ensuit que des pratiques traditionnelles entraînant un risque de destruction d'espèces animales ou végétales ou de dégradation massive de milieux naturels ne pourraient pas tomber sous le champ d'application de la Conv. Imm.

B) Le contraire se vérifie aussi (et apparaît bien plus probable), puisque la dégradation massive de milieux naturels peut entraîner la perte de certains éléments du patrimoine immatériel, notamment de ceux qui se basent sur l'interaction de communautés et groupes avec la nature. La protection de ces milieux naturels devient donc une condition nécessaire pour que la Conv. Imm. soit correctement appliquée.

Parmi les chefs-d'œuvre du patrimoine oral et immatériel de l'humanité (qui seront automatiquement intégrés dans l'une ou l'autre des listes prévues par la Conv. Imm.[33]) on trouve un certain nombre de pratiques, représentations, expressions, connaissances et savoir-faire menacés par des phénomènes de déforestation, d'urbanisation, d'acculturation ou de développement touristique, comme il est montré par les exemples qui suivent.

Le carnaval d'Oruro (Bolivie). Il représente l'héritage d'anciens rituels précolombiens qui ont perduré sous le couvert de la liturgie chrétienne avec des dieux andins transformés en saints. Il est menacé par la désertification du haut-plateau andin qui a provoqué une émigration massive[34].

La cosmovision andine des Kallawaya (Bolivie). Il s'agit d'un art de guérison procédant d'une connaissance extraordinaire de la pharmacopée animale, minérale et botanique. Elle est exercée par des guérisseurs itinérants qui connaissent des écosystèmes variés ainsi que les propriétés médicinales de 980 espèces environ. L'acculturation, le manque de protection juridique des communautés autochtones et la politique des grands groupes pharmaceutiques risquent d'entraîner la disparition de cet ensemble de connaissances[35].

Les chants polyphoniques des pygmées Aka de Centrafrique (République Centrafricaine). Ils représentent une forme complexe de polyphonie contrapuntique à quatre voix accompagnée de divers instruments à percussion et à cordes. Aujourd'hui le mode de vie des Aka est perturbé par la raréfaction du gibier due à la déforestation, par l'exode rural et par sa transformation en folklore à des fins touristiques[36].

Le patrimoine oral et les manifestations culturelles du peuple Zápara (Equateur et Pérou). Ce peuple, qui vit dans une partie de la jungle amazonienne à cheval entre deux Etats, a élaboré une culture particulièrement riche en connaissances sur l'environnement naturel (mythes, rituels, pratiques artistiques, langue caractérisée par une abondance de vocabulaire lié à la flore et à la faune). Il a été décimé par quatre siècles d'histoire marquée par l'esclavage, les épidémies, les conversions forcées, les guerres et la déforestation[37].

L'espace culturel des Bedu de Petra et Wadi Rum (Jordanie). Ces communautés nomades perpétuent une culture pastorale traditionnelle et ont préservé des connaissances concernant la faune et la flore, la médicine, l'élevage des chameaux, la fabrication des tentes, la traque et l'escalade. Depuis une cinquantaine d'années ces communautés se sont sédentarisées et leur patrimoine immatériel risque d'être dégradé par le développement du tourisme dans le désert[38].

Le savoir-faire du travail du bois des Zafimaniry (Madagascar). Cette communauté utilise vingt espèces d'arbres endémiques, adaptées chacune à un type de construction ou à une fonction décorative spéciale. Aujourd'hui elle est reléguée au rang de fournisseur d'objets d'artisanat pour le tourisme et la déforestation met en péril le travail du bois[39].

L'espace culturel du yaaral et du degal (Mali). Les festivités des communautés pastorales pratiquant la transhumance renforcent leur cohésion sociale. Mais ces pratiques sont aujourd'hui fragilisées par l'exode des jeunes et les sécheresses récurrentes qui affectent pâturages et troupeaux[40].

L'Urtiin Duu – chants longs traditionnels populaires (Mongolie et Chine). Il s'agit de chants lyriques de groupes nomades, caractérisés par l'abondance d'ornements, l'emploi de la voix de fausset, une tessiture vocale très étendue et une composition de forme libre. Depuis les années cinquante, l'urbanisation, l'industrialisation, la désertification ont obligé de nombreuses familles nomades à opter pour une vie sédentaire[41].

Le Chopi Timbila (Mozambique). Les orchestres des communautés pratiquant cette danse sont composés de cinq à trente xylophones en bois de tailles et de tonalités différentes. La déforestation a raréfié le bois nécessaire qui permet de produire la sonorité particulière de ces instruments[42].

Le Hudhud, récits chantés des Ifugao (Philippines). Il s'agit de chants narratifs d'une communauté connue pour son système de rizières en terrasses aménagées dans les montagnes. La mécanisation de la récolte du riz a affecté aujourd'hui une tradition liée à la récolte manuelle[43].

Le Kankurang, rite d'initiation mandingue (Sénégal et Gambie). Le personnage central de ce rite est un initié qui porte un masque fait d'écorce et de fibres rouges d'un arbre appelé *faara*. Il est vêtu de feuilles et son corps est peint par de teintures végétales. La tradition est en recul à cause de l'urbanisation rapide et de la réduction des surfaces des forêts sacrées, transformées en terres agricoles[44].

Autres Dangers pour le Patrimoine Culturel Immatériel

Au-delà de la dégradation des milieux naturels, d'autres dangers peuvent porter préjudice au patrimoine culturel immatériel, dont l'origine provient surtout (si l'on s'en tient à la liste des chefs-d'œuvre précitée) de certaines catégories de personnes, comme celles notamment des touristes, des missionaires, des sécularistes et des communistes. Voici quelques exemples qui démontrent bien les racines sociales du patrimoine culturel immatériel.

Le Sema, cérémonie Mevlevi (Turquie). Les appartenants à l'ordre ascétique soufi des Mevlevi (derviches tourneurs), fondé en 1273, sont connus pour leur danse giratoire, accompagnée d'un répertoire musical particulier. Ils apprenaient leur art lors d'une retraite de 1001 jours dans les cloîtres mevlevi où ils s'engageaient dans une pratique quotidienne de la prière, musique religieuse, poésie et danse. Ils revenaient ensuite à une vie normale, tout en restant membres de l'ordre. La politique de sécularisation du gouvernement turc a conduit à la suppression de cette pratique entre 1925 et 1950. Les cérémonies ont été ensuite autorisées à nouveau, mais seulement en public (la restriction a été levée dans les années 1990). Aujourd'hui un certain nombre de cérémonies ont perdu leur signification religieuse et se déroulent devant un public de touristes.

Elles ont été raccourcies et simplifiées pour répondre à des exigences commerciales[45].

Le Gule Wamkulu (Malawi, Mozambique et Zambie). Il s'agit d'un culte secret pratiqué par les membres d'une confrérie d'hommes initiés. Les danseurs sont revêtus de costumes et de masques représentant des animaux sauvages, les esprits des morts, des marchands d'esclaves et, récemment, aussi des... hélicoptères. La pratique a réussi à survivre, malgré les efforts entrepris par des missonnaires chrétiens pour l'interdire. Mais aujourd'hui elle tend à être réduite à un divertissement pour touristes[46].

L'espace culturel et la culture orale des Semeiskie (Fédération de Russie). Les origines de cette communauté de "vieux croyants" remontent à l'époque du schisme de l'Eglise orthodoxe russe (XVII siècle). Ils durent s'exiler dans une région de la Sibérie où ils vivent encore aujourd'hui s'efforçant de préserver les éléments de leur culture d'origine (principes moraux stricts, costumes traditionnels, artisanat, nourriture, chœurs polyphoniques et autres). Cette culture fut marginalisée pendant toute la période soviétique et est aujourd'hui menacée par les bouleversements socio-économiques et par la pression de technologies nouvelles[47].

La danse Mbende Jerusarema (Zimbabwe). Exécutée par des hommes et des femmes et rythmée par le battement polyrythmique d'un tambour, cette danse populaire est caractérisée par des mouvements acrobatiques et sensuels. Dans la langue shona 'mbende' signifie 'taupe', animal qui symbolisait la fertilité, la sexualité et la famille. Puisque les missonnaires désapprouvaient cette danse, elle a été rebaptisée 'Jerusarema' (adaptation du nom de la ville de Jérusalem), afin de lui donner une connotation religieuse. Elle fut un objet d'identité dans la lutte contre le colonialisme. Aujourd'hui elle est souvent exécutée comme un divertissement exotique pour touristes et les tambours, hochets et sifflets sont remplacés par des instruments de mauvaise qualité[48].

La tradition du théâtre dansé Rabinal Achí (Guatemala) constitue un cas particulier. Ce drame dynastique maya est un rare témoignage des traditions préhispaniques en Amérique centrale[49]. Les massacres subis par la communauté maya, en particulier dans les districts de Rabinal et de K'iche, ont presque déterminé la disparition de la danse. Comme il a été souligné par la Cour Interaméricaine des Droits de l'Homme dans la décision du 19 novembre 2004 (réparations) concernant l'affaire *Massacre de Plan de Sánchez c. Guatemala*[50], l'extermination des populations indigènes maya, dont les forces militaires guatémaltèques étaient responsables, a aussi affecté, au-delà des victimes individuelles, un bien collectif, la culture maya achí. Elle a aussi empêché sa transmission orale par les personnes âgées et les femmes aux nouvelles générations:

> «Con la muerte de las mujeres y los ancianos, transmisores orales de la cultura maya achí, sus conocimientos no pudieron ser transmitidos a las nuevas generaciones, lo que ha producido un vacío cultural. Los huérfanos no recibieron la formación tradicional heredada de sus ancestros. A su vez, la militarización y represión a la que fueron sometidos los sobrevivientes de la masacre, especialmente los jóvenes, ocasionó la pérdida de la fe en las tradiciones y conocimientos de sus antepasados.
> Las víctimas no pudieron realizar libremente ceremonias y ritos según su cultura maya, ya que las autoridades controlaban todas sus actividades»[51].

Aujourd'hui ce qui reste de la culture maya achí doit faire face à la précarité économique de la communauté ainsi qu'au risque de se voir transformer en folklore et d'être banalisé.

Notes

[1] Ci-après: Conv. Imm.

[2] Aujourd'hui (octobre 2009) 116 Etats sont parties de la Conv. Imm.

[3] Les réunions se sont basées sur un projet préliminaire préparé par un Groupe d'experts non gouvernementaux (doc. UNESCO CLT-2002/CONF.203/3 du 26 juillet 2002). Pour les positions des Etats cf. le doc. UNESCO CLT-2003/CONF.203/3Rev. du 12 février 2003.

[4] Pour une analyse approfondie des questions juridiques liées au patrimoine immatériel cf. BLAKE, *Elaboration d'un nouvel instrument normatif pour la sauvegarde du patrimoine culturel immatériel - Eléments de réflexion*, Paris, 2002.

[5] Ci-après: Conv. Patr. Mond.

[6] L'art. 1 de la Conv. Patr. Mond. donne une définition de ce qu'on doit entendre par "patrimoine culturel", constitué par des monuments, des ensembles et des sites.

[7] Le mot *immaterial* ayant dans la langue anglaise un sens différent, la notion d'"immatériel" est restituée par le terme *intangible* dans le texte officiel anglais de la Conv. Imm.

[8] Toutefois, des éléments du patrimoine immatériel peuvent être associés à des biens déclarés patrimoine mondial d'après la Conv. Patr. Mond. Rien dans la Conv. Imm. ne peut être interprété comme altérant le statut ou diminuant le niveau de protection de ces biens (cf. art. 3, a). De manière générale, une conception large du terme "environnement" ne peut qu'inclure plusieurs éléments de l'"immatériel".

[9] Cf. *infra*, par. 5.

[10] Cf. *infra*, par. 7.

[11] "Omnis definitio in iure civili periculosa est" (*Digesta*, livre 50, titre 17, fragment 202).

[12] Par contre, la définition de la culture traditionnelle et populaire donnée par la

Recommandation de l'UNESCO de 1989 précitée ne semble viser que les créations et ne pas prendre en considération les connaissances: "Au sens de la présente recommandation: la culture traditionnelle et populaire est l'ensemble des créations émanant d'une communauté culturelle fondées sur la tradition, exprimées par un groupe ou par des individus et reconnues comme répondant aux attentes de la communauté en tant qu'expression de l'identité culturelle et sociale de celle-ci, les normes et les valeurs se transmettant oralement, par imitation ou par d'autres manières. Ses formes comprennent, entre autres, la langue, la littérature, la musique, la danse, les jeux, la mythologie, les rites, les coutumes, l'artisanat, l'architecture et d'autres arts" (par. A).

[13] "Le caractère 'traditionnel' des connaissances traditionnelles ne tient donc pas à leur ancienneté mais plutôt à leur mode d'acquisition et d'utilisation" (BLAKE, *op. cit.*, p. 53).

[14] Les conséquences à tirer de cette restriction ne semblent pas claires.

[15] Pendant les négociations pour la Conv. Imm., les délégués des pays africains ont beaucoup insisté sur l'importance des formes d'éducation non formelle.

[16] D'après le *Glossaire - Patrimoine culturel immatériel*, réalisé sous la direction scientifique de W. van Zanten, La Haye, 2002, par le terme de "revitalisation" on entend la "réactivation ou réinvention de pratiques sociales et de représentations qui n'ont plus cours ou tombent en désuétude" (si le terme s'applique aux pratiques de la communauté culturelle) ou "l'encouragement et le soutien à une communauté locale, avec l'accord de cette même communauté, en faveur de la réactivation de pratiques sociales et de représentations qui n'ont plus cours ou tombent en désuétude" (s'il s'applique aux politiques du patrimoine).

[17] La tenue et la mise à jour d'inventaires sont des obligations lourdes à accomplir par les Etats disposant d'un important patrimoine culturel immatériel. L'art. 12, par. 1, prévoit qu'un Etat s'acquitte de cette obligation "de façon adaptée à sa situation".

[18] D'après l'art. 26, les Etats parties s'engagent à verser au Fonds une contribution dont le montant sera calculé selon un pourcentage uniforme décidé par l'Assemblée générale des Etats parties. Cette contribution ne pourra pas dépasser 1 per cent de la contribution de l'Etat partie au budget ordinaire de l'UNESCO. Toutefois, un Etat partie peut déclarer qu'il ne sera pas lié par l'obligation de verser une telle contribution. Dans ce cas, l'Etat en question devra verser une contribution se rapprochant le plus possible de la contribution qu'il aurait dû verser s'il n'avait pas fait la déclaration.

[19] Ci-après: Comité. Il est composé de 18 Etats parties, élus pour un mandat de quatre ans par l'Assemblée générale des Etats parties. Les nombre des Etats membres du Comité a été porté à 24 dès lors que le nombre des Etats parties a atteint 50.

[20] Ces critères ont été élaborés par le Comité en 2007 et ont été approuvés par l'Assemblée générale des parties en 2008.

[21] L'expression "Etat partie concerné" laisse ouverte la possibilité, dans certains cas particuliers, que la demande soit présentée par un Etat différent de l'Etat sur le territoire duquel le patrimoine immatériel serait situé.

[22] Ces critères ont été élaborés par le Comité en 2007 et ont été approuvés par

l'Assemblée générale des parties en 2008.

[23] Les critères objectifs des cas d'extrême urgence sont proposés par le Comité et approuvés par l'Assemblée générale des Etats parties.

[24] Les premiers 19 chefs-d' œuvre ont été proclamés en 2001: la langue, les danses et la musique des Garifuna (Belize), le patrimoine oral Gèlèdé (Bénin), le carnaval d'Oruro (Bolivie), l'opéra Kunqu (Chine), les trompettes Gbofe d'Afounkaha: la musique et l'espace culturel de la communauté Tagbana (Côte d'Ivoire), le mystère d'Elche (Espagne), l'espace culturel et la culture orale de Semeiskie (Russie), le chant polyphonique géorgien (Georgie), l'espace culturel du Sosso-Bala (Guinée), le théâtre sanscrit Kutiyattam (Inde), le théâtre de marionette sicilien "Opera dei Pupi" (Italie), le théâtre Nôgaku (Japon), la création et le symbolisme des croix (Lituanie), l'espace culturel de la place Jemaa el-Fna (Maroc), l'espace culturel du district Boysun (Ouzbekistan), les récits chantés Hudhud des Ifugao (Philippines), le rituel royal ancestral et la musique rituelle du sanctuaire de Jongmyo (République de Corée), l'espace culturel de la fraternité du Saint-Esprit des Congos de Villa Mella (République Dominicaine), le patrimoine oral et les manifestations culturelles du peuple Zápara (Equateur et Pérou). D'autres proclamations ont été faites en 2003 et 2005 (par exemple, dans le cas de l'Italie, le "canto a tenore", chant pastoral sarde). Pendant la réunion du Comité de 2009, les premiers éléments ont été inscrits sur la base de la Conv. Imm.: il s'agit de 76 éléments sur la Liste représentative du patrimoine culturel immatériel de l'humanité et de 12 éléments sur la Liste du patrimoine culturel immatériel nécessitant une sauvegarde urgente.

[25] Il s'agit de la disposition où la définition du "patrimoine culturel immatériel" est donnée. Mais, en réalité, la phrase en question concerne moins la définition que le champ d'application de la Conv. Imm.

[26] Par exemple, la Convention sur la protection du patrimoine culturel subaquatique, adoptée en 2001 à la suite d'une mise aux voix.

[27] Les différences entre les adjectifs autochtone, indigène et tribal ne sont pas claires. Par exemple, la Convention relative aux peuples indigènes et tribaux (Genève, 1989) s'applique "aux peuples tribaux dans les pays indépendants qui se distinguent des autres secteurs de la communauté nationale par leurs conditions sociales, culturelles et économiques et qui sont régis totalement ou partiellement par des coutumes ou des traditions qui leur sont propres ou par une législation spéciale" (art. 1, par. 1, a) et aux "peuples dans les pays indépendants qui sont considérés comme indigènes du fait qu'ils descendent des populations qui habitaient le pays, ou une région géographique à laquelle appartient le pays, à l'époque de la conquête ou de la colonisation ou de l'établissement des frontières actuelles de l'Etat, et qui, quel que soit leur statut juridique, conservent leurs institutions sociales, économiques, culturelles et politiques propres ou certaines d'entre elles" (art. 1, par. 1, a). En outre "le sentiment d'appartenance indigène ou tribale doit être considéré comme un critère fondamental pour déterminer les groupes auxquels s'appliquent les dispositions de la présente Convention" (art. 1, par. 2).

[28] Sur la question en général cf. BLAKE, *op. cit.*, p. 15.

[29] Cette remarque concerne le droit de l'auteur à l'exploitation commerciale de sa

création et le droit des brevets, tandis que le droit moral de l'auteur à revendiquer la paternité de l'oeuvre, les marques de fabrique et les appellations d'origine géographique peuvent mieux s'adapter aux caractéristiques du patrimoine culturel immatériel.

[30] Au cas où des brevets ou autres droits seraient attribués à des collectivités, il faudrait résoudre le problème de l'identification des ceux qui peuvent autoriser l'utilisation du patrimoine immatériel et bénéficier des avantages économiques relatifs.

[31] Les formules "dans la mesure du possible et selon qu'il conviendra" et "sous réserve des dispositions de sa législation nationale" ne peuvent que limiter fortement la portée de cette disposition.

[32] *Supra*, par. 5.

[33] *Supra*, par. 4.

[34] Cf. UNESCO, *Chefs-d'œuvre du patrimoine oral et immatériel de l'humanité*, Paris, 2006, p. 20.

[35] *Ibidem*, p. 21.

[36] *Ibidem*, p. 27.

[37] *Ibidem*, p. 39.

[38] *Ibidem*, p. 57.

[39] *Ibidem*, p. 61.

[40] *Ibidem*, p. 65.

[41] *Ibidem*, p. 68.

[42] *Ibidem*, p. 71.

[43] *Ibidem*, p. 76.

[44] *Ibidem*, p. 84.

[45] *Ibidem*, p. 90.

[46] *Ibidem*, p. 63.

[47] *Ibidem*, p. 82.

[48] *Ibidem*, p. 99.

[49] *Ibidem*, p. 43.

[50] Le 18 juillet 1982, 268 personnes, pour la plupart de l'ethnie maya achí, furent massacrées à Plan de Sánchez par les forces militaires guatémaltèques: «Entre las 2.00 y las 3.00 de la tarde llegó a Plan de Sánchez un comando de aproximadamente 60 personas compuesto por miembros del ejército, comisionados militares, judiciales, denunciantes civiles y patrulleros, quienes estaban vestidos con uniforme militar y con rifles de asalto. Reunieron a las niñas y a las mujeres jóvenes en un lugar, donde fueran maltratadas, violadas y asesinadas. Las mujeres mayores, los hombres y los niños fueron reunidos en otro lugar, quienes fueron posteriormente ejucutados, lanzando dos granadas e incendiando la casa en que se encontraban. Alrededor de 268 personas fueron ejecutadas, en su mayoria miembros del pueblo maya achí» (par. 49.2 de l'arrêt, série C, no. 116). A la suite des menaces subies, la plupart des survivants abandonnèrent la localité qui fut occupée par les militaires.

[51] Par. 49.12 et 49.13 de l'arrêt. En 2004, en audience publique devant la Cour, le Guatemala se reconnut responsable de violations à plusieurs dispositions de la Convention américaine des droits de l'homme (San José, 1969), y compris l'article

concernant la liberté de pensée et d'expression (art. 13, par. 5), et demanda pardon aux victimes.

CHAPTER TWENTY-TWO

BEYOND INDIVIDUAL RESPONSIBILITY: ENVIRONMENTAL GOVERNANCE AND THE AFFLUENZIC SOCIETY

ANDREW BRENNAN AND NORVA Y. S. LO

Abstract

Contemporary industrial society is characterized by a condition of affluenza, in which the continuous consumption of material goods sits at the centre of human life and aspirations. Through a set of positive feedbacks, affluenza contributes to and is reinforced by a dysfunctional and self-destructive economic and ecological system. One of the key drivers in this system is the transnational corporation, an entity that wields disproportionate economic and political power in relation to its accountability. While much of the emphasis in policy discussions is on how individuals can respond to the challenges posed by environmental catastrophe and climate change, we argue that a new form of constitutional corporatism is required – one that limits the freedoms of large corporations to exercise political, social, economic and environmental influence, and which will make them accountable again to people and governments. In the absence of limits to the influence and actions of the major corporations, the political and ecological future of the planet is bleak.

Affluenza

The problem of affluenza is best approached by thinking about it in relation to the two global crises that characterized the year 2008: climate change, and the major financial crisis. The difference in response to these two problems reveals a major problem in human psychology. The problem of climate change met until recently with corporate and government denial, confusion, and disorientation. Responses to scientific warnings were widely held by scientists to be inadequate to the problems faced in a world with rising sea levels, failing crops, water scarcity, drought, intensifying storms and the other damaging effects of a changing climate. Thanks to some extreme weather events (which may or may not be associated with anthropogenic climate change) and to increasing attention in the media, public opinion and beliefs on the issue have changed during the last ten years. The idea of contracting industrial countries' carbon emissions, aiming for convergence, if not equity, across the globe in per capita emissions, now appears to have achieved some credence in both the mainline media and also in the mainstream social and political movements of the rich industrial countries. Even semi-technical talk about carbon taxes, emissions trading schemes and emissions caps has entered the standard vocabulary of politics. Both of the candidates in the 2008 United States' presidential election seemed to agree that a reduction on carbon emission levels was necessary. Under the Kyoto protocol, the base year for several significant greenhouse gases is 1990, and targets for emissions reductions are set to that year. Both candidates wanted to see a target for emissions that would commit the United States to emitting at least 60 per cent less of these gases in 2050 than in 1990. The European Union, which – unlike the United States – has an agreed set of Kyoto targets, is already committed to huge reductions in greenhouse gases. As a result, Europe in 2006 released slightly less greenhouse gas than in 1990, with the 15 European Union nations having reduced their emissions to 97.4 per cent of the 1990 levels. Oil and gas rich Canada, by contrast, increased its emissions of greenhouse gases by 25 per cent over the period 1990–2005, the United States released 16 per cent more in the same period, while in Australia (where rapid and extensive land-clearing also contributes to increasing gas emissions) the increase was 22 per cent.

What do these figures tell us, in the context of international discussions, increasing media coverage, and changing popular perception of the problem of climate change? We conjecture that the formerly widespread situation of denial that human beings were responsible for climate change has now given way to one in which there is anxiety without much action. At national, regional and global conferences, human-induced climate change continues to figure as a major preoccupation, but so far relatively little has been done to address the problem of carbon emissions, and expenditure on mitigation has been relatively modest. This contrasts sharply with the talk and action related to a second crisis that hit the world in 2008. That crisis was more directly concerned with material goods, pleasure and wealth, and – perhaps unsurprisingly – action was far more immediate. This second crisis was the collapse of financial markets around the world. It started with the failure of the sub-prime loans market, which then triggered a cascade of credit freezes, market shocks, takeovers and bankruptcies that not only led to the partial nationalization of previously private companies and banks, but also to further use of the language of catastrophe. Terms like "meltdown" and "Armageddon" were used by journalists to capture the sense of panic associated with the massive downward movements in stock markets. Government injections of liquidity into national financial systems all over the world showed that even in an era of intense globalization and unparalleled corporate power, the nation state was alive and well, and capable of taking significant economic, financial and political action. National governments took major roles not only in internal political and economic matters, but also in ways that affected the global economy. This should be a lesson for those theorists who have been somewhat premature in declaring the death of the nation-state (Hobsbawm 1990, Appadurai 1996). Another lesson, much more practical, was also learned. Philosophical and political theory has laid out conditions for a healthy and flourishing industrial society. Thinkers have advocated the importance of many factors including environmental sustainability, social justice, open and democratic decision-making, appropriate social welfare arrangements and the importance of education and training. But the financial crisis drew clear attention to the importance to all advanced economies of a reliable and responsible financial sector.

In striking contrast to the lacklustre response to climate change, the funds voted to restore financial health and build up confidence in banks, finance houses and companies around the world were enormous and appeared with impressive rapidity. In the months following the 2008 crisis, the economies of the United, States and Europe paid $10.2 trillion in order to bail out banks, prop up financial institutions and stimulate recovery. This kind of expenditure puts a new perspective on climate change and the costs of its amelioration. Late in 2009, the World Bank issued a report indicating that the cost of protecting developing countries from a two degree increase in average global temperature would be in the region of $75 to $100 billion per year for the period 2010 to 2050. Before the crash of 2008, such a sum would have seemed to large to be realistic as a cost that should be born by the industrial world. By comparison with the expenditure that shored up the world financial system, the sum now seems modest: $100 billion, after all is only one per cent of $10 trillion. There is little evidence, so far, of even this amount of support being provided to control and provide amelioration for climate change.

Positive Feedback

The difference in response to the two crises shows that at both global and national levels there is an asymmetry in thought, sentiments and action about the two challenges. The financial collapses led to co-ordinated and urgent political action to try to save heavily indebted financial systems, even in the face of criticism from voters concerned about giving financial help to those who had been the agents of the collapse. Headlines complained about rewarding bad behaviour and the spotlight turned on the huge salary payments and share allocations received by chief executives. Despite these doubts, the G7 finance ministers came together to endorse huge spending that put real money into collapsing systems.

By contrast, very little by way of real resources, or real amelioration has been pumped into the declining and seriously stressed global environment. This asymmetry requires some explanation. Many scientists, conservationists and philosophers found it baffling that it was so hard to get species loss, water scarcity, pollution, climate changes and other environmental crises onto the public agenda in the first place. They have wondered why there was such a successful degree of evasion, denial and procrastination about these things. Can it be because we can see our money suddenly lose value, while we do not in the same way see the slow disappearance of species or experience a sudden reduction in environmental resources? In the case of individual people, unfitness and failing health creep up in cumulative ways that avoid triggering alarms. Bankruptcy, or massive financial losses can – like a heart attack – happen of a sudden and trigger just the panic and fear that was evident at the time of the financial crisis.

There are other possible explanations too. One is that both the ecological and the financial crises are products of a single system of positive feedback loops. Positive feedback occurs in a system when part of the output from a process is fed back into the same process, thus amplifying its effects. For example, in some models of climate change, increasing global temperatures are imagined as causing melting of the permafrost in polar and alpine regions. If melting permafrost releases carbon dioxide and methane into the atmosphere, then this may lead to further global warming. According to this conjecture, there is a positive feedback between increasing global temperature and release of methane and carbon dioxide. Unless there are negative feedbacks elsewhere in a system, positive feedback will have destabilizing and destructive effects leading to ever more rapid and sometimes unpredictable change. Such feedback damage is intensified by a feature of contemporary industrial society which appears to have favoured the development of a peculiarly single-minded kind of actor, one whose basic hedonistic features are exploited and perverted so as to feed the continual operation and growth of the system. These features are of course central features of human nature, ones which are fed and exaggerated by the system under which many people now live. Globalization has effectively strengthened this already robust structure of positive feedback loops by rapidly expanding its territory of persuasion. Common labels for such a system are "consumerism" and "capitalism". Both labels miss a crucial aspect of the system, one better captured in the popular label *affluenza* (De Graaf, Wann and Naylor 2005) and its popular definition as an dysfunctional condition characterized by addiction, anxiety, depression and *ennui* (James 2007).

There is a line of sociological theory that argues the consumer society actually feeds off individual vulnerability – this is an idea captured in Stephen Pfohl's idea that the shopping mall is a kind of "parasite café" (Pfohl 1990). According to this theory, individual consumers are infected or parasitized by the system of commodities within which they live their lives, measure their status and seek ever more elusive satisfactions. Affluenza refers to the involuntary nature of both people's exposure to, and damage by, the socio-economic structures in which they live: such structures are like tram lines along which people's lives run, and from which they find it hard to break free. The condition is suffered to a high degree by the affluent in their addictive pursuit of capital and consumer goods, at the cost of destroying social welfare, the environment and individual health.

Affluenza is a product of a system that is rich in available resources, hence one able to thrive and multiply on the addictions cultivated in consumers. At the same time, these addictions are focused by the system on a range of pleasures and satisfactions that are relatively narrow in comparison to the capacities of normal human beings. So narrow are these satisfactions that things which used to be regarded as free, essential goods for human well-being, such as communion with nature, the building of trust with others, and self-realization through meeting challenges are now marketed as "extraordinary experiences" to be obtained through commercially guided adventures such as wilderness adventures, white-water rafting and eco-tourism (see Arnould and Price 1993). Since such "extraordinary experiences" are only for a minority of consumers, the result for most people is the atrophy of higher sensitivities, a loss of the richness of lived experience and its replacement by repetitive satisfactions based on material goods. This the one of the great harms of consumer society, which has not only provided a host of gadgets and commodities on which to focus inflated and deranged desires (about which more in a moment), but has also managed to commodify (at least to some degree) the experiences that were previously some of life's freest and most shareable goods. Alongside this are symptoms of addiction – typically a failure to recognize the scale of the problems that are faced, optimism about the ability to get out of deeper and deeper problems, borrowing from the future, and an unrealistic and exaggerated sense of future profits and satisfactions.

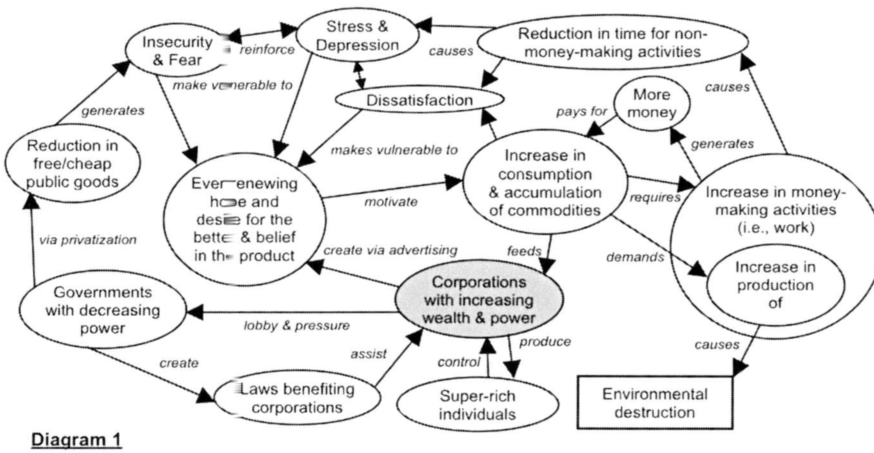

Figure 22-1. Loops in the system of reinforcing atrophy

What would the interlocking system of positive feedback be like if this model of contemporary consumer society is correct? Figure 22-1 shows several of the loops in the system of reinforcing atrophy. Corporations with increasing wealth and power create via advertising hope and desire for a better life and a belief in the products that are going to provide just such improvement for the individual. The pursuit of these desires leads to increasing consumption but at the same time also promotes discontent and dissatisfaction. To get more and more goods and services, the individual has to make more money, hence spend more time working. The ideal "citizen" of the affluenzic society is someone for whom buying and acquiring material goods and marketable services is an all-consuming passion. Of course, as Hannah Arendt repeatedly emphasized, the pursuit of private goods and the discussion of how best to obtain them is not a proper concern of the public realm in a healthy society, hence our use of quotes around the word "citizen" in the last sentence (see Arendt 1958, 1968). While we agree with Arendt's analysis, our concern in this chapter is to look at a further problem with the expansion of what she calls "the private realm", and its domination of contemporary life. When consuming is all important, then things are ranked in terms of the private preferences of individuals seeking satisfaction and public goods are diminished. Additional reinforcement to extend working hours and earn more is the natural desire to do the best for those near and dear to us, especially when governments with reducing power provide fewer public goods such as health care and education. Securing the best education for the children from kindergarten to university, the best and safest prams, keeping pace with friends' and neighbours' family expenditures – these are all lures that encourage people in working longer hour to fund more purchases in a world where more and more goods have become privatized. The imperative to consume thus gives rise to reduced time for other activities, including, paradoxically, reduced time for consumption itself.

The reduction in time available to individuals, as we shall see, in turn reinforces the drive to consume more. If consumption were an ordinary desire, focused on ordinary things, it would be unlikely to have the paradoxical effect of driving people to devote more time to work, when what they rationally want is more time for enjoying the other things they desire. However consumption is not an ordinary desire, but a pathology: it appears to be addictive, in just the way described in popular books about affluenza (see James 2007), and also in the way pointed out years ago by eco-psychologists. In work published in the 1990s, Chellis Glendinning argued that the consumption of, and dependence on, technology is a damaging addiction at the heart of industrial life (Glendinning 1995). The pleasures of such consumption seem not to be intrinsically satisfying. Consumer goods, as such, are always updateable, upgradeable, upscalable. To update, upscale and upgrade is always possible – so long as you have the money to do so. It is the fate of any consumer product to become unsatisfying and therefore need replacement. Consumer goods are not meant to last or provide long lasting satisfaction. On the contrary, the ideal commodity is something that people initially lust after, and then quickly become dissatisfied or bored with, and therefore soon seek a better replacement.

There is an interesting ontology surrounding the idea of the commodity itself. What is it for something to be a commodity? We think there are two conditions that are necessary for something to be successful commodity.

First, it must be intrinsically deficient, in that either it fails to satisfy a function completely, or it is readily made obsolete. Typical examples of incompleteness are found in cosmetics, for example, where a multiplicity of skin care products are prescribed by manufacturers for a range of skin problems and nourishing needs. Instead of applying a single product nourishes and protects, consumers are expected to experiment with different with different layers of serum, toner, moisturizer, and so forth. Planned obsolescence has been the subject of critique since the 1930s and theorists have distinguished various types of obsolescence. For example, some mechanical parts have a designed lifetime, and then fail, requirement replacement by a new part. In other cases, a fully functional piece of machinery is still in good shape after years of use, but comes into competition with newer and supposedly better models. In the computing industry, the upgrading and replacement of software by newer versions ensures that operating systems and programs that are adequate for many purposes are no longer supported because they have become incompatible with current industry standards.

There may be other forms of intrinsic deficiency apart from incompleteness and manufactured obsolescence. These two alone provide plenty of opportunity for the marketing of new products, the promoting of competition among products many of which serve extremely similar purposes and the development of brand diversity in the market place. In addition to them, however, is what René Girard has called "mimetic desire". According to this theory, marketing shows us a "model" who is associated with a commodity. A famous sporting personality advertises a shaver, or a car, a beautiful and scantily clad woman stares lovingly at a watch or a camera. The consumer wants to be like the model, or to be associated in some way with the model, and the advertising dupes him (or her) into thinking that buying the product will bring about that assimilation. Alas, after purchasing the shaver, the watch, the camera or the car, life is very much as before. The consumer does not become the model, and the acquired object loses its shine. Never mind, though, for there is always another object, associated with another model to whom the consumer can aspire, and maybe this time purchasing the object will bring about the desired change. But that, too, fails to happen. And so the cycle continues, based on the illusion of false hope and belief.

This problematic condition is represented by another loop in our complex diagram one that links discontent, desire, hope and belief in an endlessly frustrating cycle of addiction. It is consumers' dissatisfactions with what they currently have, and their desire for, and belief in, better prospective products that maintain the flow of the market. It follows that successful product marketing must not only create desires and beliefs but must also create dissatisfaction by building its seeds into the very consumer products themselves. Accordingly, we have an everlasting cycle that runs from discontent through to desire, hope, belief, and then back to discontent. For many people, the cycle of yo-yoing emotions from the low ends of disappointment and boredom to the high ends of excitement, hopefulness and optimism is a psychological equivalent to a physical addiction resulted from drug abuse. In all forms of addiction, the highs are short lived, and when they give way to the lows the addict is compelled to seek the highs again to ease the pain, the void and emptiness inside. Affluenza apparently promotes addiction. The highs it offers and the lows it generates co-ordinate and feed into each other, so that the addict is locked into the cycle.

A self-destructive system

The system of interlocking feedback loops we have described can simultaneously explain the two meltdowns, ecological and financial. The opposite of atrophy is hypertrophy, the excessive growth in the economy and the overconsumption of material goods that leads to environmental destruction. The entire project of contemporary industrial and economic developments is one big scheme of borrowing with no security. We borrow from the planet's future reserves (the ecological goods and services that would be of value to humans and members of other species yet unborn) in order to satisfy our present desires for, and addictions to, consumption and profit making. We justify these huge borrowings on the basis of capital that we hope we have in reserve – a faith in human resourcefulness and the technologies that we hope we are going to invent for paying back all our debts. Furthermore, our borrowings from future generations are security-free, meaning that if we are unable to pay back the principal (let alone the interest) and if the natural environment goes bankrupt as a result, then there is no immediate loss to us. In short, we are simultaneously the borrower, the creditor, and the broker, all of whom will be long gone by the time the ecological meltdown occurs and the contemporary industrial bubble bursts. Exactly the same kind of unsustainable borrowing and lending in finance sectors all over the world has already produced very similar kinds of consequences: abuse of credit, hugely inflated prospective returns, reckless underestimation of investment risk, and finally the bankruptcy of giants that have been previously thought to be unsinkable. While the financial system has now been bailed out, at least for the time being, the associated cycles of environmental destruction may well produce breakdowns that we have little chance of fixing.

Seen in this way the two crises are strikingly similar from a psychological point of view: the sacrifice of the future in pursuit of short-term prospective gains which will themselves fail to satisfy us, so that we make further borrowings against the future to pursue yet further illusory gains, and so on. Put concisely we seem to have described a system of *future abuse*, both ecologically and economically. Note that there is a certain cleverness to the way the future eating is carried out. People take the ecological resources now that will be lost to future generations of humans, and to future generations of plants and animals. In the economy, people at the rich end of town speculate not with their own money but with the mortgages, pension funds and savings of those at the poorer end of town. In each case, the addictions involve gambling beyond the means available, and some people gambling away wealth that belongs to other people. Gambling itself is another highly addictive form of activity. When different forms of addiction are pulled together and causally intertwined as they are in the system of affluenza, they grow on, feed off, and justify each other. When individuals in the system experience the lows of one addiction, they can quickly fix it by resorting to the highs of another addiction. There are no limits to the availability of the many forms of highs – except that every one of them is transient, intrinsically unsatisfying, and invariably brings the individual back to the lows of pain, emptiness and frustration.

Richard Sennett has coined the phrase "corrosion of character" to apply to the atrophying of values and ethics within organizations (Sennett 1998). The atrophying we have in mind, however, is widespread and occurs throughout the societies we inhabit, not just inside organizations. It involves a withering of sensitivity to everyday sources of pleasure that are characteristically human. As sensitivity withers, we become less and less able to appreciate the very joys that are central to a fully human life. Indeed, the better the consumer, the narrower the kinds of desires that person should have. Here is the real price of the consumer society. Only the kinds of desires that can be satisfied by marketed goods and services are encouraged as proper for cultivation by consumers. Once other, non-market desires and sensitivities wither away, then sources of enjoyment become increasingly limited. However, unlike the bodily changes due to ageing and disease – ones that are often irreversible, ones that we have to cope with as best we may – the changes induced by consumerism are potentially reversible and able to be resisted. Without such resistance, though, and in a world where commodities are a prime source of value, the other sources of value become increasingly reduced and increasing wealth is not accompanied by additional happiness (see chapter 4 of Frey and Stutzer 2002). To an extent, then, our diagnosis supports E.F. Schumacher's gloomy view that "the result of the lopsided development of the last three hundred years is that Western man has become rich in means and poor in ends" (Schumacher 1977, 58).

We recognize that the model of interlocking systems that lead to the two different forms of future abuse is highly oversimplified and may be wrong. The diagnosis we have suggested for the problems of the money-based consumer society is not intended to blame individuals for the problem. Addiction is normally seen as a problem for the person addicted, not something that is due to wider socio-economic realities. Individuals, we claim, act according to what is offered in their surroundings. In a society where most goods have to be purchased, and where experience itself is increasingly commodified, advertising and the desire to conform to wider social expectations themselves drive the behaviour of most people. Asking individuals to change their behaviour may actually not be very helpful when the problematic elements in the situation are due to larger forces at work. We have so far suggested that among these forces are structures within industrial societies that encourage certain kinds of individual behaviour and which prey on all-too-human vulnerabilities. There are other matters too that lie beyond individual action and control.

Constitutional Corporatism

At the end of David Guggenheim's 2006 film, "An Inconvenient Truth" Al Gore suggests a number of ways to help to save the earth. This include changing to energy-saving light bulbs, walking more and driving less, recycling more, keeping your vehicle tires fully inflated, washing clothes in cool water, planting a tree, and so on. The focus is on actions that individuals can take. These, though valuable, ignore the structural and institutional facts of environmental destruction just discussed. Think about replacing traditional incandescent light bulbs with energy efficient ones. Australia has 0.32 per cent of the global population but produces nearly 1.5 per cent of global carbon emissions, a total of 560 million tonnes of carbon dioxide equivalent per year. The country is the first in the world to force domestic consumers to purchase energy efficient light bulbs, which will reduce carbon emissions by an estimated 2 to 4 million tonnes each year – well under one per cent of emissions. Every little helps, but action on light bulbs can a distraction from action on serious inefficiencies in industrial and transport energy use, especially in a country which relies on poor quality brown coal for much of its electricity production, and which is also the largest exporter of coal in the world. In a set of institutional structures that rely on corporations, governments and advertising to control, manipulate and direct individual behaviour, each person has only a limited part to play in a wider scheme of environmental disaster.

Another reason for switching perspective, away from a focus on individuals is also that our analysis raises a problem that individuals themselves cannot solve. In policy discussions, there is frequently appeal to rationality – as if the issue were just one of making human beings behave more like some ideally rational actor. Our model shows why appeals to rationality, self-control, care for biodiversity, the triple bottom line and the other solutions usually given for the environmental ills of the age are all likely to fail. Our vision of the present global system is of a beast with a tendency to addictive consumption of both human and environmental resources, and self-perpetuating growth that leads ultimately to catastrophe. If there is to be hope for the future, then we need to find some ways of building limits into a system that threatens to run beyond control.

While the major focus of philosophical and ethical theorizing has been on individuals and their own moral responsibility, a minority of thinkers have looked at other features that set the agenda for individual action. Some philosophers have speculated about the religious and philosophical worldviews that structure thought and behaviour, and there are well-known contributions to environmental philosophy and politics by thinkers who have tried to identify features in our world views – perhaps anthropocentrism or patriarchy – which might explain our current predicaments. Chellis Glendinning's views on our addiction to technology are another minority view, as is David Korten's attack on the corporatized world (Korten 2001). Korten's claim is that the global economy has become like a malignant cancer, advancing the colonization of the planet's living spaces for the benefit of powerful corporations and financial institutions. He is not opposed either to the corporations as such or to the financial sector as a whole. Rather, he regards the global economic system overall as having made what were once useful institutions into instruments of a market tyranny that feeds on life in an insatiable quest for money. While we have speculated earlier on the compulsive nature of status-seeking, addiction to consumer pleasures and the dominance of affluenzic modes of behaviour, Korten places blame on a deregulated global economic and rogue financial systems that put making money, particularly in terms of shareholder returns, well ahead of commitment to sustainability, community building or any other large social interest.

Although we differ in emphasis from him, Korten's analysis is complementary to ours and adds depth to the suggestions already made.

Unlike Korten, however, we propose a solution that focuses on one item in the system – namely the corporations themselves. Companies, governments and non-government organizations are all important players in the system, and each of them can be seen as having important roles in amplifying or dampening some of the forces that propel the system to destruction. We propose that one way of dampening some of the positive feedback within the system we have described can be achieved by limiting the power of the corporation, and by going back to earlier understanding of corporate charters. An analogy may be helpful in this case. Think of how democracy glimmered for a while in ancient Greece, and then was tried at various times throughout Europe – but only as occasional experiments. Although the democratic ideal was known, the practice itself was lost for the most part for more than a millennium.

Something good – like democracy – is vulnerable to loss, can easily be destroyed, and can be recovered only with difficulty. But the loss of democracy was not permanent, and a recovery was made. From Magna Carta in 1215 to the passing of the United States' Constitution in 1787 there was a prolonged struggle to regain democracy, often involving a series of negotiations and struggles with the tyrannical power of monarchies. In the nearly six centuries of struggle, what was happening was a gradual limiting of the power of tyrants and the result was that in those democratic countries where a republic was not established an accommodation was made, often in the form of a constitutional monarchy, where the absolute power of the ruler was limited through government limitations on that power.

Constitutional corporatism – an accommodation that may have to be reached through prolonged struggle against corporate power and influence – would likewise seek to use legal instruments, both nationally and internationally, to limit the powers of corporations. In opening up this possibility for exploration, policy theorists, political scientists and philosophers would be recognizing that the fears expressed by earlier writers and thinkers were well founded. There is dispute about whether Abraham Lincoln himself wrote in one letter that "I see in the near future a crisis approaching that unnerves me and causes me to tremble for the safety of my country. … corporations have been enthroned and an era of corruption in high places will follow, and the money power of the country will endeavor to prolong its reign by working upon the prejudices of the people until all wealth is aggregated in a few hands and the Republic is destroyed" (quoted in Shah 2002). But there is no doubt that Adam Smith looked with concern at the possible monopoly and secrecy powers that corporations might wield (Korten 2001). More recent writers have given readable accounts of the rise of the corporation and the way in which legislation abetted the abandonment of the originally restrictive – and socially responsible – charters under which early corporations operated (Hartmann 2004).

As economic entities of enormous scale and influence, corporations are able to set their own agendas aimed at securing huge wealth for their chief executives and senior management, good returns for their shareholders, and control over their suppliers and potential competitors. It may seem to be a simple idea to limit their role in amplifying the destructive features of contemporary economic systems. But even though the idea seems simple, its implementation is likely to be very difficult. Many corporations are adept at avoiding tax, participate in the funding of political parties, bring influence to bear in the awarding of contracts, and fail to give any lasting benefits to the countries within which they operate. These all look like behaviours that need to change if environmental destruction is to stop. The direction in which change has to be made may seem clear, but the steps towards the final goal are not yet determinate. If the corporation is the problem, then we need to understand a great deal about it, and to study in detail how problematic organizations manage to operate, survive and expand. But it is not clear that there is yet agreement for undertaking research into the complex "wicked problems" of corporations and their power (Norton 2005, Brennan 2005). Governments are often unwilling to interfere with corporations given not only the huge influence of the private sector but also because of the dangers associated with changing to something new. The 2008 financial crisis did not lead to action to change the financial system in any fundamental way, but rather to shore it up and keep it running. Controlling or limiting the power of the big corporations raises the spectre of an intervention that may make things go worse rather than better – and this fear is acute in countries with a horror of socialism or communism (seen as the obvious alternatives to free market capitalism).

In many contemporary democracies, drastic change is not up for discussion. Political power operates over relatively short periods of time, and governments have to face re-election before long-term policies have a chance to be evaluated. The pressure on governments is thus to favour incremental and gradual change, change which is small rather than dramatic. As far as environmental conservation is concerned, this looks like bad news. It is likely to be business as usual as far as the voters' consumption patterns are concerned, and there will be no incentive to contemplate large-scale changes to the economic system or the massive corporations that dominate it. By contrast, in countries where capitalism is not the predominant mode of thought, there may be fertile soil for new thinking, and opportunities for exploring a greater role for government in developing new experiments in constitutional corporatism. Successful experimentation in these countries may be our best hope for a sustainable future.

Acknowledgments

An earlier version of this paper was delivered at a seminar at La Trobe University, Australia, in May 2009, and the authors are grateful to members of the audience for helpful comments and suggestions. They are also grateful to Carolyn Brennan for suggestions that changed the structure and argument of the paper in several places.

References

Andrew, Edward G. (2006) *Patrons of Enlightenment*. Toronto: University of Toronto Press.

Appadurai, Arjun (1996) *Modernity at Large: Cultural Dimensions of Globalization*. Minneapolis: University of Minnesota Press.

Arendt, Hannah (1958) *The Human Condition*. Chicago: University of Chicago Press.

—. (1968) *Between Past and Future*. New York: Viking Press.

Arnould, E. J., Price, L. L. (1993) "River magic: Extraordinary experience and the extended service encounter." *The Journal of Consumer Research* 20: 24 – 55.

Bauman, Zygmunt (2005) *Liquid Life*. London: Polity.

Brennan, Andrew (2004) "Biodiversity and Agricultural Landscapes: Can the Wicked Policy Problems Be Solved?" *Pacific Conservation Biology* 10: 124 – 143

De Graaf, John, Wann, David, and Thomas Naylor (2005) *Affluenza: The All-Consuming Epidemic,* 2nd edition. San Francisco: Berrett-Koehler.

Frey, Bruno, and Alois Stutzer (2002) *Happiness and Economics.* Princeton: Princeton University Press.

Garnaut, Ross (2008) *The Garnaut Climate Change Review.* Melbourne: Cambridge University Press.

Glendinning, Chellis (1995) "Technology, Trauma and the Wild," in T. Roszak, M. Gomes and A. Kanner (eds.), *Ecopsychology: Restoring the Earth, Healing the Mind*. New York: Sierra Club Books.

Gray, John (2003) *Al Qaeda and What it Means to be Modern.* London: Faber and Faber.
—. (2007) *Black Mass.* London: Allen Lane.
Hamilton, Clive and Richard Deniss (2005) *Affluenza.* London: Allen and Unwin.
Hartmann, Thom (2004) *The Rise of Corporate Dominance and the Theft of Human Rights.* Rodale Press.
Hobsbawm, E. J. (1990) *Nations and Nationalism Since 1780: Programme, Myth, Reality*. Cambridge, Cambridge University Press.
James, Oliver (2007) *Affluenza*. London: Vermilion.
Korten, David (2001). *When Corporations Rule the World*. Kumarian Press.
McLean, Bethany and Peter Elkind (2003) *Smartest Guys in the Room: The Amazing Rise and Scandalous Fall of Enron*. New York: Penguin.
Norton, B. (2005) *Sustainability: A Philosophy of Adaptive Ecosystem Management* Chicago: Chicago University Press.
Pico della Mirandola, Giovanni (1486) *On the Dignity of Man*. Paul J. Miller, Charles G. Wallis and Douglas Carmichael (eds. and trans.). Indianapolis: Hackett (1998).
Pfohl, Stephen (1990) "Welcome to the Parasite Cafe: Postmodernity as a Social Problem." *Social Problems*, 37:4, pp. 421–442.
Schumacher, E. F. (1977) *A Guide for the Perplexed*. New York: Harper and Row.
Sennett, Richard (1998) *The Corrosion of Character: The Personal Consequences of Work in the New Capitalism*. New York: W W Norton.
—. (2006) *The Culture of the New Capitalism*. New Haven: Yale University Press.

CHAPTER TWENTY-THREE

INTERROGATING AND RECONCEPTUALIZING NATURAL LAW TO PROTECT THE INTEGRITY OF THE EARTH

SHEILA D. COLLINS

Abstract

A foundational task for preserving the integrity of the earth is preserving its ecology of knowledges (de Sousa Santos 2008). For centuries, what has passed for universal knowledge has really been a localized and quite parochial form of knowledge: Western culture. Our ability to preserve earth's integrity is now deeply compromised by our reliance on this one form. For Westerners, the first task for preserving this ecology of knowledges is the critical interrogation of our own culture. In this chapter, one part of that interrogation is explored – an excavation of the conception of Natural Law that forms the basis of Western jurisprudence, human rights liberal democracy and international law particularly as it is exemplified in the work of John Locke. Locke's conception of Natural Law, I argue, suffers from a conceptual distortion and diminution of possible imaginaries derived from its being the only form of knowledge recognized as such by those who became the conquerors. At its heart, we also see that it is based on a relation of conflict and coercion – the right of property. The Western conception of Natural Law is contrasted with a more holistic version of Natural Law, rooted in a deep understanding of the earth system and exemplified in the knowledges of Indigenous peoples and in a variety of religious traditions going back centuries. A process of deep civilizational self-reflection is called for.

The Kogi people, modern descendants of the ancient Tairona civilization of the high Sierra Nevada de Santa Marta range on the coast of Colombia, occupy a unique eco-system. They call their mountain, "the Heart of the

World," because it contains every different climate on earth. Isolated for centuries high up in the deep folds of this 17,000 foot mountain, the Kogi are the only indigenous people to have survived the destructive impacts of the *conquista* and the subsequent modernizing forces of capitalist development with their culture intact, and they see themselves as guardians of the knowledge that is necessary to save the earth from immanent destruction. In recent years they have sought through emissaries to send messages to us, the "younger brothers," about the dangers facing the earth community and what will be necessary to save it. The Kogi, like all Indigenous peoples, are the canary in the mine. They are asking us to quit our chatter and to listen. The Kogi priests, or Mamas, are trained from birth through a process of rigid sensory deprivation, and esoteric teaching, so that when they emerge into the world at the age of about 20 they are filled with wonder and awe at the beauty and diversity of the living earth which they had never seen up to this point and are motivated to serve as its protectors.[1]

The training process of the Kogi Mamas is a heightened version of the way in which all children come into the world: filled with wonder at the colors, shapes, textures, sounds, smells, and motion they encounter and with an innate drive to observe and explore that world. Unfortunately, that wonder is all too quickly snuffed out by poverty, disease, malnutrition, war or, in the West, by our educational systems. This early childhood experience – composed of both a feeling of wonder or reverence and a heightened capacity for observation, in which the child is both the subject and the object of the observation, is part of mastering what the Kogi call the material manifestation of the Law of the Great Mother, whom they call Gaulcovang, from the root *gau* – to make, create, consecrate. For the Kogi, the Sierra Nevada is that material manifestation of this Natural Law, and the whole of their culture is based on their understanding of the structure and functioning of the Universe. According to an anthropologist who studied the Kogi,

> "Kogi cosmology is … a model for survival in that it moulds individual behavior into a plan of actions or avoidances that are oriented toward the maintenance of a viable equilibrium between Man's demands and Nature's resources. In this manner the individual and society at large must both carry the burden of great responsibilities which, in the Kogi view, extend not only to their society but to the whole of mankind."[2]

In varying degrees, Indigenous and peasant peoples around the world are the contemporary bearers of this ancient knowledge which, until quite recently, Western culture considered non-existent as knowledge, much less as saving knowledge (Shiva 1993, 12). As Boaventura de Sousa Santos has argued, Western epistemology was built upon

> "a system of visible and invisible distinctions, the invisible ones being the foundation of the visible ones. The invisible distinctions are established through radical lines that divide social reality into two realms, the realm of "this side of the line" and the realm of "the other side of the line." The division is such that "the other side of the line" vanishes as reality becomes nonexistent, and is indeed produced as nonexistent. Nonexistent means not existing in any relevant or comprehensible way of being. Whatever is produced as nonexistent is radically excluded because it lies beyond the realm of what the accepted conception of inclusion considers to be its other." (de Sousa Santos 2008, 1)

The knowledges built up over the centuries in those parts of the world, labeled by de Sousa Santos, the "colonies," are not allowed to enter into a dialogical relationship with the dominant knowledges. Not only are they considered non-knowledges, but their bearers have been subject to projects either of cooptation into the dominant culture's epistemology or, as has been more often the case, to destruction. On the visible side of the lines – roughly corresponding to the northern industrialized countries today – the tension between what de Sousa terms "regulation" and "social emancipation" are at the bottom of all human conflicts, but on the invisible side of the line, the dichotomy is between appropriation and violence (de Sousa Santos 2008, 1). On the "other side of the line," lawmakers are allowed to engage in every kind of behavior that would be illegal in their own systems. We see this today in the dumping of toxic wastes by northern and Western multinationals in developing countries or in the poorer parts of their own countries, in the bombing of civilians in places like Afghanistan, Pakistan and Iraq in order to keep the invisible parts of the world from encroaching on the visible, and in the creation of Guantánamo, Bagram, and other so-called "black sites" as places beyond the reach of any law, whose inhabitants occupy an ontology of non-existence. In addition to the lines separating human populations, however, there is another set of lines separating humans from the rest of the natural order. In this line drawing, even the flora and fauna on "this side of the line" are subject to the dichotomy between appropriation and violence.

Today the fate of humankind rests on our ability to end this line making in our thinking, actions and laws. We have come to a time when the confluence of the economic and ecological trajectories we are on demands that we develop an ethics that includes not only those peoples on the other side of the line, but the whole of the created order; but to do so requires an excavation and critical interrogation of the sources of such line making. I suggest that one of the places it may be found is in the conception of Natural Law which forms one of the sources of liberal Western jurisprudence, human rights and democracy. The forms in which Natural Law has come down to us suffers from a conceptual distortion and diminution of possible imaginaries derived from its being the only form of knowledge allowed by those who became the conquerors.

While we can be grateful to Locke for advancing the dignity of the human individual at a time of monarchical authoritarianism and providing the intellectual legitimacy for rebellion against tyranny, his philosophy, nevertheless, owes its preoccupations to its author's social location on "this side of the line." The intellectual memes that his conception of Natural Law perpetuated have left us bereft of the intellectual, ethical and political resources we need to prevent the swift dismemberment of Mother Earth.

The Origins of Line Drawing

Natural Law in the West finds its origins among the ancient Greeks, particularly Aristotle, who was both a philosopher and a scientist whose work on *phusis* (physics) anticipates in primitive form the concept of "emergence" used by modern systems theorists to describe the way complex systems arise out of multiple discrete interactions and autopoiesis, the term coined by Chilean biologists Humberto Maturana and Francisco Varela to described the self-generating quality of living systems. Aristotle's work on politics was informed by his notion of inherent form and *telos* in the natural world. Humans, he observed, seemed to have an inherent purpose or *telos* to live in community – first as families and then as collections of families that come together as the city-state. The city-state exists to satisfy basic needs, to provide for humans' need to be social, and for the common good (Aristotle, bk. 1, Part 1), and its "constitution is analogous to the natural laws governing physical organisms" (Westra 2004, 34–35). While Aristotle was onto something in his conception of the inherent forms and ends of natural phenomena and in his application of this principle to human social life, the particular nature of the observations from which he drew his philosophical principles were informed by the socioeconomic conditions of the time in which he lived as well as by the primitive instrumentation with which he made his physical observations. Thus, while his conception of the ends of the political community were laudatory, his ascription of certain social hierarchies to ontological categories – master over slave, male over female – as "natural," we now know to be inherently flawed and thus his definition of the person entitled to enjoy the rights of citizenship as he "who shares in the administration of justice and of offices" leaves out over half the population of the city-state (Aristotle, bk. III, part 1, 52). Thus, a line is drawn separating humans into those on "this side" and those on the "other side." Males of a certain social class are citizen/subjects while slaves and women are possessions of the subjects, and all barbarians are no better than slaves.

Aquinas, along with Albertus Magnus, is credited with recovering for Western intellectual history the Aristotelian notion that truth is available to all human beings through *natural revelation* or reason, which Aquinas construed as human participation in God's eternal law and in restoring the essential worth of the natural world and the value of empirical observation and experiment after their depreciation by the Neo-Platonists. But Aquinas, too, is a creature of his time and place. His conception of Natural Law owes its preoccupations to its author's social location as the son of nobility, whose life's work became one of scholarship, logical disputation, and teaching, but whose basic needs were probably supplied by subaltern others, those with knowledge of how things actually work – the hewers of water, the growers and processors of grain, livestock, herbs and garden vegetables. Aquinas inhabits a hierarchical universe where humans are still superior to other creatures by virtue of their ability to reason, where the subordination of women to men and the subjection of peasants to nobles that he saw around him were taken to be in the "natural" order of things, where the State, ruled by a monarch, and indirectly by the Pope, takes its hierarchical form from the Divine Lordship of God, where the highest truth – supernatural law – can only come through Divine, (that is, Christian, monotheistic) revelation and where the highest form of happiness is found not in the wonder of the created world, but in an otherworldly beatific vision of God.

Aquinas somewhat tempers the harsher "natural" hierarchal order of earlier Aristotelians by saying that economic or civil subjection of some human beings to others is undertaken for the benefit and good of the subjected person and not for the benefit of the superior person (Aquinas, *Summa Theologica* 1a, q. 92 "The Production of the Woman") . Nevertheless, the fact that he accepted such social hierarchies identifies him with those on "this side of the line." Despite its lofty aspirations toward the goodness of an infinitely loving God and *his* creation, Christian Europe's understanding of that goodness as expressed in human action extended only so far. When it came to positive law, the conceptual hierarchies still held. The legal community was identical with the Christian community united in the Roman Church, and although, according to Miéville, the adoption of notions of the Roman law in late medieval Christendom were predicated on the idea of equality of exchange (Miéville 2006, 70), this equality did not apply to those who fell outside the *respublica Christiana*. Thus, anyone who fell outside the authority of the Church – whose ultimate legal authority rested in the Papacy, as guardian of the even higher divine law[3] could neither be subject to its positive legal norms nor was considered to exist within its Natural Law. Medieval Christianity thus rested on a radical negation of the Other against whom it defined itself. Europe could believe that human beings were created in God's image but carry out bloody Crusades against the Infidel, pogrom against the Jews, burn heretics and witches (aka. midwives who adhered to the pre-Christian religion and peasant herbalists) at the stake, and take other people's lands and sources of sustenance by right of conquest

By the end of the fifteenth century, the spatial–political categories of the medieval era and the normative hierarchies they constituted were beginning to dissolve, and a new spatial–political order – that of the system of modern nation states was emerging. The so-called "discovery" of the New World in the late fifteenth century signaled the beginning of the end of *respublica Christiana* and made possible the emergence of modern law and, in particular, modern international law and its concept of state sovereignty. China Miéville has argued, following the Soviet legal theorist, Evgeny Pashukanis, that international law "can only be understood as those of market relations slowly generalizing globally in the transition to capitalism"(Miéville 2005,7). At the base of market relations is the commodity form, an abstract relation of supposedly equivalent values of labor time, but one which, under conditions of colonialism and evolved capitalism constitutes a relation of violence. Once labor is congealed into a commodity it becomes a thing that can be "owned."

> "In the commodity form itself, dispute, coercion and violence are inherently implied. The notion of 'mine' necessary to ownership and commodity exchange is only meaningful inasmuch as it is 'mine-not-yours'." (Miéville, 95, n. 99)

> "For a commodity to be exchanged – some forceful capabilities are implied. If there were nothing to defend its 'mine-ness,' there would be nothing to stop it becoming 'yours,' and then it would not longer be a commodity, as I would not be exchanging it." (Miéville, 126)

Miéville traces this new world order and its system of laws to "the early colonial experience of transatlantic and eastern trade. … a dialectical result of the very process of conflictual, expanding inter-polity interaction in an age of early state forms and mercantile colonialism and the abstraction of commodity exchange." For Miéville, "international law *is* colonialism, not a result of such" (Miéville, 168–169).

The competition between the Old World kingdoms to plunder the riches of the newly discovered lands led to a series of lines being drawn which were thought to separate a European sphere subjected to peace and the law of nations from an overseas sphere where neither law nor peace prevailed. Such lines may not have been formalized on paper but seem to have been discussed in several 17th century bi-lateral treaties (de Sousa Santos 2008, 3). On "this side of the line," nations were thought to enjoy equal sovereignty – a form homologous to the relations of abstract equivalence that inhered in commodity exchange and could enter into formal relations to regulate their transactions with each other and their exploitation of the non-European world (Miéville, 144). On the "other side of the line," however, rape, pillage, slavery, and wanton destruction of the natural environment could go on without either crimes being committed or punishment meted out.

John Locke and the Doctrine of Private Property

The conception of Natural Law that was to correspond to and justify this new arrangement of the world was that of John Locke far more than it was of Hugo Grotius. By the time we get to the 17th century, nature has become desacrilized, stripped of its numinous and spontaneous qualities, its complexity, its intrinsic integrity and ultimate ineffableness. Reduced to "dead matter" it can now be quantified, manipulated and used up to benefit human "progress." Locke, the Father of modern liberal democracy, now posits a "state of nature" in which humans supposedly began life. While acknowledging that God gave the world as a common inheritance to all men "for the support and comfort of their being," "there must of necessity be a means to appropriate them some way or other, before they can be of any use" (Locke 1690, Chap. 5, Sec. 26). He then goes on to explain how the labor theory of value and thus the rule of property comes into being.

> "Though the earth, and all inferior creatures, be common to all men, yet every man has a property in his own person: … Whatsoever then he removes out of the state that nature hath provided, and left it in, he hath mixed his labour with, and joined to it something that is his own, and thereby makes it his property. It being by him removed from the common

> state nature hath placed it in, it hath by this labour something annexed to it, that excludes the common right of other men: for this labour being the unquestionable property of the labourer, no man but he can have a right to what that is once joined to, at least where there is enough, and as good, left in common for others." (*Second Treatise*, Chap. 5, Sec. 27)

Notice how, in the first instance, nature has no intrinsic value but was placed here to benefit humans. Notice also how in this mythical state of nature there are only individual, atomized humans who exist in perfect freedom and who each have access to land which, by appropriating it, becomes their own. Although the "state of nature" is essentially a theological statement and not meant to convey anthropological truth (Dunn 1969, 97), Locke's exegesis of the concept inevitably falls into the trap of conflating this ahistorical account with 18th century European sociological realities. We can forgive him for being ignorant of the communal nature of actual early life and of the gift giving cultures that preceded and indeed surrounded the development of capitalism in the West where gifts developed from the land were not to be hoarded, but given away (Hyde 2007). After all, anthropological research was not then an established science. But what we cannot forgive is the system of modern economy and law that still rests on this 17th century theological myth. Finally, notice that through creating property, man "removes" a part of nature from the whole. It is a short step from this removal to the commodity form through which these amputated parts of nature are further abstracted and turned into equivalent values that can be exchanged in the market. The fictional character of Locke's "state of nature" and its function as a justification for elite appropriation is revealed in the next few lines in which he asserts that

> "God gave the world to men in common; but since he gave it them for their benefit, and the greatest conveniencies of life they were capable to draw from it, it cannot be supposed he meant it should always remain common and uncultivated. He gave it to the use of the industrious and rational, (and labour was to be his title to it;)." (*Second Treatise*, Chap. 5, Sec. 34)

Here he reveals the Western prejudice against and ignorance of the Other, for only Europeans were thought to be industrious and rational. The uncultivated land that the Indigenous hunter–gatherers lived on was, by its apparent lack of cultivation, a sign of their lack of industriousness, and their alternative ways of expressing their knowledge and wisdom were considered irrational, primitive, magical. Locke, like most Europeans, was apparently ignorant of the complex biodiverse agricultural and food processing systems that had been developed by the Indigenous of South America over centuries of observation, trial and error, or of the *milpa* or "Three Sisters" system that characterized agricultural patterns in North and Central America. To the European eye, used to seeing linear rows of one crop laid out in rectangular plots, these random forest clearings in which hillocks of corn, beans and squash grew amid the denuded trunks of trees in a seemingly wild but actually carefully cultivated symbiotic relationship that fixed nitrogen in the soil, prevented weeds and insect infestation must have seemed like land that was uncultivated and therefore, waste (Weatherford 1988, 62–64;79–97).

Locke seems to suggest in several places that the private appropriation of land is subject to some limits. Humans in the state of nature have the right to take land that their labor has made productive only to the extent that "there is enough, and as good, left in common for others" (*Second Treatise*, Chap. 5, Sec. 27) and only so much as a person can make use of without causing spoilage (Chap. 5, Sec. 31). Moreover, they are not to "harm another in his life, health, liberty, or possessions" (Chap. 2, Sec. 6). "To force or cheat a man out of his property is at all times a crime …" (Locke 1663–1664). "Is the Binding Force of the Law of Nature Perpetual and Universal? Yes." (*Political Essays*, 122). Yet these limitations on human greed are soon overridden in his acceptance of the benign role of money as the mechanism by which humans can accumulate property that is not subject to spoilage. In Locke's acceptance of property as a form for aggregating wealth we have come a long way from Aristotle's notion that property was a tool for the purpose of life (Aristotle *Politics* 1.1253b). We might compare Aristotle's notion of property with Locke's as the distinction between use value and exchange value. For Locke, humans' violation of Natural Law did not rest

> "in the largeness of his possession, but the perishing of any thing uselessly in it. And thus came in the use of money, some lasting thing that men might keep without spoiling, and that by mutual consent men would take in exchange for the truly useful, but perishable supports of life." (*Second Treatise*, Chap. 5, Sec. 46, 47)

Elsewhere Locke acknowledges that people can enter into contractual relationships to hold some lands in common, so that any individual appropriation must be consented to by the other parties to the compact.

> "It is true, in land that is common in England, or any other country, where there is plenty of people under government, who have money and commerce, no one can inclose or appropriate any part, without the consent

of all his fellow commoners; because this is left common by compact, i.e. by the law of the land, which is not to be violated." (*Second Treatise*, Chap. 5, Sec. 35)

But the compacts that people work out to hold some lands in common are not available to everyone. "And though it be common, in respect of some men, it is not so to all mankind; but is the joint property of this country, or this parish" Here we see the visible and invisible lines described by de Sousa taking shape in the development of Western ideas of property law and the rudiments of the idea of national sovereignty. The regulation of land use can take place on "this side of the line." Those who have money and are engaged in commerce can decide to have some lands held in common, which then become the property of the juridical entity created to protect them – the country or parish – but the ability to enter into a legal contract to protect some lands from private appropriation does not apply to those on the other side of the line. The Indigenous have no legal standing either as collectivities or individuals. Nor do the peasants, whose common lands were appropriated through a series of legal compacts, the Enclosure Acts, which excluded their further use, making them the exclusive property of the owners. And, apparently for Locke, neither do servants, whose labor becomes the property of the master.

"Thus the grass my horse has bit; the turfs my servant has cut; and the ore I have digged in any place, where I have a right to them in common with others, become my *property,* without the assignation or consent of any body. The *labour* that was mine, removing them out of that common state they were in, hath *fixed my property* in them." (*Second Treatise*, Chap. 5, Sec. 28)

This development of the idea of legal sovereignty is given further legitimacy by being commanded by God and is connected to the idea that humans are made to have dominion over the created order. Here is the architecture of the system of capitalism and its extension in the colonial world order.

"The law man was under, was rather for appropriating. God commanded, and his wants forced him to labour. That was his property which could not be taken from him where-ever he had fixed it. And hence subduing or cultivating the earth, and having dominion, we see are joined together. The one gave title to the other. So that God, by commanding to subdue, gave authority so far to appropriate: and the condition of human life, which requires labour and materials to work on, necessarily introduces private possessions." (*Second Treatise*, Chap. 5, Sec. 35)

Thus, private property is at the heart of the formation of the modern liberal state and its protection is the state's sole reason for being. For Aristotle, it will be recalled, the purpose of the state was the common good. Further on, Locke seems to hint at a "common good" purpose for the state when he contends that the private appropriation of land "increases the common stock of mankind" for the surplus that is produced is then available to others (through the mechanism of the market), whereas if the land had not been appropriated it would have been useless, pure waste (*Second Treatise*, Chap. 5, Sec. 37). Such a definition of the common good is a quantifiable one, not a moral one. From this concept of quantity would flow the modern measure of the common good – the Gross Domestic Product – a poor measure of any society's real quality of life.

For Locke it is human labor that produces ninety-nine hundredths of the value in the world, nature, itself, contributing a miniscule fraction (*Second Treatise*, Chap. 5, Sec. 40). Locke provides supporting evidence for his theory of private appropriation as the source of all wealth, with evidence drawn from his knowledge of life in America. Here the Eurocentric ignorance of other ways of knowing and being-in-the-world is starkly revealed:

> "There cannot be a clearer demonstration of any thing, than several nations of the Americans are of this, who are rich in lands and poor in all the comforts of life; whom nature having furnished as liberally as any other people, with the materials of plenty; i.e. a fruitful soil, apt to produce in abundance, what might serve for food, raiment, and delight; yet for want of improving it by labour, have not one hundredths part of the conveniences we enjoy: and a king of a large and fruitful territory there, feeds, lodges, and is clad worse than a day-labourer in England." (*Second Treatise*, Chap. 5, Sec. 41)

Such reasoning is just a few centuries and several levels of abstraction short of the idea that "what's good for General Motors is good for the country" – a commonly held assumption in the United States until the recent economic meltdown, but rephrased now in the idea that the private banking industry is too big to fail because what's good for banking is what's good for everyone in the world. This conception of private property, so foreign to pre-capitalist peoples, is given added authority by its being thought to be "in the very nature of things" – thus a "natural right." Individual rights, which flow from this natural right to property are thus the rights of property owners (Miéville, 145). As Miéville has pointed out, "From the right of property flows the erection of modern law whose very premise is to manage or settle disputes," implying that the right of property on which they are based is basically a conflictual one (Miéville, 44).

In the early legal–political systems that emerged from this conception of natural law those without property – women, slaves, servants, indigenous peoples – had neither contractual nor political rights. The early formation of systematized international law can be seen in the lines or *raya* that were drawn by the Pope to divide the New World into zones controlled by the Spanish and Portuguese at a time when the nascent state was still evolving. But these lines could not hold and so a series of so-called "amity lines" were drawn up that delineated a European "law of nations" which was undergirded by a zone of anarchy. According to Miéville, "Where they operated, overseas colonial possessions were not recognized, but were up for grabs: beyond the line, might made right" (Miéville, 182). Such lines lasted only until about 1634 when they were undermined by the rising powers of England and the Dutch East India Company each of which sought to have their juridical–political right to colonies recognized. But what is important in the drawing of these lines is the assumption that legality and legitimacy, the right to possession lie on "this side of the line" and anarchy and anything goes lie on the other side. It is a dichotomous way of thinking that would not entirely disappear in the history of colonialism and has resonances down to today. The lines drawn to define nation state borders are remnants of the amity lines. Within these borders, law and peace prevail, but outside them the law does not exist. Only those defined by the law as "citizens" have rights. The compact between the Soviet Union and the United States to divide up the world and the doctrine of mutually assured destruction is a form of the amity lines, the two superpowers tacitly agreeing not to attack each other, but feeling free to carry out any kind of atrocity beyond the East/West line in the zones of the world that were up for grabs.

While over the centuries, and especially in the last sixty years, the international community has enlarged its understanding of human rights, codifying them in a series of international agreements, it has not been able to operationalize this understanding in any but the most elemental ways, since the enforcement of these rights rests in the hands of national sovereigns. As Miéville argues, sovereign states are posited as legal subjects, but with no overarching authority. Thus, interpreting law becomes a never ending series of arguments which can never be settled except through the violence of an imperialist state. Perhaps, as Miéville argues, western law, and by extension international law, is at base unworkable, lodged as it is in the violence of the commodity form which, in turn, is based on the foundation of a Natural Law that is most "unnatural." Not only is law's concern only with the human subject – the earth and its non-human inhabitants have no enforceable "rights" – but it is based on a series of dichotomies, sometimes recognized, sometimes covert that are foreign to the way in which the earth system actually works as an intricately connected system of systems that is intelligent, self-creating, spontaneous, contingent, infinitely varied, yet with form (Capra 1996).

Such dichotomous thinking was developed by men who, by virtue of their education and social location were cut off from the lessons that attachment to the earth teaches. While modern science has given us a picture of how the earth system works, it provides no meaningful way for us to appropriate its lessons. That must come through our cultural institutions; but modern Western culture, with its rapacious capacity for accumulation has so far drifted from this elemental understanding of the earth system that it is fast cutting us off from even the reality on which such an understanding must be based.

We have drifted so far from our original "home" that we take pride in the creation of "virtual communities" and "virtual cities" – electronic pixels moved around on a screen. We produce "creative designers" who spend all of their waking hours dreaming up the next breakthrough in toothbrushes and spend hours watching "reality shows," purporting to be "real life," but actually carefully scripted prisoner's dilemmas in which only one person is allowed to win. Our scientists are put to work creating "smart bombs" that kill their faceless, nameless, history-less, victims at a distance and then wonder why those victims come back to haunt us. Our "best and brightest" can't see that a financial system floating so far above the real economy of production and negating so thoroughly its dependence on the natural environment from which it began was bound to come cascading down like a house of cards, and when it did, they were so lacking in an understanding of the interrelations that constitute the real world, that they continue to pour money into the same failed system. When that system utterly fails, as it is bound to if we continue on our current trajectory, we will have no knowledge of how to survive, because we will have destroyed the vocabulary – the flora and fauna – from which that knowledge arises.

Toward an Ecology of Knowledges

Throughout the history of accounts of encounters between Native Americans and European or Euro-American settlers there is a constant refrain. Native Americans contrast their own knowledges and ethical understandings to those of the white men, but almost always they are simply not heard. Red Jacket, a 19th century Native American, said of the Western clergy

The black coats tell us to work and raise corn; they do nothing themselves and would starve to death if someone did not feed them. All they do is to pray to the Great Spirit; but that will not make corn and potatoes grow; if it will why do they beg from us and from the white people.[4]

In 1744, the commissioners from Maryland and Virginia negotiated a treaty with the Indians of the Six Nations, the contemporary Haudenosaunee, who were invited to send boys to William and Mary College. The Indians turned the offer down with these words:

> "We know that you highly esteem the kind of learning taught in those Colleges, and that the Maintenance of our young Men, while with you, would be very expensive to you. We are convinced, that you mean to do us Good by your Proposal; and we thank you heartily. But you, who are wise must know that different Nations have different Conceptions of things and you will therefore not take it amiss, if our Ideas of this kind of Education happen not to be the same as yours. We have had some Experience of it. Several of our young People were formerly brought up at the Colleges of the Northern Provinces: they were instructed in all your Sciences; but, when they came back to us, they were bad Runners, ignorant of every means of living in the woods. … neither fit for Hunters, Warriors, nor Counsellors, they were totally good for nothing. We are, however, not the less oblig'd by your kind Offer, tho' we decline accepting it; and, to show our grateful Sense of it, if the Gentelemen of Virginia will send us a Dozen of their Sons, we will take Care of their Education, instruct them in all we know, and make Men of them."[5]

Needless to say, the Native Americans' offer was never accepted. Today, Indigenous peoples around the world are making that offer again. They are asking us to go back to school in the great university of Life. They are pointing out that what the West has taken for a universal system of knowledge is really a universalized version of a very parochial and localized system constituted by and constituting a particular historical culture, class and gender (Shiva 1993, 9; 1995). They are asking us to consider that their own cultures might have something valuable to teach us, steeped as they are in a detailed knowledge of the earth system. They inform us that their system of communal ethics, rooted in a deep relationship to local biomes might be better able to guide us through the dilemmas that now confront us, such as the apparent tradeoff between reducing greenhouse gas emissions and ensuring an adequate life for the millions of malnourished people in the so-called developing world; or the apparent conflict between the needs of the contemporary population locked in an economic meltdown and the needs of future generations. They might just suggest that looked at through a different lens these apparent conflicts may not in fact be conflicts at all.

While ecologists have been concerned about protecting and conserving the physical ecology of the earth, an equally important, if not foundational task is to preserve the ecology of knowledges. I argue that this is foundational because without a change in consciousness we cannot possibly build the communities of resistance and reconstruction that will be needed to save the earth. As de Sousa has pointed out, due to the dominance of the monocultures of the West, "a massive epistemicide has been under way for the past five centuries, whereby an immense wealth of cognitive experiences has been wasted" (de Sousa Santos, 16).

Preserving the ecology of the knowledges that are left is, we now realize, essential to the preservation of the integrity of the earth. To be engaged in this process means that we in the West must support the many efforts of Indigenous and peasant peoples around the world to create seed banks, to create and preserve lexicons of their languages, and medicinal knowledges, and to keep their ancestral lands – which are the basis of their knowledges and without which their cultures cease to exist – free of the ravaging effects of fossil fuel production, biopiracy, and developers intent on taking land for private interests. We in the West must support the demand of Indigenous peoples, articulated in their "Anchorage Declaration", that the "inherent and fundamental human rights and status of Indigenous Peoples" affirmed in the UN Declaration on the Rights of Indigenous Peoples "be fully recognized and respected in all decision-making processes and activities related to climate change," that their right to self-determination be respected with regard to any program or project that affects their environments (including the right to say "no"), that "the [United Nations Framework Convention on Climate Change] UNFCCC's decision-making bodies establish formal structures and mechanisms for and with the full and effective participation of Indigenous Peoples", and that the Annex I countries pay the historical and ecological debt owed for their disproportionate contribution to climate change.[6]

Although the doctrine of national sovereignty has questionable ethical roots and serves as a barrier to the adoption of enforceable international laws to protect the earth and its human species from violation by individuals, states and corporations, preserving the ecology of knowledges in a world of nation states may require us to use this doctrine selectively, as a kind of ju jitsu tactic to defend the sovereignty of countries like Ecuador and Bolivia whose governments are now seeking to embed

Indigenous understandings of ecological diversity and earth integrity into their constitutions and legal systems and to give "national" or community title to lands and the genetic resources that exist on the lands that Indigenous peoples occupy.[7] The concept of national sovereignty, it seems to me, can only be redeemed if we use the mechanism in this way to protect those political–juridical units that see themselves as guardians of the natural environment.

Preserving the ecology of knowledges also requires trying to find ways of translating not only the words, the logic, of subaltern cultures, but their myths into a form that is comprehensible to the Western mind but that does not damage their integrity.[8] The facile adoption of Indigenous lore by the New Age movement is not what I am suggesting. Such a practice has rested too easily within the bowels of the Western appropriationist project. Rather, I am calling for a genuinely humble intercommunicative process of dialogue and learning. This is a tall order and admittedly may not be possible as the languages may not be commensurable, but it is an effort worth trying. The project, however, raises all kinds of procedural questions, as de Sousa points out:

> "What types of relationships are possible between the different knowledges? How can we distinguish between incommensurability, incompatibility, contradiction, and complementarity? Where does the will to translate come from? Who are the translators? How should we choose translation partners and issues? How can we form shared decisions and distinguish them from imposed ones? How can we make sure that intercultural translation does not become the newest version of abyssal thinking, a soft version of imperialism and colonialism?" (de Sousa Santos, 18)

And further:

> "How can we identify the perspective of the oppressed in real world interventions or in any resistance to them? How can we translate this perspective into knowledge practices?"(de Sousa Santos, 18)

Conclusion

In summary, we must learn to see our own culture in dialogical relationship to those other diverse and equally meaningful cultures in the world. Diversity is at the heart of true Natural Law; indeed, we may say that it is one its foundational premises. There are deep lacunae in our culture that might be filled with insights from other cultures. Indigenous cultures and Buddhist teachings tell us that human life is connected to non-human life and that non-human species have sentience, a kind of moral consciousness, and are imbued with "spirit." The Koyukon people of Alaska and the Yukon, for example, have stories that relate a Distant Time when animals had the form of human beings and spoke like humans and at some point when they died became the plants and animals that populate the territory the Koyukon live in today (Kinsley 1995, 37–38). In many Indigenous cultures animals, plants, insects and even rocks are spoken of as relatives. Unlike the Western conception of time and development as a linear process, most Indigenous cultures see the universe and the interactions of living systems as a circular process of birth, growth, decay, regeneration, which are simply phases in an ongoing continuum. Ancestors speak to the living' and trees, eagles, coyotes and stones have messages that can be "read." Humans participate in the natural world, not as creatures set apart, but as "two-leggeds" among four leggeds, six leggeds, etc. Their languages and cultural traditions are drawn from the ecology of the area in which they live. Such seemingly mythic understandings actually have much more in common with what contemporary scientists are discovering about the nature of living systems. Maturana and Varela, for example, describe living systems

> "as units of interactions; they exist in an ambience. From a purely biological point of view they cannot be understood independently of that part of the ambience with which they interact: the niche; nor can the niche be defined independently of the living system that specifies it. ... This circular organization constitutes a homeostatic system whose function is to produce and maintain this very same circular organization by determining that the *components* that specify it be those whose synthesis or maintenance it secures." (Maturana and Varela 1980, 9)

The cultural teachings, stories, rituals and practices of many Indigenous peoples set moral limits on human destructiveness of the non-human world and thus function to help maintain the homeostasis of the entire life system, not as something acting on that system from outside, but as component parts of the living system itself. The ritual of asking forgiveness for taking the life of the animal or fish that is needed for food keeps the community from killing more than is needed for their survival. Traditional agricultural methods play a key role in ensuring food security, preserving soil and water, and resisting natural disasters (Nabhan 2002). The agroforestry practiced by the Amazonian Guaraní, for example, exploits diverse ecological niches and uses small areas of forest for a short time before moving on to use another area of forest, so that no single resource will be over-harvested and the forest will be allowed to regenerate (Reed 1997, 65). These methods are infinitely superior to the wholesale clear-cutting and mono cropping associated with industrial agriculture and we could learn much from them about meeting the world food crisis in a time of climate change. As Shiva has demonstrated, the vaunted Green Revolution which was said to have saved millions of people from starvation in the developing world provided only a short-term benefit, in the long run contributing to the erosion of the planet's genetic diversity and destroying in a few short years what it had taken farmers working with nature millennia to develop (Shiva 1993, 84–85).

Our Western conception of health is far too narrow, focused as it is mostly on curing disease. Most Indigenous people view health as not just the physical well being of the individual, but the social, emotional and cultural well being of the entire community. Scores of studies in medical journals attest to the effects of poverty, inequality, pollution, and stress in degrading human health, yet our Western health systems for the most part treat individuals, ignoring the larger socio-economic and environmental contexts in which they are situated. Western education is carried out in a place set apart from daily life. It is seen as a preparation for life, which is to be understood theoretically before it is lived. Indigenous peoples, on the other hand, have developed their knowledges from a deep and continuous dialogue and interaction with nature, drawing their lessons from observation and experimentation with the materials nature offers and then passing on this knowledge through apprenticeship systems. Their knowledge, as one observer has put it, comes from deep within the materials worked with, in the way we say a sculptor discerns the form inside the stone (Crickmay 2002, 46).

There are things about our culture that may need to be rooted out because they are destructive: the excessive individualism, the sanctity of economic growth as a measure of well-being, and the win/lose dichotomy fostered by Western capitalism; the linear idea of "progress" and the technological fixe to name a few. There are, however, ways in which some kinds of Western technology and medical practices may be used to complement traditional Indigenous practices to achieve ecological protection and human health at a time when the options for action are narrowing swiftly. There are also subversive knowledges buried within our Western monotheistic religious systems that have resonances with the spiritual teachings of Indigenous peoples and need to be recovered: the teachings of St. Francis and the experiences of the mystics, for example. According to Nasr, the refusal of Islamic society "to surrender to the dicta of the [Western] machine" is testament to the stubborn survival of an alternative understanding of nature found in the Quran's lack of a clear line of demarcation between the natural and supernatural, between man and nature.

> "... according to the Islamic perspective God himself is the ultimate environment which surrounds and encompasses humanity. .... The environmental crisis may in fact be said to have been caused by the human refusal to see God as the real 'environment' which surrounds us and nourishes our life." (Nasr 1992, 89)

The writings of the Romantics, poets like Walt Whitman and American environmentalists like Henry David Thoreau, John Muir, and Aldo Leopold are now being resurrected. Quantum mechanics, evolutionary biology, astrophysics and neuroscience, among other scientific fields, are confirming much of what Indigenous cultures and mystical traditions have known for centuries.

As Westra has observed, Natural Law as found in Aristotle, Aquinas, Grotius and others has given us a set of foundational doctrines, establishing for posterity such principles as the rights of humans to subsistence and physical security, to a safe environment, to live in a community that fosters the common good , to the acquisition and pursuit of knowledge, to form with each other a government that protects these rights as well as the right to rebel against one that does not (Westra 2004, 45). But we must also recognize that their conception of rights was limited by the framework of the cultural and physical repertoires in which they lived. Our task today if we are to preserve the integrity of the earth, is to broaden that framework with the knowledges that were for so long excluded from recognition as knowledge by the West. Thanks to the discoveries of modern science, augmented by the Internet and global travel we now have the possibility of broadening our cognitive framework. Indeed, this has become an evolutionary necessity. Such a project of deep civilizational self-reflection and cross-cultural learning is currently occurring in many parts of the world. Participants in the World Social Forums, the Global Ecological Integrity Group, the Earth Democracy Project, the Toda Institute for Global Peace and Policy Research conferences and those who have been involved in writing the Earth Charter represent such a process. But this process must become much more widespread, requiring vast changes in our educational systems, our religious practices, our production and exchange systems, and our approach to scientific inquiry and technological innovation.

References

Aristotle, *Politics*, trans. Benjamin Jowett. Kitchener, Ont.: Batoche Books, 1999.

de Sousa Santos, Boaventura. 2008. "Beyond Abyssal Thinking: From Global Lines to Ecologies of Knowledge," *Eurozine* (18 February) http://www.eurozine.com.

Capra, Frijof. 1997. *The Web of Life: A New Scientific Understanding of Living Systems*. New York: Anchor Books.

Crickmay, Lindsey. 2002. "Transmission of Knowledge Through Textiles: Weaving and Learning How to Live." In *Knowledge and Learning in the Andes: Ethographic Perspectives*, Stobart, Henry and Rosaleen Howard, eds. Liverpool: Liverpool University Press, 40–55.

Dunn, John. 1969. *The Political Thought of John Locke: An Historical Account of the Argument of the 'Two Treatises of Government.'* Cambridge: Cambridge University Press.

Hyde, Lewis. 2007. *The Gift: Creativity and the Artist in the Modern World*. New York: Vintage Books/Random House.

Locke, John. 1690. *Second Treatise of Civil Government*. The Project Gutenberg ebook. John Locke. *Two Treatises of Government* http://www.gutenberg.org/dirs/etext05/trgov10h.htm

—. "Is the Binding Force of the Law of Nature Perpetual and Universal? Yes." Essays on the State of Nature VII. In *Locke: Political Essays*. Ed. Mark Goldie. Cambridge Texts in the History of Political Thought. Cambridge: Cambridge University Press, 1997.

Maturana, Humberto R. and Francisco J. Varela. *Autopoiesis and Cognition: The Realization of the Living*. Dordecht: D. Reidel Publishing Co., 1980. Vol. 42: Boston Studies in the Philosophy of Science.

Miéville, China. 2005. *Between Equal Rights: A Marxist Theory of International Law*. Leiden, The Netherlands: Koninklijke Brill NV.

Nabhan, Gary Paul. 2002. *Enduring Seeds: Native American Agriculture and Wild Plant Conservation*. Tempe, AZ: University of Arizona Press.

Nasr, Seyyed Hossein. "Islam and the Environmental Crisis." In *Spirit and Nature: Why the Environment is a Religious Issue*, eds. Steven C. Rockefeller and John C. Elder. Boston: Beacon Press, 1992.

Reed, Richard. 1997. *Forest Dwellers, Forest Protectors: Indigenous Models for International Development*. Needham Heights, MA: Allyn and Bacon.

Shiva, Vandana. 1993. *Monocultures of the Mind: Perspectives on Biodiversity and Biotechnology*. London: Zed Books.

—. Ingunn Moser, eds. *Biopolitics: A Feminist and Ecological Reader on Biotechnology*. London: Zed Books.

Westra, Laura. 2004, *Ecoviolence and the Law*, Transnational Publishers, Ardsley, NY

Notes

[1] For more about the Kogi culture, see: Tairona Heritage Trust http://www.eremite.demon.co.uk/Tairona/1pages/secb/b9mamased.html; Alan Hereira, *The Heart of the World* (Jonathan Cape/Random House, 1990); *The Elder Brother* (New York: Alfred A. Knopf, 1992).

[2] Tairona Heritage Trust, *ibid.*

[3] During the long disintegration of the late Medieval order, Aquinas's distinction between natural law and supernatural law seemed to give the Pope, the "vicar of Christ" supreme law-making power. In fact, argues Miéville, it did not always work out this way as the Holy Roman Empire was a "corollary of the relatively atomized, shifting polities of vassal feudalism." Nevertheless, the legal categories continued to be based on those of *respublica Christiana*. Miéville, ibid., 168-9.

[4] Red Jacket, quoted in: Roger Hammer, ed. *The People: Native American Thoughts and Feelings*, Summertown, TN: The Book Publishing Company, 1990, p. 41.

[5] *Ibid.*, 31.

[6] "Anchorage Declaration," adopted by the Indigenous People's global Summit on Climate Change, 24 April, 2009.

[7] Various proposals to protect the rights of communities to their biological resources and knowledges have been offered and some have made it into the national legislation of their countries. Most define the rights bearing entity as "local community" rather than "nation," but the underlying purpose is the same--the legal sovereignty of such communities to protect their resources and innovations from outside predators and the right to negotiate their use by outsiders. Examples include The [Philippine] Republic Act 8371 or the "Indigenous People's Rights Act of 1997; The Community Intellectual Rights Act proposed by the Third World Network and the Biodiversity Related Community Intellectual Rights Act proposed by groups in India. See: The Role of Intellectual Property Rights in Protecting Traditional Knowledge (The Philippine Experience) http://www.unctad.org/trade_env/test1/meetings/delhi/Countriestext/Philipinestext.doc; Nijar G. Singh, "Community Intellectual Rights Protect Indigenous Knowledge," *Biotechnology and Development Monitor*, No. 36, (1998): 11-12; Shiva V. S., H. Afsar. G.B. Jafri, and R. Holla-Bhar. *The Enclosure and Recovery of the Common: Biodiversity, Indigenous Knowledge and Intellectual Property Rights*. Research Foundation for Science, Technology and Ecology. New Delhi.

1997, 163-174. For a full discussion of the issues see: Martin Khor, *Intellectual Property, Biodiversity and Sustainable Development*. London: Zed Books, 2002.

[8] De Sousa Santos argues that "through these knowledges it is possible to nurture an enhanced value or concept of commitment that is incomprehensible to the positivistic and functionalist mechanisms of modern science. From such nurturing will develop a new capacity for wonder and indignation, capable of grounding a new, non-conformist destabilizing, and indeed rebellious theory and practice." Boaventura de Sousa Santos, "Beyond Abyssal Thinking: From Global Lines to Ecologies of Knowledge," *Eurozine* (18 February), 17, http://www.eurozine.com

CHAPTER TWENTY-FOUR

GLOBALIZATION AND EXTENDED ECO-FOOTPRINTS: NEO-COLONIALISM AND (UN)SUSTAINABILITY

WILLIAM E. REES

Abstract

The world is currently being driven by a model of world development and poverty alleviation based on the alleged efficiencies associated with economic globalization, open markets and expanding trade. The theory of "comparative advantage" argues that if all countries specialize in and export those goods that they can produce at a lower relative price than other countries and import those good for which their relative costs are higher, total global output and incomes per unit input will rise. This implies that trading nations invariably benefit from material exchange because their citizens can consume more goods of all kinds. In this chapter I argue that despite its apparent economic advantages, the current form of trade-oriented globalization is a socially vacuous and ecologically empty concept that threatens future of humankind. Growth-based global commerce under current WTO rules actually represents institutionalized *inefficiency* and hypocrisy masquerading as economic vision. To begin with, globalization fails on its own terms. The economic benefits of trade-induced growth go primarily to the already wealthy who don't need them while the 'collateral' social costs weigh most heavily on the poor. Second, by extending the ecological footprints of trading nations, expanding trade accelerates the loss of ecological integrity. Half or more of the ongoing ecological degradation in some countries can be attributed to economic production dedicated to satisfying foreign consumers. This degradation threatens future productive capacity and thus the security of both domestic and import-dependent populations. Trade-dependent countries put themselves at additional risk from global ecological change exacerbated by trade itself. In short, the continuous expansion of inter-regional trade creates a pattern of *potentially fatal interdependence*, an inherently unstable "entanglement of nations", that contributes significantly to the unsustainability of techno-industrial society. No individual nation can "fix" this situation on its own. However, mutual interdependence implies that individual and national self-interests have for the first time converged with humanity's collective interests. The world community thus has an unprecedented incentive to work cooperatively to create a new model of global sustainability for the common good.

Globalization and Conventional Trade Theory

The unchallenged objective of national economic policy virtually everywhere is to maximize gross domestic product and, therefore, per capita incomes (i.e. consumption). The implicit belief is that welfare is invariably associated with income growth, even in rich countries. Most tellingly, most governments, the United Nations, World Bank and other mainstream development organizations see economic growth as the only practical means to reduce poverty.

Economists argue that so-called "globalization" can contribute significantly to accelerating economic growth. Globalization theory says that we can maximise gross *world* product (GWP), and thus individual national "products", by taking advantage of the efficiency gains available through trade. This would require that each country specialize in those goods and services that it can produce most efficiently and trade for everything else. This, in turn, implies the effective dissolution of national boundaries and the horizontal integration of national economies into one. In a fully globalized economy, there are no formal legal-institutional barriers to trade; import and export tariffs are non existent.

Everybody wins! (Or do they?)

According to conventional or "Ricardian" trade theory, nations should contemplate trading when the relative prices of various domestic goods differ – that is, when the cost of inputs to produce a unit of good "A" is considerably more or less than the cost of inputs to produce a unit of good "B". More specifically, the "law of comparative advantage" states that a country should export those goods for which it has a lower relative price than other nations and use the foreign currency thus earned to import things for which its domestic relative price is higher.[1]

Suppose, for example, that Petroland can produce oil at a cost of $70 dollars a barrel and fine wine for $35 a bottle. This means, in effect, that to produce an extra barrel of oil, the country would have to give up two bottles of wine. For this example, we say that the "relative price" of oil in Petroland is 2 and the relative price of wine is ½. Now assume another country, Vinland, that produces oil for β36 and wine for β12. Vinlanders place a relatively higher value on oil than do Petrolanders. They have to give up 3 bottles of wine to produce a single barrel of oil, i.e., in Vinland, the relative price of oil is 3, that of wine 1/3.

Because Petroland and Vinland have different relative prices for oil and wine, each country stands to gain from trade. Petroland has a "comparative advantage" in producing oil because its relative production costs are lower. If Petroland redirects resources from producing two bottles of wine to producing an additional barrel of oil, it can sell that oil to Vineland for β36 and use the cash to buy three bottles of Vinland's fine wine (worth $105 in Petroland!). Petroland gains a bottle of wine in the exchange. Similarly, Vinland has "comparative advantage" in producing wine – its relative wine production costs (1/3) are less than Petroland's (1/2). By redirecting inputs from oil production to wine, and selling three bottles of wine to Petroland for $105, Vinlanders are able to purchase a barrel and a half of oil from Petroland (or a barrel of oil plus $35 worth of something else). Again, this more than compensates for the one barrel decline in domestic oil production.

It is easy to see why conventional trade theory is so seductive. Economic resources are scarce, factors of production expensive and everyone is looking for ways to reduce production costs. Globalization seemingly provide a means of achieving this from which all parties can only benefit. Free trade enables each country to specialize in those products that it produces most efficiently (i.e., at least cost), so output per unit input goes up in all trading countries. As GDP and income per capita increase, the citizens of trading nations can consume more of just about everything.

Let's return to our earlier example. Before trade Petroland might produce a single barrel of oil and two bottles of wine for a certain value of inputs; similarly Vinland produces a barrel of oil and three bottles of wine. Thus, pre-trade global output would be two barrels of oil and five bottles of wine. However, in a "globalized economy", with both economies working more efficiently, total production *for the same inputs* would rise to the equivalent of 2.5 barrels of oil and six bottles of wine. Note that even if Petroland produces both oil and wine more efficiently (fewer inputs per unit output) than Vinland,[2] it still stands to gain from trade. (It is pre-trade *relative* prices that determine whether a country stands to gain from trade.)

There is an additional bonus. With higher total output and competitive markets, average prices will likely fall below pre-trade prices so realizable demand increases. This positive feedback may increase GWP even further. As the rising incomes of trading nations enable their people to maximize the consumption of goods from all over the world, everyone seems to emerge as a winner!

If only things were that simple. The reality is that conventional trade theory is based on flawed assumptions and a simplistic model of the economy that is abstracted from both the society it supposedly serves and the ecosystems in which it is embedded. The purpose of this chapter is to make the case that conventional logic overstates the gains from trade, blinds analysts to the hidden social costs of structural change and accelerates ecosystems degradation. Moreover, by fostering excessive material interdependence, globalization further exacerbates the current unsustainable state of techno-industrial society.

Practical Flies in the Theoretical Ointment

The alleged productivity (efficiency) and welfare gains from global trade depend on several critical assumptions whose violation compromises the entire theory. For example, basic trade theory assumes that:

1. *Investment capital is essentially immobile.* For neoclassical trade theory to work its magic, *capital must stay home*. Ricardo believed that, with the opportunity to trade, capitalists would invest in those parts of their domestic economies in which their nations had a comparative advantage. This would increase GDP and foster local employment. However, in today's world of increasing capital mobility, capital flows are determined mainly by opportunities to maximize profit wherever those opportunities may lie. Capital knows no loyalties to people or place. Indeed, if capital is as freely mobile as goods, then trade and investment "will be governed by the labor theory of value rather (absolute advantage in terms of labor costs) than by comparative advantage" (Daly and Cobb 1994, 214; also, Daly 1999). Investment thus flows to countries with the cheapest labour and access to cheap resources (think China and India). Since labour is relatively immobile, employment declines elsewhere and some countries may lose previously important sectors of their

economies (think US rust-belt states). In short, so-called free trade is a double-edged sword that grows some economies while simplifying and shrinking others.

2. *Higher income (consumption) always increases personal well-being.* Economists tend to view human beings as self-interested utility maximizers with fixed preferences and unlimited material wants. Continuous economic growth is therefore necessary to enhance welfare in poor and rich countries alike. From this perspective, economically rational people will always strive to maximize their personal consumption regardless of any impact on other people or "the environment". This caricature assumes a *Homo oeconomicus* empty of emotion and unconstrained by non-material values. It reduces people to selfish, atomistic consuming machines devoid of personal relationships and abstracted from wider community (for an expansion of this critique, see Daly and Cobb 1994, chap. 4). As critically, the model ignores the evidence that beyond a certain income level, there is no correlation between rising incomes and either objective indicators of population health (e.g., longevity, post-operative survival [Siegel 2006]) or subjective expressions of felt well-being (e.g., the proportion of people reporting themselves as happy or very happy [Lane 2000]). I.e., the reality is that, in high-income countries, well-being is no longer tied to income growth *per se*. The presumed universal need for income growth therefore cannot be used to justify globalization .
3. *Growth in GWP will reduce or eliminate poverty.* Poverty alleviation remains a major rationale for trade-induced economic growth. In theory, if the economic pie continues to grow at the margin, then even those with the thinnest slices will eventually enjoy sufficiency. (It is rarely acknowledged that this theory provides a convenient excuse to avoid the politically difficult alternative of income redistribution.) No doubt economic growth has raised millions out of poverty in recent decades. However, despite the fact that GWP growth has exceeded population growth by a factor of two to three in recent decades, the absolute number of poor has never been greater. About 1.2 billion people, particularly in the impoverished parts of Africa, Asia and Latin America, still lack access to potable water and 2.6 billion have no sanitary or sewage facilities; close to a billion people live on less than a dollar per day and most of these are at least calorically deprived; almost half the world – over three billion people – live

> on less than $2.50 a day and most are nutrient deficient; over 26,000 children die every day from poverty (meaning hunger, water-borne and other preventable illnesses) (Shah 2009, World Bank 2008). Clearly, growth in GWP isn't working as advertised. These billions of people, ostensibly the beneficiaries of growth, would certainly gain from even modest income increases. However, most of the gains from growth are going to the already rich who don't really need them while the poor fall further behind – much of Africa is actually in retreat as incomes fall in absolute as well as relative terms.

It seems that the principle structural assumption of trade theory (capital immobility) has long since collapsed in the real world; the principle rationale for continuous growth (a continuous association with well-being) is betrayed by the data; and the principal moral objective of globalized trade (poverty alleviation) is not being achieved. The falsification of the major *explicit* assumptions and goals of globalization should, in itself, be enough to undermine the allure of mainstream trade theory. But economic models make two additional *implicit* assumptions that that are potentially even more fatal to the cult of globalization:

1. *The economy has no important connections to the environment.* Much neo-liberal economic theory, including trade theory, is inspired by a conceptual model that sees the economy as a self-perpetuating "circular flow of exchange value" floating disdainfully above real communities or the natural world. In this master model, value in the form of goods and services (defined as "national product") flows from firms to households and is exactly balanced by a counter-current flow of value in the form of factors of production – e.g., labour, finance capital – from households to firms (defined is "national income"). As Herman Daly points out, studying the economy in this way is similar to studying human physiology solely in terms of the circulatory system with no reference to the digestive tract and material throughput. It is impossible to understand the economy's relationship to or impact on ecosystems from the perspective of the circular flows model because the circular flow is conceived as "an isolated, self-renewing system with no inlets or outlets, no possible points of contact with anything outside itself" (Daly 1991, 196).

2. *There are no biophysical impediments to continuous economic growth*. It follows from the previous point that mainstream economists recognize few ecological constraints on the economy (Christensen1991). If significant environmental externalities (e.g., pollution) do emerge, economists assume that these can be "internalized" by politically determined pollution charges or taxes. This should induce waste treatment (to avoid paying the tax) and ensure that subsequent prices more closely reflect the true cost of production. Problem solved. Similarly, rising prices will signal incipient resource scarcity and thus encourage greater "factor productivity" (efficiency) or the substitution of some other resource. In the absence of significant ecological constraints, there are no limits to growth. From this perspective, the increasing interdependence of nations is a good thing and such questions as: "What happens if critical trade flows are disrupted by climate change, ecosystems collapse and subsequent geopolitical unrest?" never comes up.

The rest of this chapter focuses mainly on the largely unaccounted ecological and geopolitical impacts of globalization that flow from violations of these unspoken assumptions. My basic premise is that, far from floating free of nature, the economy is actually a fully embedded dependent subsystem of the ecosphere. The loss of vital natural capital and associated life support functions thus poses a threat to geopolitical stability on a finite planet. In this light, ecological sustainability is a *prerequisite* for economic sustainability (it is a *necessary* if not sufficient condition). Any form of globalization based on growth-based economic models that do not adequately represent the ecosystems that contain the economy exacerbates both the environmental and political risks already confronting global civilization.

The Ecological Costs of Globalization

The real world is obviously vastly more complex than the simple two-country model used to rationalize free trade among nations. One obvious complexification is that there are actually more than 190 independent nations than might potentially engage in trade. This ensures that numerous countries simultaneously try to exploit their assumed comparative advantages in the same products and, since each is vitally interested in increasing its market share, intense competition develops among exporting nations.[3] Transnational trading corporations naturally exploit this competition by playing exporters off against each other, thus driving down export prices. Of course, lower prices stimulate greater demand which producing countries rush to satisfy – they need the enhanced sales to maintain profitability at the lower prices. Matters are further complicated if national governments assist domestic firms through direct or indirect subsidies that prop up domestic production (i.e., increase resource consumption) and, by further lowering prices, distort global markets. (Subsidies to the fisheries and farm sectors are chronic problem areas.)

The dissipation of natural capital

This roiling boil of competitive relationships has serious adverse implications for ecosystem integrity and prospects for sustainable development. The downward pressure on prices has at least three direct negative ecological consequences. First, as noted, it increases demand and thus the direct consumption of resources.[4] Even in the case of renewable and replenishable resources (e.g., forests, fisheries, agricultural soils, ground-water, etc.), this can lead to the depletion of essential natural assets. Second, low prices reduce the producer surpluses (revenues minus production costs) that exporters might otherwise use to maintain their natural capital. This means that prospects for restoration are grim if critical stocks *are* seriously depleted. Third, the same low prices and desire for profit encourage firms to pollute. The more production costs that can be "externalized" onto both local ecosystems and the global commons, the greater the money profits to producer-exporters. World Trade Organization rules actually encourage developing nations to exploit their ecosystems in this way to subsidize the growth of their economies.

There is another ultimately maladaptive economic factor at work here. Basic resources – food, water, energy, etc. – are absolutely vital for survival. Nevertheless, the modern world places the primary resource sectors (agriculture, fisheries, forestry, water, mining, etc.) at the bottom of the economic pecking order well below manufacturing and not even visible from the high-end service sector. To most economists and politicians, "development" implies ascending the structural hierarchy from resource dependence to knowledge-based wealth creation,[5] oblivious to the facts that the latter can never be divorced from the former and that the foundation of any structure is ultimately more important than the penthouse.

Within this conceptual framing, prevailing terms of trade undervalue natural capital, particularly slowly- reproducing resource stocks.

Consequently, as Colin Clark revealed as early as 1973, even the *owners* of biological resources have an incentive to liquidate their assets if the return to capital from alternative forms of investment – e.g., manufacturing – exceeds the natural growth rate of their natural capital (and the latter may be as low as one or two percent). In short, it is economically rational for an individual or firm, motivated by the desire to maximize profits and responding to high discount rates, to obliterate privately-owned ecosystems (risking the extinction of local species populations) and invest the proceeds in something more profitable (Clark 1973).[6] The increasingly global and the sometimes overheated resource commodity markets fostered by globalization exacerbate this tendency.

Eco-footprints and trade in biocapacity

Globalization has clearly increased the flows of goods and services across national boundaries – trade as a percentage of GDP has more than doubled since 1970. However, while the world celebrates the contribution of trade to the increase in money wealth, it ignores the fact that expanding material flows (of both economic goods and wastes) also represents trade biological potential or biocapacity.[7]

We can monitor trade in biocapacity using ecological footprint analysis (EFA) (Wackernagel and Rees 1996; Rees 2006). The "ecological footprint" of a specified population is the area of productive ecosystems required by that population, on a continuous basis, to produce the resources that it consumes and to assimilate its wastes, *wherever on Earth those ecosystems may be located*. To facilitate comparison among nations, analysts usually express biocapacity in terms of standardized global hectares (gha) where each gha represents a hectare of global average productivity.

Many countries now depend largely on international trade and on the global commons for the biocapacity necessary to support prevailing levels of resource consumption and waste production (Princen 1997, Rees 2006). For example, most high-income western European countries are running substantial "ecological deficits" with the rest of the world: the Netherlands has a *per capita* eco-footprint (EF) of 4.0 gha, or almost four times its domestic biocapacity; similarly, the UK requires 3.3 times its domestic biocapacity to maintain current material standards; even France has a deficit equivalent to two thirds the area of its domestic productive ecosystem area. And there are even more extreme examples – the Japanese consume eight times the biocapacity of their home territory, much of it obtained through trade (EF data from WWF 2008).[8]

A few countries have apparent large ecological surpluses. These are only "apparent" because they are generally traded away or appropriated through natural flows (e.g., carbon assimilation) to cover the eco-deficits of other countries. (Global trade is a zero-sum game in biocapacity terms.) Canada, for example, has a large per capita eco-footprint of 7.1 gha but a small population and large land area. Canada's biocapacity is about 20 gha *per capita*, so Canadians need the equivalent of only 35 per cent of their domestic productivity to sustain themselves – the rest is available for trade.

And most of it is, in fact, traded away. For example, Kissinger and Rees (2009a) analysed the allocation of Canada's agricultural land-base between domestic and foreign consumers, focusing on the Canadian prairies which represent about 85 per cent of the country's farmland. On average, between 1989 to 2006, 62 per cent of crop production and more than 45 per cent of meat production by weight was exported. This corresponds to an average of 28.3 million of the total 55.1 million *actual* hectares, or about 51.4 per cent of the prairie agricultural land-base.[9]

The "exported" land was widely disbursed – about 37 per cent went to the US, 26 per cent to Asia and 11 per cent to Latin America. The rest went to the European Union and other countries (Figure 24-1). The total foreign food eco-footprint on Canada's prairies increased significantly over the period studied, reaching almost 34 million actual hectares, an area equivalent to the area of the United Kingdom, the Netherlands and Denmark combined (Kissinger and Rees 2009a).[10] No doubt, similar data could be developed for Canada's fisheries, forest and carbon sink capacities.

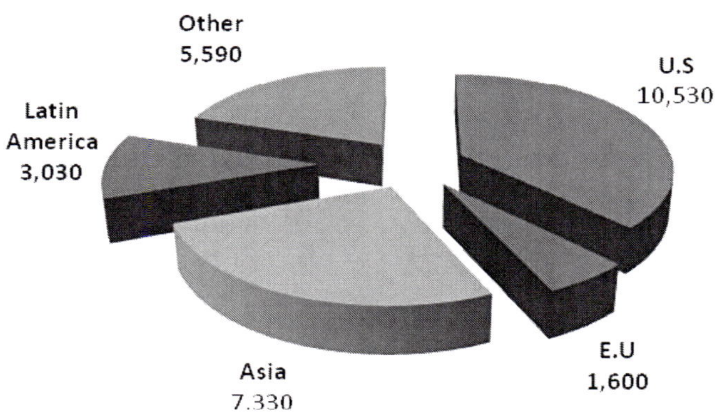

Figure 24-1. The average area of agricultural land (1000s ha) in the Canadian prairies devoted to foreign consumers. Source: Kissinger and Rees (2009a)

Farming takes a heavy toll on the land including the loss and fragmentation of natural habitat and its consequent impacts on biodiversity. In the case of the Canadian prairies, 99 per cent of the tall grass prairie, 80 per cent of fescue prairie, and 67 per cent of mixed-grass prairie habitat has been lost to agricultural development and the remaining natural prairie is broken into small islands surrounded by non-native vegetation. Approximately 70 per cent of wetlands have been drained. These impacts have seriously reduced wildlife and waterfowl populations. Meanwhile, an estimated 36–50 per cent of the organic matter and natural nutrient content of Canada's prairie soils have been eroded, oxidized or otherwise dissipated by agriculture activities. Such "primary" degradation has led to increases applications of fertilizers which further contributes to soil degradation, water contamination, GHG emissions and the continuing erosion of the natural prairie ecosystem (Kissinger and Rees 2009a). Since over half of the agricultural activity and land-base of the Canadian prairies is dedicated to export markets, it seems reasonable to attribute half of the biodiversity loss, soil degradation and general depreciation of ecosystem integrity to export markets and offshore consumption.

We can examine the same issue from the perspective of an importing country using the United States (US) as an example (Kissinger and Rees 2009b). The US is, of course, a major food and fibre producer and still a net exporter. Yet, as globalization proceeds, US imports, and therefore its dependence on other countries, are steadily increasing (Figure 24-2). From 1995-98 to 2004-05 the US imported an average of 122,768,000 m^3 of wood products annually and imports as a fraction of total wood consumption increased from 23 per cent to 30 per cent. During the same period, the US imported 1,338,500 tonnes of beef; 53,170 tonnes of lamb; and 2,078,500 live cattle per year and meat imports as a fraction of consumption increased from 8 per cent to 14 per cent. Imports of other agricultural commodities averaged 56,550,300 tonnes annually and total import weight increased by factor of 1.5. While total crop commodity imports were stable at 12 per cent of market share over the study period (but up from only 8 per cent in 1980s), the market share of some imported crops increased significantly: fruits from 28 per cent of consumption in 1995 to 41 per cent in 2005; legumes from 5 per cent in 1995 to 12 per cent in 2005; stimulants from 23 per cent in 1995 to 30 per cent in 2005; and agricultural fibres from 43 per cent in 1995 to 49 per cent in 2005 (Kissinger and Rees 2009b).

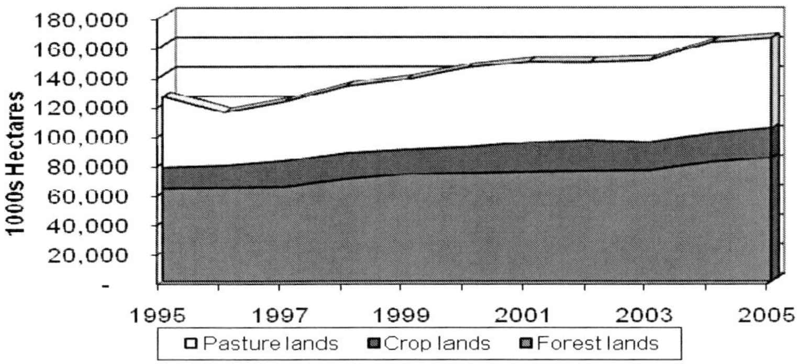

Figure 24-2. Biocapacity "imported" by US consumers, 1995–2005
Source: Kissinger and Rees (2009b)

Clearly, a large quantum of biocapacity is involved in producing the import share of US consumption. Using productivity data from the various exporting countries, Kissinger and Rees (2009b) estimated that, on average, US consumption of imported food-crops, meat and forest products effectively "appropriated" 141×10^6 actual hectares of productive land annually during the 1995 to 2009 study period. This is equivalent to 18 per cent of the area of the lower forty-eight states or approximately the area of Germany, Italy, Spain, and the United Kingdom combined. Annual US demand for extra-territorial biocapacity embodied in food and forest products increased from 120×10^6 hectares to over 160×10^6 hectares during the study period (Kissinger and Rees 2009b; Figure 24-2). Canada is the US largest supplier, but dozens of countries, mostly in the Americas and Asia, but also Europe, Australia and New Zealand, export portions of their biocapacities to satisfy US import demand (for full country listings by individual product, see Kissinger and Rees 2009b).

Discussion and Conclusions: Can the World Afford "Free" Trade?

There is little doubt that globalization and international trade have bettered the lives of millions of people and contribute significantly to the growth in GWP. However, neoliberal economics and trade theory are essentially divorced from the social and ecological contexts within which the economy operates. In particular, existing models pay little attention to the distributive effects of trade-induced growth and do not account for ecological externalities.

Failing on its own terms

As noted earlier, *de facto* capital mobility (a violation of classical trade theory) ensures that a disproportionate share of investment flows to countries with an absolute advantage in low-cost labour and access to cheap resources. The former is a natural "asset" of low-income developing countries with young, rapidly growing populations, and the latter is assured by price competition in the global marketplace. Thus while trade-oriented investment may provide low-paying jobs in parts of the third world, the flight of capital often abandons existing infrastructure, destroys established communities, and uproots countless families in previously industrialized countries, including North America and Europe. (How efficient is that?) Moreover, the narrow specialization demanded by globalization ultimately leaves restructured economies on both sides of the equation increasingly vulnerable to the vagaries of the global marketplace[11] and ecological change (more on this below).

In addition, the conceptual disconnect between economic trade models and the ecosphere positively encourages the over-exploitation of both vital resource stocks and the overfilling of waste sinks. "The environment" is often willingly traded off by development agencies and national governments so as not to inhibit investment and rapid economic growth. Even if a country has adequate anti-pollution legislation, lax enforcement takes its toll. China, the world's preferred manufacturer of consumer goods, has among the most polluted air, waters and soils on Earth (Figure 24-3).

Figure 24-3. Sunrise in Suzhou: much of China's eco-degradation represents the externalized eco-footprints of wealthy consumers halfway around the world

In all such cases, the costs to health and well-being go largely unaccounted and the value of long-term losses to biodiversity and other indices of ecological integrity are inestimable. What is obvious is that these "external" (outside the market) costs are born disproportionately by innocent parties both within and outside the producer/exporter countries (e.g., consider the impacts of trans-boundary water or air pollution, including CO_2 emissions). Moreover, because such negative externalities represent real costs, the goods and services inflicting them enter the marketplace at prices below their true cost of production. Such under-pricing leads to over-consumption, inefficient resource use and pollution, all classic symptoms of market failure.

Regrettably there is little incentive to correct the failure. Developing countries, desperate for investment and growth, are reluctant to act unilaterally for fear that corrective pollution charges would price them out of global markets. Meanwhile, wealthy consumers, the principal beneficiaries of the underpriced goods, are disinclined, also out of self-interest, to press their own governments for necessary reforms to the trading system.[12] We may brag about being a market-based society but happily tolerate what is in effect, ecological and social dumping. This undermines the fundamental assumption of full-cost pricing upon which efficient functioning of the market depends. From this perspective, globalization under prevailing rules and terms of trade actually represents institutionalized *inefficiency* – exploitation and hypocrisy masquerading as economic vision. Meanwhile, resource depletion and ecosystem degradation accelerate, making mockery of claims to that we are serious about sustainable development.

In these circumstances, it is hardly surprising that the global growth machine is not meeting its oft-stated primary goal of eliminating poverty. The structure of the economy and prevailing terms of trade ensure that the greatest share of national and global income growth from trade flows to the owners of capital and other upper income groups who need it the least – indeed, who may not benefit at all. In 1966 the world's wealthiest countries with one billion people – 15 per cent of the world population – accounted for 76 per cent of gross world product ($36.6 trillion out of $48.2 trillion). The richest 20 per cent of the world's population take home 76.6 per cent of the world's income; the poorest 20 per cent subsist on 1.5 – and the gap is increasing both within countries and between rich and poor countries (Shah 2009, UNDP 2007). Clearly, "the time is ripe for our measurement system to *shift emphasis from measuring economic production to measuring people's well-being*. And measures of well-being should be put in a context of sustainability" (Stiglitz, et al., p.12).

Global security and the increasing "entanglement of nations"

> "The natural effect of trade is to bring about peace. Two nations which trade together render themselves reciprocally dependent." Montesquieu (1758, cited in Martin *et al.* 2007).

Globalization and trade help stimulate both population and economic growth but also blind the seeming beneficiaries to potentially risky long-term consequences. Most significantly, the exchange of biocapacity enables all trading partners unwittingly to exceed their domestic carrying capacities, even as they draw down resource stocks elsewhere – high income European countries and Japan are good examples. This potentially dangerous situation is completely invisible to standard abstract economic trade models.

The case of the Canadian prairies shows that millions of people, in arguably overpopulated countries around the world, have become at least partially dependent for their daily bread on the output of Canadian prairie farmland. However, in just a century, industrial high-input agriculture has displaced most of the native prairie ecosystem, eliminated whole species populations and dissipated up to half the soil upon which the same export-oriented agriculture depends. Soils that only a few decades ago produced high yields of grain of outstanding quality without artificial inputs now must be fertilized to maintain both quality and quantity. Excess fertilizer, together with pesticides and mechanization, accelerate the degradation of the prairie agro-ecosystem itself. Those millions of foreign consumers (to say nothing of Canadians) are dependent on a form of industrial agriculture that is inherently unsustainable.

Even the US, the world's greatest agricultural powerhouse, is becoming more dependent on possibly fragile "offshore" biocapacity as the country's food footprint extends into terrestrial ecosystems far beyond its borders. US consumption now has a significant but invisible (from the US) impact on the long-term integrity of productive ecosystems in many exporting countries, including Canada. As similar scenarios unfold in different contexts around the world, they are driving the depletion of many kinds of critical natural capital, from fish and tropical forests to groundwater and petroleum. The end result is that all nations are exceeding *global* carrying capacity simultaneously (Rees 2006; WWF 2008) – and the dynamics of the conundrum ensure that there is no fall-back position.

As the world's nations become increasingly specialized and trade-dependent, some have learned the hard way that they are also increasingly vulnerable to supply or price shocks. This should serve as a warning shot across their free-trading bows. Globalization fosters an irreversible entanglement of nations that could be disastrous for everyone if underlying assumptions of ecological and geopolitical stability are seriously violated (particularly in a world is at or beyond carrying capacity). Trade is generally considered to have a pacifying effect between mutually dependent partners (Martin *et al.* 2007) but this assumes ecological stability and the absence of external shocks. Consider the effect on the national security of import-dependent nations if human-induced climate change should:

- Directly undermine the productivity of exporting nations and thus eliminate major sources of supply (e.g., drought may become chronic or permanent; mountain glacier or ground water sources essential for irrigation may dry up and disappear; soil and landscape degradation may accelerate).
- Induces sea-level rise. permanently flooding coastal plains (and much of the world's prime agricultural lands).
- Displaces millions of people from their homes and generates mass migration, local strife and geopolitical chaos.

People who have been too long smugly insulated by their wealth from unfolding global realities may think such scenarios unlikely, in which case they might consider *The Age of Consequences...*, the 2007 report of Washington's Center for Strategic and International Studies (SCIS). According to CSIS:

> "In the case of *severe climate change*, corresponding to an average increase in global temperature of 2.6°C by 2040 [now considered by some analysts to be unavoidable], massive nonlinear events in the global environment give rise to massive nonlinear societal events. ...nations around the world will be overwhelmed by the scale of change and pernicious challenges, such as pandemic disease. The internal cohesion of nations will be under great stress, ...as a result of a dramatic rise in migration and changes in agricultural patterns and water availability. The flooding of coastal communities around the world, ...has the potential to challenge regional and even national identities. Armed conflict between nations over resources, ...is likely and nuclear war is possible. The social consequences range from increased religious fervor to outright chaos" (CSIS 2007, p.7).

In the press release announcing the study, contributing author Leon Fuerth remarked that wealthy countries could "go through a 30 year process of kicking people away from the lifeboat as the world's poorest face the worst environmental consequences". Prospects for sustainability with equity and the pacifying effect of trade would clearly dissolve in the event of global crisis.

The latest threat: Colonization by legal means

Food is the resource most likely to raise international tensions. Concerns about future food supplies were heightened in 2008 when various pressures (including the diversion of corn and other grain crops from food to bio-ethanol production) doubled the prices of various staple foods over those of only a year or two earlier. Some governments implemented food export bans to maintain adequate domestic supplies at affordable prices. These events rattled food importers into realization that in future they might not be able to depend on world food markets and trade for essential staples. The shock helped accelerate a new wave of mainly government-to-government land deals in which "countries that export capital but import food are outsourcing farm production to countries that need capital but have land to spare" (*Economist* 2009).

Out-sourcing in this context means leasing or the outright purchase of other countries' "surplus" arable land, a trend that is potentially explosive. For example, while Saudi investors spend $100m raising wheat, barley and rice on land leased to them by the Ethiopian government for export back home, the World Food Program is spending even more ($116m) to provide "…230,000 tonnes of food aid between 2007 and 2011 to the 4.6m Ethiopians it thinks are threatened by hunger and malnutrition" (*Economist* 2009). One such land transfer has already felled the government that brokered it. Madagascar's president was overthrown by popular uprising in March 2009, just months after South Korea's Daewoo Logistics announced (in November 2008) that it had signed a 99 year lease on 3.2 million acres of land in Madagascar, to produce corn and palm oil for food and bio-fuel production in Korea.

While there has long been limited investment in foreign farmland, the scale of the latest land rush is unprecedented. For example, South Korea has signed deals for 690,000 hectares and the United Arab Emirates for 400,000 hectares in the Sudan (a country apparently willing to reserve a fifth of its arable lands for Arab countries). China has acquired the rights to 2.8m hectares of Congo to grow palm oil for biofuel and is negotiating for two million ha in Zambia (where it already has extensive land investments) (*Economist* 2009). The United Nations reports that the acquisition of farmland in poor countries by rich countries and international corporations is accelerating, with an area half the size of Europe's farmland targeted in the last six months [to mid-2009] – at least 30 million hectares of food-land is about to change hands (*Guardian* 2009).

Transferring biocapacity

In effect, the world is witnessing the legal transfer of biocapacity from poorer countries to richer nations that have exceeded their domestic carrying capacities and cannot produce enough food to feed themselves. Both parties to may well benefit in the short term and if population growth and global change were of no concern, one might be sanguine about such exchanges. However, in a world expecting an additional two billion people at the table in coming decades, where the demand for food may well double even as climate change and peak oil threaten production, they are potentially disastrous. Access to foreign biocapacity facilitates further growth and deepens the dependence of importing nations while potentially alienating the population of the "surplus" country (who might well soon need the land and food themselves).

The growing competition for surplus biocapacity throws a darkening shadow over the world's largely uncritical commitment to globalization. As resources become scarcer in an inequitable world, globalization might well be interpreted as a process by which rich countries use their paper wealth to acquire the real wealth of developing and resource-rich countries. At first, trade was enough, but more recently outright acquisition has begun to emerge as a significant alternative. In either case, it is clear that rich, powerful nations can now achieve through the instruments of globalization what used to require territorial occupation.
The emerging picture raises several questions that should have been asked long ago but that at least may serve to temper enthusiasm for the further "entanglement of nations":

- What are the economic, ecological and moral implications of purposefully creating irreversibly dependent material relationships among nations if those relationships are unlikely to sustainable in the face of accelerating global change and resource scarcity?
- Is it wise for any nation to develop such that its future well-being and security depend significantly on vulnerable imports or extra-territorial biocapacity located half a planet away?
- What risks does a nation assume by committing substantial portions of its limited land-base to foreign consumers?
- At what point are the gains from globalization and trade cancelled by the loss of the economic diversity, resilience and security associated with greater self-reliance?

Seriously addressing these questions may show that for all the theoretical benefits that might accrue to globalization in a stable world, the excessive entanglement among nations may well be maladaptive in a world of tightening ecological constraints.

Epilogue: Toward mutual sustainability

We are at the point where no region or country can achieve sustainability on its own. The UK or Canada – any country – could become an exemplar of sustainability, even to the point of reducing its per capita eco-footprints to a globally equitable 2 gha, but to no avail. It would be swept away by global change if the rest of the world continued on a business as usual path.

Regrettably, we share a finite planet effectively governed by obsolete economic concepts that are abstracted from biophysical and social reality. As a result, the entire world is entrained in a form of globalization that has created a potentially fatal interdependence of nations and undermines the sustainability of techno-industrial society.

This truly global crisis means that for the first time in the evolutionary history of *H. sapiens*, short-term individual and national (i.e., "tribal") self-interest has all but converged with humanity's long-term collective interests. The world community thus has an unprecedented incentive to work cooperatively to create a new model of global sustainability for the common good. The question is whether humans have the intelligence and (political) will-power to abandon deeply entrenched patterns of individual and political behaviour. Will we allow short-term interests and competitive relationships give way to long-term concerns and international cooperation in order to salvage a future for civilization? There can be no greater motivation than enlightened self-interest. If this fails, then… (I'll let the reader contemplate the alternatives.)

Acknowledgements

I would like to thank Meidad Kissinger, whose doctoral studies provided the data on Canada and the US. This research was supported by grant #410-2004-0786 from the Social Sciences and Humanities Research Council of Canada.

References

Christensen, Paul. 1991. Driving Forces, Increasing Returns, and Ecological Sustainability. Chap. 6 in Robert Costanza (ed.), *Ecological Economics: The Science and Management of Sustainability*. New York: Columbia University Press, 75–87

Clark, C.W. 1973. The Economics of Overexploitation. *Science* 181 (4100): 630–634

CSIS (2007). *The Age of Consequences: The Foreign Policy and National Security Implications of Climate Change*. Washington: Center for Strategic and International Studies. Available at www.csis.org/media/csis/pubs/071105_ageofconsequences.pdf

Daly, H.E. 1991. The Circular Flow of Exchange Value and the Linear Throughput of Matter-Energy: A Case of Misplaced Concreteness. Chap. 10 in H. E. Daly, *Steady-State Economics* (2nd ed.) Washington: Island Press, 195–210

—. 1999. *Five Policy Recommendations for a Sustainable Economy*. Sophie Prize acceptance lecture, Oslo, Norway (15 June 1999). Available at www.sophieprize.org/Articles/49.html

Daly, H.E. and J.B. Cobb, jr. 1994. *For the Common Good*. Boston: Beacon Press

Economist, The. 2009. Outsourcing's third wave. Print edition, 21 May 2009. Available at www.economist.com/world/international /displayStory.cfm?story_id=13692889

Guardian, The. 2009. Fears for the world's poor countries as the rich grab land. (4 July 2009) Available at www.ethiomedia.com/adroit/2555.html

Kissinger, M. and W.E. Rees. 2009a. Footprints on the prairies: Degradation and sustainability of Canadian agricultural land in a globalizing world. *Ecological Economics* 68 (2009) 2309–2315

Kissinger, M. and W.E. Rees. 2009b. Importing Terrestrial Biocapacity: The US Case and Global Implications. *Land Use Policy* (in press), doi:10.1016/j.landusepol.2009.07.014

Lane, R. 2000. The Loss of Happiness in Market Democracies. New Haven: Yale University Press

Martin, P., T. Mayer, M. Thoenig. 2007. Does globalisation pacify international relations? *Vox*. Available on line at www.voxeu.org/index.php?q=node/354

Princen, T. 1997. The shading and distancing of commerce: When internalization is not enough. *Ecological Economics* 20 (3): 235–253

Ricardo, D. 2005. *Principles of Political Economy and Taxation*. Vol. 1 of *The Works and Correspondence of David Ricardo*, Piero Sraffa (ed.) with the Collaboration of M.H. Dobb (Indianapolis: Liberty Fund,), 11 vols

Rees, W.E. 2006. Ecological Footprints and Bio-Capacity: Essential Elements in Sustainability Assessment. Chapter 9 in Jo Dewulf and Herman Van Langenhove, eds. *Renewables-Based Technology: Sustainability Assessment*, pp. 143–158. Chichester, UK: John Wiley and Sons

Shah, A. 2008. *Poverty Facts and Stats*. GlobalIssues.org (Last updated: Wednesday, September 03, 2008) Available at www.globalissues.org/article/26/poverty-facts-and-stats

Siegel, C. 2006. *The End of Economic Growth*. Berkeley: Preservation Institute. Available at www.preservenet.com/endgrowth/EndGrowth.html

Stiglitz, J., A. Sen, J-P. Fitoussi. 2009. Report by the Commission on the Measurement of Economic Performance and Social Progress. Available at www.stiglitz-sen-fitoussi.fr/documents/rapport_anglais.pdf

UNDP. 2007. *Human Development Report 2007*. United Nations Development Program. New York: United Nations.

World Bank. 2008. World Bank Development Indicators 2008. Available at http://web.worldbank.org/WBSITE/EXTERNAL/DATASTATISTICS/0,,contentMDK:21725423~pagePK:64133150~piPK:64133175~theSitePK:239419,00.html

WWF. 2008. Living Planet Report 2008. Gland, Switzerland: World Wide Fund for Nature.

Notes

[1] The law of comparative advantage was originally formulated by David Ricardo (2005, orig. 1817).

[2] In this case, Petroland would have an *absolute* advantage over Vinland in producing both oil and wine.

[3] E.g. Australia, Russia, Indonesia and other major coal producers compete for markets in such major coal importing countries as Japan, South Korea, Taiwan and the UK; the US, Canada, the EU, Australia and Argentina and others go head to head to satisfy demand for wheat in Egypt, Brazil, the EU, Japan, Algeria, etc.

[4] Most analysts see this as a good thing. After all, lower prices and increased consumption are the explicit goals of globalization. Consumers everywhere should benefit.

[5] Some observers are deluded that this ascent means the economy is decoupling from nature.

[6] I.e., contrary to conventional wisdom, the privatization of nature is no guarantee of conservation.

[7] Biocapacity is related to carrying capacity. Carrying capacity is the human population that could be supported by a given geographic area and is thusdependent on the subject area's biocapacity.

[8] Note that there are only about 2 gha of biocapacity per capita on Earth (WWF 2008). This represents each person's equitable share of global biological output.

[9] The purposes of this study were better served using actual hectares rather than gha.

[10] This figure includes a 2.1 million hectare "carbon footprint", the ecosystem area required to sequester the carbon dioxide emitted producing products for export.

[11] For example., Alberta, Canada, went from a seven-billion dollar provincial budget surplus to an eight billion dollar deficit with the collapse of world petroleum prices and the withdrawal of investment from its oil-sands in 2008-09.

[12] This assumes, of course, that they are even aware of the damage most of which is occurring at great distance.

APPENDIX

Hanfried Blume joined the Global Ecological Integrity Group in 2006 when he attended his first GEIG meeting on Samos. My friendship with him goes back to 1985, when he translated Christopher Stone's legendary "Should Trees Have Standing?" into German. I was instantly captured by his warm, engaging personality, his modesty and sharp, investigative intellect. No one personified the ideals of personal and ecological integrity more than Hanfried. An environmental law practitioner and lecturer, a writer, translator, philosopher, poet, artist and farmer/gardener, he brought many talents to his course as an eco-activist. An early champion of the eco-integrity approach to law, he was among the founders of the Institute of Environmental Law (*Institut für Umweltrecht*) in 1987, the first of its kind in Germany. He famously represented the seals of the North Sea in a case against the German Government (*"Robbenklage"*). He argued that the anthropocentric character of federal environmental laws completely failed to protect ecosystems, animals and plants, in this case, maritime life of the North Sea. This landmark case had a major impact on public awareness and helped paving the way to various legal and constitutional reforms.

A few weeks before his sudden death, he gave a paper "Ecology and Romantic Thought" to the 2009 GEIG meeting in Florence. For its insights into the history of German and American ecological thinking, but equally for its personal, one might say "romantic" tone, the paper is included here.

For those unfamiliar with the German 19th century romanticism (*'Romantik'*) – in arts, music, science, philosophy and political theory –, it was the protest movement against utilitarianism and rationalism of the late 18th century. These were among the philosophical strains, that brought about the global ecological crisis. The modern ecological movement, by contrast, draws from the entire experience of human existence, i.e. rational thought *and* emotions, rights *and* responsibilities, respect for the integrity of human beings *and* the natural environment, and so on. The romantic tradition – from 'romancelike' – nurtures an adventurous, holistic attitude that embraces all aspects of live. Exactly this may be needed to overcome the narrowness and shallowness of current environmental policy, law and governance. Far from being escapists, many romantic thinkers were highly political and continue to be relevant for today. This is one message of Hanfried Blume's paper.

The other message is very personal, especially now after Hanfried's death. The integrity we seek and respect in ecological systems, must also be sought in ourselves. A person of integrity is someone who is able to 'integrate' feelings, thoughts and actions in a coherent and ethical manner. Through his paper Hanfried reveals himself as one such person. When he speaks of the difference between human artifacts (recordings of singing birds, plastic trees) and natural things he evokes integrity that only the natural has. When he positions the romantic worldview as opposed to the dominant Cartesian view of nature, he insists on the validity of arts and poetry for understanding nature. And when he shows the difficulty that Germans of his generation had to embrace such romantic sentiments, he again reveals his own integrity. Germany's Green movement has, for a long time, struggled to accept such diverse philosophers as Kant, Hegel, Marx, Heidegger, Marcuse, Bahro or writers like Novalis, Goethe, Hesse and American counterparts such as Thoreau, Waldo Emerson, Leopold, Bateson or Snyder as *all* belonging to their ideological ancestry. Yet, they are, and people who, like Hanfried, insist on such ancestry were often dismissed as 'romantic' or, worse, as idealistic radicals (*Fundamentalisten*).

The history of the global ecological movement has yet to be written. When it eventually happens, we may well see how rich the Western tradition of ecological thought has, in fact, been. When we leave the - rather outdated - distinction between (progressive) left and (conservative) right behind, we may detect how powerful political ecology really is. It can penetrate and transform the entire political spectrum. In this way we can be confident that humanity may have a future.

Hanfried Blume was one of those who saw this possibility and lived for it. He wrote this paper to remind us of this precious part of Western thought. His life is a reminder of the incredible strength that lies in the human spirit.

—Klaus Bosselmann

ECOLOGY AND ROMANTIC THOUGHT1

HANFRIED BLUME

I think I am a romantic, or let me say: neo-romantic. And I think some of us here are, too.

This isn't so much a confession but having read and thought a little about this time in history it's a statement: yes this seems to be so – the similarities are strong.

And I am quite sure that what we call engagement in ecology, in all the questions of ecology have a good deal of that romantic impregnation.

If we look at the movement of environmental consciousness – our brothers and sisters in spirit are those romantic thinkers of the time of the early 19th century. Yes!

Look for instance at one of the key texts of the environmental movement – the wonderful and powerfully inspired book "The Silent Spring" of Rachel Carson written in the 1960s: What is the motivation behind such an approach? Why being concerned in the Silence in the gardens and groves and in the meadows when the birds of spring are not there anymore? – We'll miss the birds' songs; this wonderful unfolding vulnerable beauty of the singing bird which enchants. A place….

Without this, the world is a different place.

Granted, we do have recordings of the singing birds – one could play this in Dolby stereo surround sound in those places: this sounds wonderful, it is even better than the real bird's song, we could take the best bird singers (there are good and bad singers[2]! We have a nightingale in our farm garden[3] it is a nightingale "cock" as we call the male, the "hens" do not sing at all, and our singer is "mediocre" when I analyze the modulations, "worse" than the average cocks performance. So let us replace this with a CD recording of an Italian nightingale super singer?! (I have a digital recorder in my suitcase – to record a nightingale in Italy and to bring this sound back over the Alps seems as a truly romantic project!)

I remember the famous essay "Ways not to think about plastic Trees" by Lawrence H. Tribe[4] – how funny it was to discuss the difference between a plastic tree and a real one. For a "drive by" in a convertible car on some boulevard with its plastic palm trees, there is almost no difference. These pretty durable, washable wonderfully man-made low cost tree factory products. But: It also reminds us of "Brave new world": "There is something hidden behind." (Nietzsche's strange dark guest at the door).

Comparing plastic and natural tress it is not quite the same: The old rumpled imperfect Trees feel different somehow. And it is a bit spooky: there seems to be a kind of giant "shadow" as Carl Gustav Jung[5] would have put it: „und man sieht nur die im Lichte, die im Dunkeln sieht man nicht" (and you see only those in the light, those in the dark you will not see). It is like taking pills against a depression: Everything appears quite fine. – But as I said there is something hidden "behind" – and one is afraid that this ugly side is the real one!

And strangely enough – the romantics also wanted to beautify the world, to "enchant" it with a magic wand, with art and poetry (*Zauberstab*) (Novalis). And they were not immune from trying to make the sky bluer, the rose a little bit redder to be satisfied. And we say: a rose is a rose is a rose is a rose because we went through phenomenology, existential philosophy, Buddhism and through Gertrude Stein.

There is this strong New Romanticism which came to life in North America in the 20th century. Think of the poets of the West coast and what William Everson called the "archetype West".

My thesis is that the romantic tradition which originally was a European phenomenon waved over to North America in two big waves, survived there and came back to Europe in the sixties. And this was extremely helpful to develop our environmental movement.

So thank you, America, you were a good biotope to let the best of the European soul survive, one can say "it wintered" on the West Coast, in California, thank god.

Our terrible European 20th century probably would have killed this delicate vulnerable seedling. A kind favour of history waved it back over the big pond and it had a chance to grow and to blossom again.

You know, America and especially the American West coast kept the flame alive.

In Germany we had Adorno's verdict that after Auschwitz one could not write poems anymore – and much less nature poems, of course. And words like Bertold Brecht's poem "what are these times alike when a conversation about trees is almost a crime – Because it includes being silent about all the atrocities that happen?"[6] These words have dealt a hammer blow to our collective consciousness, and in my understanding this was true not only for Germany.

In my generation "friends of nature", folksong lovers, hikers were suspicious – and I think you know why: This all seemed to be contaminated, it had also been a big thing in our "twelve years"[7]. The Nazis had used all this in their propaganda, they have spoiled the nature movement. And this meant to us: "Hands off, strictly!"

Well, I was at the university from 1967 to 74. Klaus Bosselmann is a contemporary of mine, he knows what I am talking about, and we saw what that "campus revolution" was all about then. And we were quite radical in these years. For example, we were strictly left: Marx, Engels, Lenin etc., but also Marcuse Adorno, Levi-Strauss were daily reading matter. On that side we were sure that this was antifascism: these were the deadly enemies of Hitler, and this seemed to be enough for a while.[8]

I remember how astonished we were when the first signs of environmental consciousness came from America. I remember the phrase "to be pollution minded" and we were mocking about it all the time.

And now this strange new "romanticism" in California – they were reading Hermann Hesse! Millions of readers! Siddharta, Der Steppenwolf, Die Morgenlandfahrt! Hesse was "out" in the German speaking countries – his books they were "stale, sedate, out". And his enthusiasm for India was "escapism" etc… For us, he was one of those "*Jugendbewegten*[9]" – but he also was an emigrant, he was a pacifist, he was a moral person, and full of integrity: that's also true.

When we heard about Rachel Carson's "(The) Silent Spring" our first reaction was: "Oh, that's a clever move, to criticize capitalism from that side!" All those worries about the environmental crisis – just a new aspect of anti-capitalism!

Well that was not the whole story, as we know. That approach is deeper, more profound, and more "radical" in the sense of radix, the roots: to uncover the roots, to ask fundamental questions.

So it is good to go back to the roots of that romantic civilization critique which began at the beginning of modern history, i.e. the time of the French Revolution, around 1800.

Let us first look at the historic era around 1800. At this time, Romanticism was seen as a modern revolutionary force which revolted against the "classical" heritage of the 18th century. Goethe always used this term ("*Romantik*", "*romantisch*") in that sense, with a pejorative undertone: non-classical, not serious, a little bit crazy, "drivel" ("*Spinnerei*"), however, he was one of the main sources for a kind of a generation conflict. In fact, he was the main influence to the new wild generation of poets and philosophers who lived in a commune in Jena, less than 30 miles from his hometown Weimar. I am speaking about the intellectual scene of that time. Look at Goethe's poetry, his sensitivity towards nature. There are countless quotes that became household chores in German culture. One can say that nature meant to Goethe what history meant to Hegel: it was the "category" – the "lens" - through which he perceived reality.

And in our context the most powerful change of epochs was the new *Naturphilosophie* which revolutionized the conscience of the world, and the being-in-the-world (Meister Eckart's famous word). The 18th century had been dominated by the French rationalists and by Newton's world mechanics. Descartes/Newton ruled the intellectual Europe. They were modern. They represented the age of enlightenment then. The ruling theory was a mechanistic concept of nature, instrumental for man's purposes, and the world was seen as a machine, quite complex but obeying the universal laws of physics and logically/rationally completely explainable.[10]

Nature and animals and plants are totally marginal, no subject/theme of real interest. The only thing that made a difference was the human mind – a "thing that thinks" (res cogitans). – And a thing that can think about god.

This construct had given way to the mechanistic materialism of the French encyclopedists, which still dominates in the various fields of science as the scientific world view.[11]

This concept, the Romanticist revolted against – the world view of the scientist, of the engineer, and of the finance investor –, this engineer's attitude of feasibility is called "western thinking". It changed the world totally, until today as we know. This concept is so strong; it is mainstream, and a model for success, which made the Western countries so rich and so powerful. And it seems to be nearly self-evident, common sense and very convincing, of course! It is even a way of life that we do not even recognize it as a special attitude – there are parallels to "Freedom and Democracy". It is the strongest element in the world, and it rules in all countries, from China to Latvia, From Finland to Patagonia. – Maybe some Amazonas tribes didn't join this world wide consensus yet.

One is afraid that what was called "Eastern Thinking" – Japan, China, Tibet and also ancient India – almost totally withdrew into monasteries and places like Kyoto (the Kyoto School of Philosophy, the "philosophy of the place" very interesting *Philosophie des Orts* which has very strong links to Heidegger). But this kind of "Eastern thought" also became a strong undercurrent in the Western Culture as we know, namely in the US. There was no "Eastern Thought" in the era of communism! Their attitude towards nature was super-materialistic, super mechanistic, super pro industrialism. Stalin spoke about the necessity of the "class war against nature" and we remember Lenin's word "communism is Soviet power plus electrification". Industrialization would bring prosperity and progress. But we also remember that Karl Marx as a young man has written his doctor's thesis about Schelling's *Naturphilosphie* –and this was, in our context, *the* turning point in the history of philosophy. Marx' early writings were passed over in silence in the Soviet Union, but they were not by the New Left in the West. Look at Marcuse's "One Dimensional Man" – how he dealt with Marx' term "alienation" and turned it to modern man's/woman's alienation from Nature. But he wrote this in America – in Germany he would not have been able to think like this after the war I am afraid.

And there was a time when there was another force that felt like an opponent or even as a rival to that "Western thought" – something that called itself "German Mind/Spirit" ("*Deutscher Geist*"). It saw itself as an alternative way of being with strong roots in the Romantic Era – a phenomenon which perished in the 20th century almost totally.

It is quite interesting why the romantic era was so strong in Germany: Here the romantic thinkers and poets were a coherent group which wanted to make a revolution of mind, of culture – in contrast to the French revolution which was observed very, very closely and intensely. Like: "We are making the revolution within"; here we do not have the chance to change the politics. And the Romanticists saw how the French revolution could end in chaos and blood.

Romanticism was a countermovement and it was a new world view. The Romantic thinkers felt dissatisfied with that world view of the 18th century which appeared to them trivial, shallow, trite. And all those sharply dualistic concepts:

Mind - matter;
Human being - nature;
Subject - object (observer and observed);

didn't seem to be true or, at least, they seemed only half true.

The poets were the first to see this. There was no place for beauty, no place for marvelling, for amaze at nature's miracles no room for wonder ("*Staunen*"). What a shallow mind that does not see the beauty of the world! That does not have words for nature's wonders! These brains build machines, industries - but they miss life's meaning!

In this perspective, the romantic mind is the poetic mind.

Remember Novalis' word "*Toujours en état de la poésie*" (Always in a poetic mood.)

In his unfinished novel, *Heinrich von Ofterdingen* (1802), Novalis supplied German Romanticism with a symbol, "the blue flower," which stood for the unknown and unattainable goal of the melancholy yearning pervading the romantic life. A leading Romantic poet and philosopher of "Magic Idealism," Novalis expressed the Romantics' idealization of the Catholic Middle Ages in the essay, "Christendom or Europe" (1797, not published until 1826), and explored romantic emotions, spirituality, and the fusion of love and death in his "Hymns to the Night" (1800), and "Spiritual Songs" (1802).

The mind of unity: "In the human mind, Nature becomes conscious of itself. So they are both pervaded by the same spiritual principle" (Novalis). And here we remember Gregory Bateson's telling book title: "Nature and Mind. A Necessary Unity." And let me tell the other one: "Steps to an Ecology of Mind." Bateson tells us here the two main mottos of Romanticism. And I could give you dozens of quotes by Arne Naess which say exactly the same – Arne Naess that wise man from Norway, who died in January aged 96 years. In his very old age he still lived in his Tergastein hut in the woods half the year. He was born in 1912, so he was so lucky to breathe the air before the first world war for some time that war which was the beginning of the 20th century's disaster.

And there had been dissidents in the old 18th century to whom the Romantics owed a lot:

This was Rousseau's word "Back to Nature".

This was a hundred years earlier Spinoza's pantheism: God is in nature, and nature is god's revelation (*Offenbarung*) (Please excuse me, this a simplification in both cases but seems to be tenable in this short form here).

Both had an immense impact on Goethe who, as I have said, was the green soul in the Germany of these days.

One thing is important here: this was not just an intellectual attitude, it was not just thought. There is this "holistic" approach which also seems to fit quite well with the world view of modern philosophy.

Some people get suspicious when they hear the word "holistic" or "holistic" approach – wholeness, totality etc. (and having in mind the nearness of totality and totalitarianism). However, these words seem to be sometimes a kind of joker that is played when one does not want to be really concrete.

One of the critics is my good friend Chris Stone.[12] He asked me in a discussion: "Please tell me now: What do you mean with holism, holistic approach etc?" I did not find a real good answer, I am afraid. – And later I said to him that all ends up in one mind which tries to see this all (in the sense "wholeness" is the seeing-all-at-once).

But it is an irrefutable fact that human awareness is in its nature non-dualistic, that inside and outside are inseparable together. Well, thought can divide this up, which can be very useful to observe a single phenomenon, a reality fragment very closely – but the being-in-the-word is a holistic phenomenon. Arne Naess for example always talked about this: nature and humans standing on the same ground, being of the same kind, being the same etc.

For our romantic precursors the main thing was that this unity naturally appears in the everyday's practice – in the rite of daily life. Novalis also speaks about the holy – close to the word wholeness – what Gary Snyder[13] from a buddhist point of view calls "the real work" of simple, profound "household chores".

Yes here we have such a new vital source of new Romanticism: Gary Snyder the poet who lives in the Californian Mountains. Another man living in a wooden hut who is called the "Thoreau of our time" – a wonderful poet and a great thinker in the field of ecology. And Novalis in his fragment "*Die Lehrlinge zu Sais*" (The Apprentices of Sais) says in a very plain way that every human essentially is an artist – the everyday life with its gestures and simple grasps and grips as a performance of art. And very modern: the sculpturist Joseph Beuys was a reader of Novalis, that's well known.

The romantic poets especially Novalis were not unworldly, though, they did not live in an ivory tower: Novalis was an inspired poet and a mining engineer seeing both: the science of geology and the mystery of the dark depths of the earth.

In romanticism we have a pattern of civilization critique which still is alive until today. But the mainstream of Western civilization followed another track, as we know. As we know, the 19th century saw an explosion of knowledge in the natural sciences, and an immense increase in engineering skills, and this is progress and there is no end in sight. This is the other element, which appears to be stronger. However, romantic thought survived in various forms as an undercurrent to reappear strongly. And this is quite a miracle, but our culture seems to have that strong side aspect to allow different concepts of reality – of world views – existing together – or we should say better side by side.

"America, you have got it better than our old continent" this is a word of Goethe. How true this would be!

Goethe's thinking and the new romantic philosophy was received in England and the young United States very strongly. There in particular by Ralph Waldo Emerson and Henry David Thoureau.

Let's look at America in the first half of the 19th century:

I am an admirer of the great Henry David Thoreau and of Ralph Waldo Emerson – astonishing how fresh their ideas still are! Nietzsche saw this: he quoted Emerson for a motto to "The Gay Science" ("*Fröhliche Wissenschaft*") and he appreciated that new American tone in philosophy which wasn't academic, which was so much closer to practical life – that is how he thought and wrote himself (He was the first of the "life philosophers" (*Lebensphilosophie*[14]) who appeared later in the 20th century.). And there are countless quotes of Emerson and from Thoreau's hut at Walden Pond which show how much they had studied the new European approach to nature. This was really a fertile transfer of ideas.

These two men laid the foundation for a revolutionary new understanding of nature and ultimately an ecological awareness. Emerson emphasized the importance and ground-braking nature of Goethe's views when he stated that, „He, Goethe, has pronounced the best things about nature that were ever said“[15] and „His love of Nature seems to give a new meaning to the world“[16]. For Emerson, Goethe was the major force behind the revolt against the 18th century mechanistic world view (Descartes, Hume, Newton): "He revolted against the science of the day, against the French and English science ... [and] declared war against the great name of Newton."[17] And to Goethe, Schelling's nature philosophy was a liberation in the field of philosophy what he felt as a poet was not founded in the "new philosophy".

But there is another tone in their writing: Very much down to the earth, and an unbelievable sense of practical personal freedom which was unthinkable in the feudal states of middle Europe of that time. For example: the name "Novalis" was a pseudonym, a "nom de guerre" of the poet Friedrich von Hardenberg, to avoid censorship and political prosecution (novalis means "the peasant settler, the new colonist"). There was not that mysterious speculative attitude resorting to an inner world of fantasies which also was characteristic of the European romanticism.

These romantics of the New World who called themselves "transcendentalists" (according to the transcendental philosophy of the German idealism) laid the foundation of a new practical understanding of nature, and of a different life of freedom. This sense of freedom – unthinkable in middle Europe of that time – seems to have been a birth gift of the American democracy (Hannah Arendt wrote about this "On Revolution"), and Europe had reasons to be glad about it several times. Thoreau would not have had to say a single word: His life itself was a radical fundamental critique of our civilization. But fortunately, he was not staying silent.

In Europe and in America the mainstream of civilization did not follow these prophets; they were left behind as a footnote of history.

Were they?

In America and Europe, namely in Germany, there remained a strong undercurrent of that critique of civilization – one could see this in poetry, in art, in philosophy. Like: "Beware, there is something missing, our culture is going to loose its soul we are losing our sense of beauty, of wholeness..." – This sounds like the prophets in the Old Testament, a constant reminder that is ignored again and again.

And this undercurrent is still alive today as we know.

From time to time there were eruptions. One was around 1900. In all European countries we had this movement of life reform, in Italy e.g. the vegetarismo movement. In the German-speaking area it was strongest: Maybe because "the iron cage of rationalism and industrialization here weighed heaviest. As in nature the antidote to a poisonous plant often exists within the same eco-system."[18] There was a real outburst of counter-culture. The motto romantic. It was "Back to Nature": Hundreds of utopian groups of "wandering birds", anarchists, nudists, folk dancers, hikers, health food freaks, artists communes, advocates of a sexual revolution etc. And again, there was a modern philosophy – life philosophy (*Lebensphilosphie*) which became an international phenomenon (Henri Bergson in France e.g.).

But this here takes an astonishing turn:

Some of these people saw the signs of the times: years before 1914 they saw that Europe was going to end up in a disaster. A considerable number of these people left Europe, most of them to the United States. Here they continued their utopian ways of life: they did not change their minds. And they had a chance to do so in the US: here – especially in California – they found fertile soil for their ideas.

For instance, all the practitioners of natural healing methods ("nature heals"), the health food shops, supporters of the anthroposophic movement and their branch of bio-dynamic farming; the followers body culture and of the modern expressive dance etc. Jack Kerouac wrote about the long-haired, long-bearded "Nature Boy" freaks of the 50s looking like Hippies – in "On the Road". Gary Snyder told me that in the San Francisco Bay area there were a lot of leftist anarchist groups around 1900 influenced by that life reform movement. Some were hiking groups who did the same as in the old continent; they had names like *"Wandervogel", "Wanderlust", "Naturmensch"* etc. And Hermann Hesse, of course…

The "San Francisco Renaissance" of poetry in the 50s was purely of that neo-romantic spirit! Their precursors were either directly romantics like William Blake (admired and followed by Allen Ginsberg) or writers like D. H. Lawrence and Robinson Jeffers.

In his novels and stories D. H. Lawrence fights against a stiffened rigid mechanic bourgeois life – longing for an organic holistic life and world-view including sexual liberation. He was strongly influenced by the life philosophy and the life reform movement. – His wife Frieda von Richthofen had been part of that scene in Munich and Heidelberg. She was a congenial kindred intellectual partner to him. Her sister Else was married to Alfred Weber, brother of Max Weber who both were involved in the development of that life philosophy.

D. H. Lawrence lived for several years in New Mexico. He was a writers' writer. His novels and his philosophy – his new tone to the English literature found enthusiastic admirers among American writers like Henry Miller, Hemingway, and Robinson Jeffers who as a boy had been at school in Europe and studied in Germany and had read Nietzsche in German language. He lived at the Californian West coast where he had built his own house from stones and rocks he found at that coast. Gary Snyder one of the mayor beat poets and environmental thinkers says that he was influenced by Jeffers strongly.

If I'll have the opportunity to talk again in this wonderful group here I'll talk about this man, Gary Snyder, who is a constant strong inspiration to humankind. I just translated his best book, "Practice of the Wild" into German, it will come out next spring when Gary will celebrate his 80th birthday. This is our birthday gift to him.[19]

Notes

---

[1] Hanfried Blume left this text as notes for his presentation at the GEIG meeting in Florence, June 2009. For the publication it was slightly modified in the intention to keep its originality. Footnotes were inserted by Michael Schröter, Berlin.

[2] Alfred Brehm, a German zoologist and author of the 19th century, speaks of this classification.

[3] Hanfried Blume owned a farm arrangement in Dreileben, a small village close to Magdeburg (former East Germany). In the 1950th his family was expropriated, but he regained property rights after the German reunification in 1990. See also Hanfried's personal remark at Fn. 8.

[4] Laurence H Tribe, Ways Not to Think about Plastic Trees: New Foundations of Environmental Law, Yale Law Journal 1974 (83), p. 1315-1348.

[5] Carl Gustav Jung was a Swiss medicine and psychologist at the beginning of the 20th century and the founder of the "analytical psychology".

[6] Bertolt Brecht: "Was sind das für Zeiten, wo ein Gespräch über Bäume fast ein Verbrechen ist - Weil es ein Schweigen über so viele Untaten einschließt?"

[7] 1933-1945.

[8] "To me, there was a little blemish in that story – I was born in East Germany, my parents had been expropriated by the communists, our farm became a state farm, they fled to the West, I grew up in the West and as a schoolboy I visited my grandparents in East Germany, so I wasn't an enthusiastic follower of that state socialism – I did know that I would do everything to get out of this cage if had to live there."

[9] The "Jugendbewegung" in Germany was active especially in the beginning of the 20th century. In contrast to urban life, they embraced a kind of natural living and accented romantic aspects, especially the German folk songs (e.g. the Wandervogel).

[10] Note Descartes' sharp distinction of spirit and matter, *res cogitans and res extensa*, the famous Cartesian dualism. Here the world, the nature is seen as a giant complex machine, even animals were. Remember: To Descartes, an animal's life was morally totally irrelevant: He saw an animal's cry of pain in a torture like a purely mechanic process – released by a machine-like mechanism in the animal's brain without any reason to feel any pity or to worry about.

[11] It is not my intention at all to criticize or blame Newton and Descartes – what fool I would be! – but the world-view that was developed in their wake, namely by the French encyclopedists, became a little bit narrow, restricted. And this was felt by the new generation of poets and thinkers in almost all European countries.
[12] Hanfried Blume translated Christopher D. Stones, Should Trees Have Standing? Southern California Law Review 1972 (45), p. 450-501, into German.
[13] Hanfried Blume translated several poems of Gary Snyder into German.
[14] This movement was established in France by Henri Bergson and in Germany by Wilhelm Dilthey in 19th century. In contrast to natural sciences and rationalism they emphasized that life is more and to catch its wholeness non-rational and creative terms and categories were needed.
[15] The Collected Works of Ralph Waldo Emerson. Edited by Alfred R. Ferguson et al. 4 vols. to date. Cambridge: Harvard University Press, 1964, Vol. 4, p. 158.
[16] The Complete Works of Ralph Waldo Emerson. Edited by Edward Waldo Emerson. 12 Vols. Boston: Houghton Mifflin Co., 1903 -1904, Vol. 12 p. 323.
[17] Op. cit. ed. E.W.E. Vol. 10 p. 338.
[18] Gordon Kennedy, Children of the Sun. A Pictorial Anthology. From Germany to California 1883-1949.
[19] The publisher decided, despite the death of Hanfried Blume, to keep the project going, thus this last present of him to Gary Snyder will be there in spring 2010.

CONTRIBUTORS

Donald A. Brown
Associate Professor, Environmental Ethics, Science, and Law,
Penn State University

Klaus Bosselmann, PhD
Professor of Law and Director of the New Zealand Centre for Environmental Law at the University of Auckland. Chair of the IUCN Commission on Environmental Law Ethics Specialist Group. Web Address: www.nzcel.auckland.ac.nz

Andrew Brennan
Philosophy, La Trobe University
Melbourne, Australia
Website: http://www.latrobe.edu.au/philosophy/andrewb.htm

Valerie A. Brown AO, BSc MEd PhD (Human Sciences)
Emeritus Professor of Environmental Health, University of Western Sydney: Director, Local Sustainability Project, Fenner School of Environment and Society, Australian National University;
Adjunct Professor, Institute for Sustainable Futures, University of Technology, Sydney.

Ramy Bulan, Phd
Associate Professor, Faculty of Law, UM
Director, Malaysian Centre for Indigenous Studies, University of Malaya
ramy@um.edu.my

David N. Cassuto
Professor of Law
Director, Brazil-American Institute for Law & Environment
Pace University School of Law
http://www.pace.edu/page.cfm?doc_id=23145
http://baileinstitute.org

Sheila D. Collins
Professor of Political Science and Director, Masters program in Public Policy and International Affairs, William Paterson University.
Email Address: collinss@wpunj.edu
Web Address: www.wpunj.edu/cohss/polisci/faculty/collins/collins.htm

Michael Coyle
Associate Professor in the Faculty of Law
University of Western Ontario
London, Canada

Joseph Dellapenna
Professor of Law, Villanova University, Director of the Model Water Code Project, American Society of Civil Engineers, Rapporteur of the Water Resources Law Committee of the International Law Society.

J. Ronald Engel, Ph.D.
Senior Research Fellow, Center for Humans and Nature; Professor Emeritus, Meadville/Lombard Theological School
(University of Chicago affiliate)
Web Address: www.humansandnature.org

Prof. Dr. Franz-Theo Gottwald
Theologian, philosopher, author and economic ethicist.
Director of the Schweisfurth-Foundation Munich, honorary professor at Humboldt University Berlin, associate lecturer at the Munich University for Political Science.
www.schweisfurth-stiftung.de

Dr. phil., lic. theol. Johannes Heinrichs
Former Prof. of Philosophy and Social Ecology at Humboldt University Berlin, now free philosophical writer
In den Kämpen 13
47169 Duisburg (Germany)
Phone/Fax: 0049-203-578 99 520
www.johannesheinrichs.de www.stenobooks.com/deutsch

Vicky Karageorgou
PhD in European Environmental Law (University of Cologne)
Elected Lecturer in European Administrative and Environmental Law,
Panteion University, Athens Greece
e-mail: vkaragiorgou@yahoo.gr

Norva Y. S. Lo
Philosophy, La Trobe University
Melbourne, Australia
Website: http://www.latrobe.edu.au/philosophy/norval.htm

Owen McIntyre
Faculty of Law
University College Cork
National University of Ireland

Jack P. Manno
Associate Professor and Director of Graduate Studies
Department of Environmental Studies
SUNY College of Environmental Science and Forestry
www.esf.edu/es/faculty/manno.htm

Agnès Michelot
Maître de conférences HDR en droit public
Laboratoire d'Analyses de l'Action Publique et de l'Europe (LASAPE)
Faculté de droit, de sciences politique et gestion
Université de La Rochelle
45, rue Vaux de Foletier
17000 La Rochelle

Bradford W. Morse
Dean of Law, University of Waikato School of Law in New Zealand and
Professor of Law, University of Ottawa Faculty of Law (on leave).

William E. Rees, PhD, FRSC
Professor
University of British Columbia
School of Community and Regional Planning
6333 Memorial Road
Vancouver, BC, Canada V6T 1Z2
Web Address: http://www.scarp.ubc.ca/

Tullio Scovazzi
Professor of International Law
University of Milano-Bicocca
Milan, Italy

May Sim
The College of the Holy Cross, Department of Philosophy
Worcester, MA, U.S.A
Email Address: msim@holycross.edu

Colin L. Soskolne, PhD
Professor (Epidemiology) and Accreditation Coordinator
Department of Public Health Sciences, School of Public Health;
Academic Coordinator
Office of Sustainability, Office of the Provost (Academic);
University of Alberta, Edmonton, Alberta, Canada
Web Address: www.ualberta.ca/~soskolne

Prue Taylor
Senior Lecturer, University of Auckland, New Zealand,
Deputy Director, New Zealand Centre for Environmental Law.
Private Bag 92019
Auckland, NZ
Email Address: prue.taylor@auckland.ac.nz

Valentina Vadi
Lecturer in international law
Maastricht University
Email Address: v.vadi@maastrichtuniversity.nl

Laura Westra, Ph.D., Ph.D.(Law)
Professor Emerita (Philosophy)
University of Windsor
Sessional Instructor, Faculty of Law
Web Address: www.ecointegrity.net
GEIG: www.globalecointegrity.net

Michelle Zackrison
LL.B. candidate at the University of Ottawa, Faculty of Law in Canada.